CORROSIVE-CONTAINING WASTES

CORROSIVE-CONTAINING WASTES

Treatment Technologies

by

L. Wilk, S. Palmer, M. Breton

Alliance Technologies Corporation
Bedford, Massachusetts

NOYES DATA CORPORATION
Park Ridge, New Jersey, U.S.A.

Library of Congress Catalog Card Number 88-17965
ISBN: 0-8155-1180-9
ISSN: 0090-516X
Printed in the United States

Published in the United States of America by
Noyes Data Corporation
Mill Road, Park Ridge, New Jersey 07656

10 9 8 7 6 5 4 3 2 1

Library of Congress Cataloging-in-Publication Data

Wilk, L.
Corrosive-containing wastes

(Pollution technology review, ISSN 0090-516X ; no. 159)
Bibliography: p.
Includes index.
1. Corrosive wastes. 2. Factory and trade waste. I. Palmer, S. II. Breton, M. (Marc) III. Title. IV. Series.
TD898.8.C67W54 1988 628.5'4 88-17965
ISBN 0-8155-1180-9

Foreword

Treatment technologies for corrosive-containing wastes are described in this book, which will be a comprehensive source of information for those involved in the evaluation of available waste management options.

Corrosive acids and alkalis are widely used by all segments of American industry and result in the generation of approximately 40% of all Resource Conservation and Recovery Act (RCRA)-regulated hazardous wastes. Improper management of these wastes can result in altered pH of surface waters to the detriment of aquatic organisms. Land disposal of these wastes can also lead to the solubilization of toxic (e.g., heavy metal) constituents of codisposed wastes, thereby enhancing the potential for their transport into the environment. To combat the potential negative effects associated with current disposal practices, the 1984 RCRA Amendments directed EPA to ban corrosive wastes from land disposal to the extent required to protect human health and the environment.

The land disposal ban excludes acidic corrosive wastes (pH less than or equal to 2.0) from land disposal units (excluding underground injection), effective July 8, 1987. Treatment standards for corrosives which are currently managed through underground injection were to be promulgated on August 8, 1988. Finally, alkaline corrosive wastes (pH greater than 12.5) will be banned from disposal effective May 8, 1990. In addition, standards for hazardous constituents which are commonly present in corrosive wastes, such as heavy metals and toxic organics, are also being promulgated under the 1984 RCRA Amendments. Thus, prior to land disposal, corrosive wastes will also have to meet these standards as they are promulgated.

All potentially viable technologies for treating corrosive-containing wastes are identified and discussed in the book; however, emphasis is placed on proven technologies. Each of the technologies is described in terms of actual performance in removing constituents of concern, associated process residuals and emissions, and those restrictive waste characteristics impacting the ability of a particular technology to effectively treat the wastes under consideration. Cost and capacity data are also provided to help assess the applicability of the technologies to specific waste streams.

The information in the book is from *Technical Resource Document: Treatment Technologies for Corrosive-Containing Wastes,* prepared by L. Wilk, S. Palmer, and M. Breton of Alliance Technologies Corporation for the U.S. Environmental Protection Agency, December 1987.

The table of contents is organized in such a way as to serve as a subject index and provides easy access to the information contained in the book.

Advanced composition and production methods developed by Noyes Data Corporation are employed to bring this durably bound book to you in a minimum of time. Special techniques are used to close the gap between "manuscript" and "completed book." In order to keep the price of the book to a reasonable level, it has been partially reproduced by photo-offset directly from the original report and the cost saving passed on to the reader. Due to this method of publishing, certain portions of the book may be less legible than desired.

NOTICE

Contents and Subject Index

1. Introduction

Section 3004 of the Resource Conservation and Recovery Act (RCRA), as amended by the Hazardous and Solid Waste Amendments of 1984 (HSWA), prohibits the continued placement of RCRA-regulated hazardous wastes in or on the land, including placement in landfills, land treatment areas, waste piles, and surface impoundments (with certain exceptions for surface impoundments used for the treatment of hazardous wastes). The amendments specify dates by which these prohibitions are to take effect for specific hazardous wastes as shown in Table 1.1. After the effective date of a prohibition, wastes may only be land disposed if: (1) they comply with treatment standards promulgated by the Agency that minimize short-term and long-term threats arising from land disposal; or (2) the Agency has approved a site-specific petition demonstrating, to a reasonable degree of certainty, that there will be no migration from the disposal unit for as long as the waste remains hazardous.

Liquid acidic wastes (i.e., pH less than or equal to 2.0) will be banned from land disposal effective July 8, 1987 with an exception granted for wastes which are disposed via underground injection. Restrictions for these wastes will be enacted by August 8, 1988. Alkaline waste (i.e., pH greater than or equal to 12.5 or wastes which are strongly corrosive to steel) and non-liquid waste disposal restrictions will be promulgated by May 8, 1990. However, the 1984 RCRA Amendments authorize the Agency to extend the effective dates of prohibitions for up to 2 years nationwide if it is determined that there is insufficient alternative treatment, recovery or disposal capacity.

TABLE 1.1. SCHEDULING FOR PROMULGATION OF REGULATIONS BANNING LAND DISPOSAL OF SPECIFIED HAZARDOUS WASTES

Waste category	Effective date[a]
Dioxin containing waste	11/8/86
Solvent containing hazardous wastes numbered F001, F002, F003, F004, F005	11/8/86
California list:	
- Liquid hazardous wastes, including free liquids associated with any solid or sludge containing:	
- Free or complex cyanides at $\geq$1,000 mg/L	
- As $\geq$500 mg/L	
- Cd $\geq$100 mg/L	
- Cr^{+6} $\geq$500 mg/L	
- Pb $\geq$500 mg/L	7/8/87
- Hg $\geq$20 mg/L	
- Ni $\geq$134 mg/L	
- Se $\geq$100 mg/L	
- Ti $\geq$130 mg/L	
- Liquid hazardous wastes with:	
- pH $\leq$2.0	7/8/87
- PCBs $\geq$50 ppm	
-Hazardous wastes containing halogenated organic compounds in total concentration $\geq$1,000 mg/kg	7/8/87
Other listed hazardous wastes (§§261.31 and 32), for which a determination of land disposal prohibition must be made:	
- One-third of wastes	8/8/88
- Two-thirds of wastes	6/8/89
- All wastes	5/8/90
Hazardous wastes identified on the basis of characteristics under Section 3001	5/8/90
Hazardous wastes identified or listed after enactment	Within 6 months

[a]Not including underground injection for which land disposal restrictions will be promulgated by 8/8/88.

PURPOSE AND SCOPE

This Technical Resource Document (TRD) for corrosive RCRA wastes identifies recovery and treatment alternatives to land disposal for these wastes and provides performance data and other technical information needed to assess potentially applicable alternatives. This document is one of a series of documents designed to assist regulatory agency and industrial personnel in meeting the land disposal restrictions promulgated by the 1984 RCRA Amendments. To minimize redundancy, emphasis has been placed on treatment technologies (i.e., neutralization) which specifically address the corrosive nature of RCRA wastes. Similarly, discussion of recovery practices has been restricted to methods which are capable of achieving adequate performance at extreme conditions of pH. Although emphasis is placed on performance data for these processes, cost data and technical factors affecting performance (e.g., restrictive waste characteristics) are discussed to assist in the evaluation of alternative approaches to land disposal.

DOCUMENT ORGANIZATION AND CONTENT

The following section (Section 2) will identify the hazardous wastes of concern which meet the RCRA definitions of corrosive wastes. Available information concerning waste stream characteristics, generation, and management practices will be provided in Section 3. Following sections (Sections 4 and 5) will discuss neutralization and recovery practices, which are available as alternatives to land disposal. Each process will be reviewed with regard to the following four factors:

1. Process description, including design and operating parameters, applicable waste types, pretreatment requirements, and post-treatment and disposal of residuals;
2. Case study and performance data which identifies the range in potential applications, processing equipment, and system configurations;
3. Cost of treatment; and
4. Present status of the process.

Virtually all corrosive wastes will have to undergo some form of neutralization as part of the treatment/disposal process. Thus, handling of non-corrosive waste constituents will be discussed as pre- or post-treatment (neutralization), as appropriate. Treatment and disposal alternatives for these non-corrosive constituents and treatment residuals will be identified. However, the reader is referred to related Technical Resource Departments for detailed performance data since this was beyond the scope of this document.

A final section (Section 6) provides approaches to identifying and selecting appropriate technologies for corrosive waste streams. Although emphasis is placed on technical approaches, economic and institutional concerns are also discussed to assist in process selection.

2. Identification of RCRA Corrosive Wastes

As specified in the EPA regulations for identifying hazardous waste, a waste is defined as a RCRA corrosive waste if it meets either of the following criteria:[1]

- It is aqueous and has a pH less than or equal to 2.0 or greater than or equal to 12.5; or
- It is a liquid and corrodes steel (SAE 1020) at a rate greater than 0.25 in./yr at a test temperature of 130°F.

Thus, wastes which are all solids are not subject to restrictions applicable to RCRA corrosive wastes. However, wastes which contain both liquids and solids may be classified as RCRA corrosive wastes depending upon the characteristics of the liquid fraction. EPA has established a test protocol called the paint filter test[2] which is designed to separate free liquids by gravity from the waste matrix. If the recovered liquid meets the above criteria for corrosive waste, the entire waste is considered to be corrosive. The term "solids" is used throughout this document. It is assumed that corrosives which were characterized as "solids" in the literature, actually contain residual quantities of liquid which meet the definition of corrosive waste.

Corrosiveness was determined to be a hazardous characteristic because improperly managed highly acidic or alkaline wastes can present a danger to human health and the environment through the following mechanisms:[3]

- Harm to transporters and other persons coming into contact with the waste;
- Solubilization of toxic constituents of solid wastes thereby enhancing their transport into ground and surface water;

- Chemical reactions with co-disposed wastes which can result in generation of heat or toxic fumes; and
- Altered pH of surface waters to the detriment of aquatic organisms.

Although these effects occur below pH of 3 and above pH of 12, EPA ultimately promulgated less restrictive standards (i.e., pH less than or equal to 2.0 or greater than or equal to 12.5) in order to exclude certain materials from regulation. Specifically, these include otherwise non-hazardous wastes such as lime stabilized sludges (pH 12.0 to 12.5), which can be put to agricultural or other beneficial uses, and substances such as cola drinks and many industrial wastewaters (pH 2.0 to 3.0).[3] EPA felt that these less restrictive limits would still encompass those wastes which are most likely to involve damage to the skin, solubilization of toxic substances or harmful chemical reactions.

In addition to pH, EPA has elected to express corrosiveness in terms of the metal corrosion rate since many RCRA wastes are stored, transported and land disposed in steel containers. Therefore, the metal corrosion rate is indicative of a compound's ability to escape from its container or corrode other containers thereby increasing the potential for adverse chemical reactions or hazardous substance release. The EPA standard for metal corrosion rate was adopted from the Department of Transportation classification for compounds which exhibit severe rates of corrosion at temperatures which may be encountered during handling of hazardous materials.

EPA has listed two wastes from specific sources (K062, K111) and five discarded commercial chemical products and associated off-specification materials, containers, and spills (U006, U020, U023, U123, U134) as materials which meet the criteria for corrosive hazardous wastes.[1] As shown in Table 2.1, certain other listed hazardous wastes have also been reported which were characterized by the generators as corrosive. Finally, the waste code D002 has been assigned to wastes which also meet the definition of corrosive wastes but are not identified as hazardous in Part 261, Subpart D. These non-toxic corrosive compounds and other compounds which can react with water to form corrosive wastes are also listed in Table 2.1.[4]

TABLE 2.1. CORROSIVE AND POTENTIALLY CORROSIVE WASTE MATERIALS

A. RCRA Wastes Listed Due to Corrosivity

D002 - Any waste not listed in Subpart D which has either: 1) 12.5 ≤pH≤2, or 2) corrodes SAE1020 steel at a rate greater than 0.25 inch per year at 130°F

K062 - Spent pickle liquor generated by steel finishing operations of plants that produce iron or steel

K111 - Product washwaters from the production of dinitrotoluene via nitration of toluene

U006 - Acetyl chloride

U020 - Benzenesulfonyl chloride

U023 - Benzotrichloride

U123 - Formic acid

U134 - Hydrogen fluoride; hydrofluoric acid

B. Potentially Corrosive RCRA Listed Wastes

ACIDS -

K039 - Filter cake from the filtration of diethylphosphorodithioic acid in the production of phorate

K050 - Heat exchanger bundle cleaning sludge from the petroleum refining industry

K100 - Waste leaching solution from acid leaching of emission control dust/sludge from secondary lead smelting

U008 - Acrylic acid
P010 - Arsenic acid
U136 - Cacodylic acid
U052 - Cresylic acid
U088 - Diethyl phthalate
U102 - Dimethyl phthalate
U103 - Dimethyl sulphate
U113 - Ethyl acrylate
U119 - Ethyl methanesulfonate
U114 - Ethylenebis (dithiocarbamic acid)
P058 - Fluoroacetic acid, sodium salt
P063 - Hydrocyanic acid, hydrogen cyanide
U147 - Maleic anhydride
U156 - Methyl chlorocarbonate
U162 - Methyl methacrylate
P043 - Phosphorofluoric acid, bis (1-methyl-ethyl ester)
U204 - Selenium dioxide

(continued)

TABLE 2.1 (Continued)

BASES -

K060 - Ammonia still lime sludge from coking operations

U012 - Aniline	U133 - Hydrazine
U092 - Dimethylamine	U167 - 1-Napthylamine
U098 - 1,1-dimethylhydrazine	U174 - N-Nitrosodiethylamine
P046 - Ethanamine, 1,1-dimethyl-2-phenyl-	U194 - n-Propylamine
P053 - Ethylene diamine	U196 - Pyridine
	U221 - Toluenediamine

C. Potentially Corrosive Non-Listed Wastes

ACIDS -

Acetic acid	Hydrogen chloride
Acetimidic acid	Hydrogen peroxide
Adipic acid	Isocyanic acid
Butanoic acid	Methanesulfonic acid
Carbamic acid	Nitric acid
Chlorosulfonic acid	Oxalic acid
Chromic acid	Phosphoric acid
Citric acid	Phosphorofluoric acid
Dithiocarbamic acid	Propionic acid
Fulminic acid	Sulfuric acid
Hydrochloric acid	Vanadic acid

BASES -

Aminoethanolamine	Morphaline
Ammonium hydroxide	Potassium hydroxide
Caustic potash Solution	Sodium carbonate
Caustic soda Solution	Sodium hydrosulfide
Cyclohexylamine	Sodium hydrosulfite
Diethanolamine	Sodium Hydroxide
Diethylenetriamine	Triethanolamine
Diisopropanolamine	Triethylamine
Dimethylformamide	Triethylenetetramine
Hexamethylenediamine	Trimethylamine
Hexamethylenetetramine	Urea
Methylethylpyridine	
Monoethanolamine	
Monoisopropanolamine	

(continued)

TABLE 2.1 (Continued)

D. Chemicals that React in Water to Give Acids

Compound	Resulting Acid
Acetic anhydride	Acetic acid
Aluminum chloride	Hydrogen chloride + aluminum hydroxide
Benzoyl chloride	Hydrogen chloride + benzoic acid
Bromine	Hypobromous acid
Chlorosulfonic acid	Hydrogen chloride + sulfuric acid
Maleic anhydride	Maleic acid
Nitrogen tetroxide	Nitric acid + nitric oxide
Nitrosyl chloride	Hydrogen chloride + nitrous acid
Oleum	Sulfuric acid + sulfur trioxide
Phosphorus oxychloride	Hydrogen chloride + phosphoric acid
Phosphorus pentasulfide	Hydrogen sulfide + phosphoric acid
Phosphorus trichloride	Hydrogen chloride + phosphoric acid
Polyphosphoric acid	phosphoric acid
Sulfur monochloride	Hydrogen chloride + sulfuric acid + others
Sulfuryl chloride	Hydrogen chloride + sulfuric acid
Titanium tetrachloride	Hydrogen chloride + tri-hydroxyhalides

E. Chemicals that React with Water to Give Bases

Compound	Resulting Base
Anhydrous ammonia	Ammonium hydroxide
Ethylene imine	Monoethanolamine
Lithium aluminum hydride	Lithium hydroxide + hydrogen + aluminum hydroxide
Sodium	Sodium hydroxide + hydrogen
Sodium amide	Ammonia + sodium hydroxide
Sodium hydride	Sodium hydroxide + hydrogen

Sources: References 1, 4, 5 and 6.

Ultimately, the pH or metal corrosivity of a waste should be determined experimentally. Test methods for these properties are uncomplicated, inexpensive to perform and are accurate provided that representative waste samples have been collected. Estimates of pH and corrosivity are provided in the literature for pure solutions at various concentrations.[7,8,9,10] However, many wastes will be contaminated with interfering compounds which will limit the usefulness of these approximations.

The presence of chloride and sulfate ions accelerates corrosion by interfering with the development of protective films and contributing to the breakdown of passive films already in existence. Dissolved oxygen and elevated temperature also act to stimulate the corrosion process. Alternatively, calcium and bicarbonate ions tend to limit attack. Solutions with high pH cause limited corrosion but, in practice do not usually corrode steel at a rate which would classify them as hazardous under the RCRA definition. Thus, acidic solutions which are very low in pH, contain chloride or sulfate ions and/or contain dissolved oxygen should be tested for corrosivity.[4] EPA suggests using a simple immersion test (NACE Standard TM-01-69) to make this determination.[1]

In most industrial applications of acids or alkalis, the presence of contaminants will generally act to create a more neutral solution. The magnitude of this effect will be dependent on the specific type and concentration of the contaminants. If representative waste samples are not available for testing, the pH for a pure component solution can be estimated provided that the solute concentration and dissociation constant are known. The latter can be obtained from standard engineering texts for most acids and bases.[7,8,9,10] Examples of pH values for compounds with varying dissociation constants and concentrations are provided in Table 2.2. Calculational methods for pH are summarized below.

For monoprotic acids:

$$pH = -\log X = -\log \left[\frac{-K_a + (K_a{}^2 + 4K_aM)}{2} \right]$$

TABLE 2.2. DISSOCIATION CONSTANTS AND pHs OF ACIDS AND BASES IN AQUEOUS SOLUTIONS AT 25°C

Compound	Formula	K_a (1st step)	Concentration	pH
Acids				
Hydrochloric acid	HCl	10^3	1 N	0.1
			0.1 N	1.1
			0.01 N	2.0
Sulfuric acid	H_2SO_4	10^3	1 N	0.3
			0.1 N	1.2
			0.01 N	2.1
Sulfurous acid	H_2SO_3	1.3×10^{-2}	0.1 N	1.5
Oxalic acid	$C_2H_2O_4$	5.9×10^{-2}	0.1 N	1.5
Orthophosphoric acid	H_3PO_4	7.5×10^{-3}	0.1 N	1.5
Bases				
Sodium hydroxide	$NaOH$	10^3	1 N	14.0
			0.1 N	13.0
			0.01 N	12.0
Potassium hydroxide	KOH	10^3	1 N	14.0
			0.1 N	13.0
			0.01 N	12.0
Lime	$CaOH$	3.74×10^{-2}	(Saturated)	12.4

Source: Reference 5.

where X equals the hydronium ion concentration, K_a is the acid dissociation constant and M is the molar concentration of the undissociated acid (i.e., the solution formality). These equations can be applied to bases by substituting pOH and K_b where appropriate in the above equation and using the following relationship:

$$pH = 14 - pOH$$

For strong acids (bases) which are almost completely ionized, pH becomes the negative logarithm of the product of the formal acid (base) concentration and the number of moles of hydronium (hydroxide) ions formed per formula weight of acid (base).

REFERENCES

1. Federal Register. 40 CFR Part 261. Environmental Protection Agency. Regulations for Identifying Hazardous Wastes.

2. Federal Register. February 25, 1982. Volume 47, p. 8311.

3. USEPA. Identification and Listing of Hazardous Waste Under RCRA, Subtitle C, Section 3001 - Corrosivity Characteristics (40 CFR 261.22). U.S. Environmental Protection Agency. Washington, D.C. PB81-184319. May 1980.

4. Camp, Dresser & McKee Inc., Boston, MA. Technical Assessment of Treatment Alternatives for Wastes Containing Corrosives. Prepared for USEPA under Contract No. 68-01-6403. September 1984.

5. Sax, I. N. Dangerous Properties of Industrial Materials. Van Nostrand Reinhold Company. New York, N.Y. 6th Edition. 1984.

6. Versar, Inc. Springfield, VA. National Profiles for Recycling - A Preliminary Assessment. Draft Report prepared for USEPA under Contract No 68-01-7053. July 8, 1985.

7. Weast, R. C. CRC Handbook of Chemistry and Physics. CRC Press, Inc. Boca Raton, Florida. 1984.

8. Dean, J. A. Lange's Handbook of Chemistry. McGraw-Hill Book Company, New York, N.Y. 12th Edition. 1979.

9. Lyman, W. J. and D. H. Rosenblatt. Handbook of Chemical Property Estimation Methods. McGraw-Hill Book Co. New York, N.Y. 1982.

10. Green, D. W. Perry's Chemical Engineers' Handbook. McGraw Hill Book Company, New York, N.Y. 6th Edition. 1984.

3. Corrosive Waste Sources, Generation, and Management

3.1 CORROSIVE WASTE SOURCES

The primary industrial applications for acids and bases which result in generation of corrosive wastes are: (1) use as chemical intermediates in the inorganic and organic chemical manufacturing industries; (2) use as a metal cleaning agent in metal production and fabrication industries; and (3) use in boiler blowdown and stack gas treatment, primarily in electricity generating facilities. Other significant corrosive waste sources include refining processes in the petroleum industry and pulping liquor in the paper industry.

Table 3.1.1 summarizes industrial consumption of primary acid and base chemicals by end use. As shown, production of fertilizers consumes nearly half (48.3 percent) of all domestic production of acids and bases, including sulfuric acid, phosphoric acid, urea and nitric acid. The second largest industrial use category is the production of primary chemicals (16.6 percent) followed by various uses as a chemical intermediate in the production of glass, explosives, soaps and detergents, synthetic fibers, resins, coatings, plastics and adhesives (7.0 percent).

Other industries which consume large quantities of corrosive materials include primary metal, metal fabricating and equipment manufacturing industries. Corrosive chemicals are primarily used as fluxing agents in the production of steel and to clean or prepare metal surfaces prior to annealing or coating operations. The latter uses are nonconsumptive and often involve highly concentrated solutions. Therefore, a high quantity of waste is generated from these industries relative to the amount of raw material consumed.

TABLE 3.1.1. INDUSTRIAL CONSUMPTION OF PRIMARY ACID AND ALKALI CHEMICALS BY END USE (10^3) TONS/YEAR)

Chemical species	Production	Fertilizers	Primary chemical manufacturing	Chemical intermediate[a]	Petroleum industry	Pulp and paper	Water and stack gas treatment	Food mfg. or products	Metal production, cleaning, preparation, & finishing	Mining	Explosives	Soaps and detertiles	Textiles	Other[b]
Sulfuric acid	33,354	21,255	2,616	981	2,616	654	1,308	-	400[c]	1,308	-	327	-	1,889
Phosphoric acid	9,215	8,455	-	-	-	-	-	570	95	-	-	-	-	95
Sodium hydroxide	9,141	-	5,484	-	366	1,828	-	-	366	-	-	-	274	823
Sodium carbonate	8,468	-	1,524	3,387[d]	-	254	593	-	-	-	-	593	-	2,117
Urea	7,610	4,870	76	533	-	-	-	761	-	-	-	-	-	1,370
Nitric acid	7,340	4,698	1,174	-	-	-	-	-	0	-	1,101	-	-	<367
Muriatic acid	2,580	-	650	-	650	-	-	370	650	-	-	-	-	260
Acetic acid	1,469	-	1,264	-	-	-	-	-	-	-	-	-	29	176
Adipic acid	570	-	68	492	-	-	-	-	-	-	-	-	-	10
Acrylic acid	375	-	330	30[c]	-	-	10[c]	-	-	5[c]	-	-	-	>0
Methyl methacrylate	362	-	-	307	18	-	-	-	-	-	-	-	1	36
Hydrofluoric acid	330	-	116	-	20	-	-	-	135	17	-	-	-	42
Potassium hydroxide	220	20	99	-	>0	-	-	-	-	-	-	55	-	<46
Citric acid	101	-	10	>0	-	-	-	71	>0	-	-	13	-	< 7
Sodium hydrosulfite	60	-	-	-	-	38	-	>0	-	-	-	>0	>0	<22
Chlorosulfonic acid	45	-	42	-	-	-	-	-	-	-	-	-	-	3
Chromic acid	36	-	0	-	-	-	-	-	18	-	-	-	>0	<18[e]
Propionic acid	33	-	19	6	-	-	-	>0	-	-	-	-	-	< 8
Ethylene diamine	29	-	17	>0	-	-	-	-	-	-	-	-	>0	<12
Formic acid	24	-	12	3	>0	-	-	-	>0	-	-	-	3	< 6
Sodium hydrosulfide	15	-	2	-	-	-	-	-	2	-	-	-	-	11
TOTAL	81,377	39,298	13,503	5,739	3,670	2,774	1,911	1,772	1,666	1,330	1,101	988	307	7,318
PERCENT	100	48.3	16.6	7.0	4.5	3.4	2.4	2.2	2.0	1.6	1.4	1.2	0.4	9.0

[a] Includes production of glass, fibers, resins, coatings, plastics and adhesives.
[b] Includes net exports.
[c] Estimated.
[d] Use in glass production.
[e] 16,000 tons used as a wood preservative.

Source: Adapted from References 1 and 2.

Other high volume industrial uses of strong acids and bases include use as an alkylation catalyst in petroleum refining, chemical bleaching in the paper industry, textile treating and finishing, food products manufacturing, the recovery of nonferrous minerals, (e.g., copper and uranium ores in mining applications), and oil and gas well acidizing for enhanced recovery. Some of these uses are nonconsumptive and result in the generation of high concentration, high volume corrosive waste; e.g., textile treatment baths, and spent acid alkylation catalysts. Conversely, food industry and oil and gas drilling generate little waste relative to quantities of acid/alkali consumed. These uses are either consumptive or result in waste which is exempt from RCRA regulation; e.g., acid mine drainage.

The following is a brief summary, by industry, of acid/alkali consumption, corrosive waste streams generated, and their general waste characteristics.

3.1.1 Chemicals and Allied Products

The Chemicals and Allied Products Industries (SIC 28) represent the bulk of acid/alkali production (90 to 95 percent)[1], consumption (80 percent or more)[1,2], and waste generation (72 percent).[3] These industries include the inorganic chemical, fertilizer, and organic chemical manufacturing industries as discussed below.

Inorganic Chemicals Industry--

The inorganic chemicals industry is responsible for most of the production of sulfuric acid, phosphoric acid, sodium hydroxide, nitric acid, some hydrochloric acid, hydrofluoric acid and other acids/alkalis. Many of these processes generate wastes in the form of off-specification products, wastewaters, and sludges (e.g., filtration residues) which meet EPA criteria for corrosive wastes. However, due to a high degree of waste recycle, these industries only generate a modest fraction of overall chemical industry corrosive waste.[4]

Hydrofluoric acid is generally produced by the action of sulfuric acid on fluorspar which generates a calcium sulfate precipitate. Other substances such as silicon tetrafluoride, fluosilicic acid, hydrogen sulfide, carbon dioxide and sulfur dioxide can be generated due to impurities in the fluorspar.[3] The kiln discharge stream is typically below pH of 2.0 and is treated by neutralization with soda ash and clarified.[5] Combined wastewaters have a total suspended solids (TSS) range from 220 to 16,400 mg/L and are typically between pH of 2.0 to 3.0. However, values in the corrosive acid range have been reported.[6] Heavy metal concentrations (Cr, Ni, Zn) tend to be below levels which require treatment as a hazardous waste. Thus, treatment of these wastes via lime neutralization and dual media filtration provides a stream suitable for discharge.[6]

Phosphoric acid is primarily produced by sulfuric acid acidulation of phosphate rock generating a $CaSO_4$ precipitate which is removed by filtration. Gypsum and contaminants such as fluosilicates, and aluminum and iron phosphates are precipitated from the acid stream as the solution is concentrated through evaporation. The precipitates are generally recycled and used in dry fertilizer manufacturing. Wastewater sources include cooling tower blowdown, containing ammonia and chromates, and boiler blowdown.[7]

Nitric acid is primarily produced by ammonia oxidation in a converter containing a platinum-rhodium catalyst which generates water as a by-product. This is followed by further oxidation of nitric oxide to nitrogen dioxide which is absorbed in a portion of the by-product water to form nitric acid. Major contaminants in the waste by-product water include unreacted ammonia, nitric acid, and nitrogen oxides (e.g., NO, NO_2, N_2O_4).

Sodium hydroxide is produced as a co-product with chlorine during the electrolysis of sodium chloride in diaphragm or mercury cells. The reaction product is evaporated to concentrate the caustic soda solution producing sodium chloride brine muds. However, due to the buffering action of the salt, these wastes typically have pH levels below 12.5 and thus may not qualify as RCRA corrosive wastes.[5]

Most sulfuric acid is produced by the contact process in which sulfur is burned to sulfur dioxide which is then catalytically (vanadium pentoxide impregnated on diatomaceous earth) oxidized to sulfur trioxide. This is absorbed in concentrated sulfuric acid and diluted to the desired concentration. The principal waste stream generated is acid plant blowdown (i.e., off-gas scrubber liquor) which has a typical pH of 1.8 to 2.5 and contains high levels of Cu, Pb, Zn, and Fe.[8] When by-product sulfur dioxide from smelter gases is used as the source of sulfur, impurities such as arsenic and fluorine are also commonly present.[3]

Other documented sources of corrosive wastes from production of primary inorganic chemicals include combined raw wastewaters from the production of chloride (pH ranges from 1.5 to 9.0, TSS averages 32 mg/L, Al averages 44 mg/L) and production of sodium fluoride (pH in excess of 12, TS averages 167,500 mg/L, F averages 16,000 mg/L).[5]

Fertilizer Industry--

Potentially corrosive fertilizer industry wastes originate from production of their intermediates (primarily sulfuric acid, phosphoric acid, and urea), production of fertilizer products and boiler blowdown. Ammonia effluents will not reach a high enough pH to be RCRA corrosives. However, urea process condensates may exceed pH 12.5. Condensate from flashed gases from the urea concentration step result in a solution containing urea, ammonium carbonate, ammonia and carbon dioxide. The waste quantity has been reported to range from 417 to 935 l/kkg of product with ammonia concentration of 9,000 kg/1,000 kkg and urea concentration of 33,500 kg/1,000 kkg of product.[7]

In the phosphate fertilizer industry, significant discharges result from water treatment and blowdown streams associated with the sulfuric acid process, and gypsum pond water. Gypsum pond water is typically recycled in a closed loop, requiring treatment and discharge only when pond holding capacity is exceeded. The gypsum slurry, resulting from the filtration step in phosphoric acid production, has a pH of less than 2.0 with soluble phosphates (e.g., phosphoric acid) and fluoride (e.g., sodium silicofluoride) in the 8 to 15 g/L concentration range.[7] Gypsum pond waste characteristics are summarized in Table 3.1.2 for a phosphate fertilizer plant.[9]

Organic Chemicals Industry--

The organic chemicals industry generates corrosive wastes in the production of petrochemicals, polymers, organometallics, detergents, pesticides, explosives, dyes, and pigments. Corrosive raw materials used in high volumes include sodium hydroxide, sulfuric acid, sodium carbonate, hydrochloric acid and acetic acid. In addition, hydrochloric acid is produced in large quantities (90 percent of domestic production) as a by-product from manufacture of vinyl chloride monomers, isocyanates, chlorofluorocarbons and other chemical products.[1]

The production of petrochemicals generates effluents which are usually the result of by-product water and water used for chemical transportation.[3] These are typically heavily contaminated with organic substances. For example, acetic acid manufacturing wastewater contains high levels of any or all of the following: formic acids (175 to 89,400 ppm), acetic acid (30 to 177,100 ppm), propionic acid (0 to 1,200 ppm), alcohols (260 to 30,000 ppm), ketones (0 to 2,000 ppm) and aldehydes (0 to 92,000 ppm).[10]

The EPA has recently added product washwater from the production of dinitrotoluene via nitration of toluene (K111), to the list of hazardous wastes from specific sources (40 CFR Part 261.32). These washwaters exhibit a pH less than 2.0 and contain toxic concentrations of 2,4-dinitrotoluene (800 ppm) and 2,6-dinitrotoluene (200 ppm).[11]

Organic chemical wastes from the polymer industry (including plastics, resins, rubbers, and fibers) can be acidic or alkaline, and contain a variety of contaminants.[3] Effluents from reclaimed rubber manufacture have a very high pH as well as large amounts of suspended solids (usually escaped bits of rubber) and chlorides. Synthetic rubber production also yields a high concentration of solids, but they are suspended in an acidic, rather than alkaline, saline solution. Cellulosics can be produced by the xanthate process which generates a waste stream containing H_2SO_4, NaOH, heavy metals, cellulosic materials, and sulfates.[3] Polyester and alky resins are formed by condensation polymerization of a dibasic acid and a polyfunctional alcohol. The main wastes yielded by this process are caustic wash solutions contaminated with unreacted volatile fractions of raw materials.[3]

The pesticide industry routinely recycles a large fraction of its spent acid wastes. However, potential sources for corrosives include:[12]

- Waste products from acid recovery units;
- Nitric acid and other acidic wastewater from fractionation columns in the production of halogenated aliphatic pesticides;
- High pH caustic scrubber water which typically contains dissolved solids and low organic levels;
- Spent acid from settling tanks containing moderate organic content; and
- Wet scrubbers for acidic solutions containing low organic concentrations.

TABLE 3.1.2. PHOSPHATE FERTILIZER INDUSTRY: GYPSUM POND WASTE CHARACTERISTICS

Parameter[a]	Range
pH, standard units	1.6 - 2.1
Total acidity, as $CaCO_3$	20,000 - 60,000
Fluoride, as F	4,000 - 12,000
Phosphorus, as P	4,000 - 9,000
Silicon, as Si	1,000 - 3,000
Total solids	20,000 - 50,000
Total suspended solids	50 - 250
Chlorides, as Cl	50 - 500
Sulfates, as SO_4	2,000 - 12,000
Sodium, as Na	50 - 3,000
Calcium, as Ca	50 - 1,500
Magnesium, as Mg	50 - 400
Aluminum, as Al	50 - 1,000
Chromium, as Cr	0.2 - 5.0
Zinc, as Zn	1.0 - 5.0
Iron, as Fe	100 - 250
Manganese, as Mn	5 - 30
NH_3 - N, as N	0 - 1,200
Total organic N, as N	3 - 30
Color, APHA units	20 - 4,000

[a]All values expressed as mg/L unless otherwise noted.

Source: Reference 9.

An example of an acidic process waste from the production of organo-phosphorus pesticides was provided by EPA as follows: pH ranging from 1.0 to 2.5, TSS of 36 mg/L, phenol concentration of 0.005 mg/L, total chlorine 5,730 mg/L, total phosphorus 35.1 mg/L and total pesticides of 0.014 mg/L.[12]

Phosphate compounds are a principal ingredient in detergents as are alkalies (e.g., sodium carbonate), silicates, neutral soluble salts, acids (e.g., sulfuric acid), insoluble inorganic builders (e.g., clays such as kaolin) and miscellaneous organic compounds; e.g., ethylenediamine, tetra-acetic acid and sodium carboxymethylcellulose.[3] Effluents from detergent manufacture will contain a fraction of these and other contaminants in the acid and alkali wastes.[3]

The explosives industry generates wastes from three major processes: the manufacture of TNT, the manufacture of smokeless powders, and the manufacture of small-arms ammunition.[3] TNT production results in nitric and sulfuric acid wastes, contaminated with ammonia and nitrogen oxides, and alkaline wash water containing sulfur impurities. Smokeless-powder wastes are similarly concentrated in nitric and sulfuric acids, and generally contain a large amount of waste sulfates and nitrogen oxides. Wastes from the manufacture of small-arms munitions include waste pickling acids contaminated with copper, zinc and grease from cutting oils and detergents, and wastewater from equipment washout. An example of a caustic waste ($pH \geq 12.5$) is a desensitized washout solution of RDX($(CH_2N_2O_2)_3$) and lead styphnate ($PbC_6HO_2(NO_2)_3$) which has been treated with caustic. This typically contains high lead levels (225 mg/L), high TDS (17,000 mg/L), and low TSS (35 mg/L).[13]

The most comprehensive source of information regarding hazardous waste characteristics and handling methods in the organic chemicals industry is the Industries Studies Data Base (ISDB).[14] This data was compiled by EPA through actual chemical analysis of residual streams or from organic chemical manufacturer responses to RCRA 3007 questionnaires. Wastes meeting the definition of RCRA corrosives were classified by residual type, physical/chemical properties, waste quantity and constituents, constituent concentrations, and management practice. Nearly 37 million metric tons of waste was characterized which is representative of approximately 72 percent of all corrosive wastes generated.[3]

A summary of data analyzed in August 1985 is presented below.[14] This discussion is generally qualitative since little data were available to establish aggregate waste volumes which exceeded various constituent concentration levels.

Chemical industry corrosives tend to be very low in metal and oil concentrations. The small percentage of waste streams which may require specialized treatment for metals (less than 3 percent) tended to be small volume streams. However, wastes are frequently contaminated with organics at concentrations which require treatment, in addition to neutralization, prior to discharge. The most frequently reported noncorrosive organics which may require treatment are methanol, toluene, and chlorinated solvents.

The most significant corrosive constituent on a waste stream volume basis is sodium hydroxide followed by sodium chloride, glutaric/adipic acid (low concentration, high volume aqueous waste), sulfuric acid, and sodium carbonate. The most highly concentrated of these are sulfuric and hydrochloric acids. On average, sulfuric acid and sodium hydroxide streams are generated in large volumes whereas sodium carbonate and the more expensive hydrochloric acid wastes are generated in smaller volumes.

The vast majority of organic chemical industry corrosive wastes are aqueous (93.3 percent) with roughly an even split between acidic and alkaline wastes.[5] Sludges constituted 4.2 percent of the total waste volume, but nearly all of this was fluid enough to be disposed via deep well injection. Organic liquids accounted for only 2.2 percent, collected gas streams (predominately HCl) contributed 0.1 percent, and solids were only 0.01 percent of the total waste generated. Solid corrosives include materials such as spent adsorbents, filters, dried sludges, waste containers, and catalysts. An example of these is spent phosphoric acid catalyst used in the production of cumene from the alkylation of benzene with propylene.

3.1.2 Petroleum Refining Industry

The Petroleum Refining Industry (SIC 29) is a large consumer of sulfuric acid (most of which is produced onsite), hydrochloric acid and, to a lesser extent, sodium hydroxide and hydrofluoric acid.[1] Concentrated sulfuric acid (98 percent solution) or hydrofluoric acid are used to catalytically react light hydrocarbon molecules in alkylation units to yield higher octane products. Spent acid sludge, which is contaminated with water (about 3 percent) and hydrocarbons (about 10 percent), is typically recovered on or offsite through thermal and other treatment technologies.[15] Sludges which result from neutralization of alkylation waste will also contain heavy metals such as nickel, copper, lead, selenium and arsenic.[16] Other potentially corrosive wastes from petroleum refining include acid/alkali washes, acid sludges from sulfonation, sulfation and acid treatments, spent acid catalysts (e.g., phosphoric acid catalyst used in polymerization processes), boiler blowdowns,-and caustic solutions from chemical sweetening and desulfurization.[16]

Re-refining of used oil by the acid clay process generates acid sludge and dehydration wastewater which have pH less than 2.0.[16] Sulfuric acid treatment of used oil dissolves metal salts, aromatic and asphaltic compounds, organic acids, water and other polar compounds. In addition to these compounds, the resulting sludge will contain particularly high levels of lead (2,000 to 10,000 ppm) and lesser, but significant levels of aluminum, barium, magnesium, and calcium.[17] As much as 10 percent of used oil can remain in the spent acid sludge[9] which contributes to its reported heating value of 9,000 to 11,000 Btu/lb.[17] A sample of acid clay dehydration process water showed a pH of 1.6 with high iron (197 mg/L), phenol (99 mg/L), boron (54 mg/L), silicon (52 mg/L) and lead (40 mg/L) content.[17]

Caustic sludge is generated in small quantities from emulsion breaking of used oils which results in a waste containing caustic, sodium silicate, oil, lead and other metals.[17] However, as a result of waste disposal problems, the use of both acid and caustic clay processes have declined in favor of various forms of vacuum distillation. Thus, these processes are no longer as significant a source of corrosive wastes as they were prior to the implementation of RCRA regulations.

3.1.3 Primary and Secondary Metals Industry

The Primary and Secondary Metals Industries (SIC 33) generate large volumes of corrosive wastes from spent metal cleaning, preparation and pickling solutions. Other sources include certain rinses following acid baths, fume scrubbers and small quantity residuals generated from treatment of bath solutions.

The iron and steel industry accounts for approximately one-fourth of all domestic hydrochloric acid demand for use in acid pickling baths.[1] This acid is used most frequently by large primary steel producers while sulfuric acid is used in roughly one-half the amount by smaller, secondary steel finishers. Smaller quantities of nitric and hydrofluoric acid are used by specialty steel finishers.[18]

The choice of acid is dependent on the base metal and desired surface characteristics. Sulfuric acid penetrates oxide scale and reacts with the base metal to form hydrogen which aids in removing the scale. Scale eventually dissolves in the acid solution forming ferrous sulfate. Hydrochloric acid reacts vigorously and directly with the oxide scale forming soluble ferrous and ferric chloride. Combination acid pickling (hydrochloric, sulfuric, nitric, and hydrofluoric acids) is reserved for specialty steels. Wastewaters from these operations contain sulfates, fluorides and nitrates in addition to iron salts. They also contain higher levels of toxic metals (e.g., Cr, Ni) since specialty steels contain a wider range and higher levels of alloying elements than do carbon steels.[19] Pickling baths are highly concentrated with acid (e.g., 5 to 15 percent H_2SO_4 by weight) and, as a result, are frequently recycled.[20]

Other sources of acidic corrosives from the iron and steel industry include pickling bath rinse waters, pickling bath scrubber systems and scrubber effluent from cleaning absorber vent gases emitted from (hydrochloric) acid regeneration systems.[21] Rinse waters can be below pH of 2.0 if counterflow or high pressure spray systems are used to minimize water consumption. Out of 43 rinse discharges sampled by the EPA, 22 had recorded pH values which were less than or equal to 2.0. Combination acid rinses had the lowest tendency to qualify as corrosive wastes (27 percent).[21]

Wet fume scrubbers generate wastes with similar constituents as the pickling baths but much lower concentrations of solids, oil and grease.[22, 23] Much of this waste can be recycled or eliminated altogether by replacing wet systems with dry systems such as acid demisters. Similarly, wet scrubber wastes from spent pickling bath regeneration fumes are commonly recycled along with regenerated acid. A summary of constituent types and concentrations for wastes associated with acid pickling is provided in Table 3.1.3.

Alkali cleaning is practiced at over 60 carbon and specialty steel facilities to remove dirt, oil, and grease prior to other finishing steps. Typical baths are composed of caustic and have pH of 12 to 13, oil and grease concentration of 1,500 mg/L, iron concentration of 100 mg/L and high solids content (TDS of 25,000 mg/L, TSS of 1,000 mg/L).[23] Since alkali solutions are not as aggressive as acid pickling baths, the spent solution will contain smaller quantities of most toxic metal pollutants. These baths are very amenable to recycling through processes such as ultrafiltration since the major contaminants are easily removed solids and oils. Low molecular weight alkali and builders will readily pass through membranes. Rinse waters generally do not have high enough pH to be considered RCRA corrosive wastes.[23] They are typically treated through oil skimming or flotation, blended with acidic wastes and then treated for solids/metals removal.

The nonferrous metals industries, including primary and secondary copper, lead and zinc, also generate significant quantities of corrosive wastes. Primary copper smelters generate sulfuric acid from treatment of sulfur dioxide off-gas in contact sulfuric acid plants. Dilute sulfuric acid waste is generated from blowdown of conditioning towers which are designed to remove dust from the metallurgical operations off-gas. Sampling data at three facilities showed pH ranges of 1.8 to 2.5 with the following concentration averages: TDS - 244,000 ppm, TSS - 1920 ppm, Zn - 218 ppm, Pb - 89.8 ppm, Fe - 38.2 ppm, As- 59 ppm, Cd - 9.7 ppm and smaller amounts of Cu, Ni, and Se[8]. Several firms recycle this waste as a coolant to hot ESP units following neutralization, thereby returning the metals content to metallurgical processing.[8]

Copper smelters also generate sulfuric acid waste from the dimethylaniline absorption process which involves cleaning particulate matter from SO_2 gas streams. This waste is similar to the sulfuric acid blowdown waste described above. Other acidic wastes include spent sulfuric acid from electrolyte solutions or sludges from recovery (e.g., dialysis, evaporators) of copper from these solutions. These wastes contain high concentrations of

TABLE 3.1.3. SUMMARY OF CONSTITUENT CONCENTRATIONS FROM SAMPLED IRON AND STEEL PLANTS (mg/L)

	Pickling baths			Rinse waters				
	Sulfuric acid	Hydrochloric acid	Combination acid	Sulfuric acid	Hydrochloric acid	Combination acid	Raw fume scrubber-all acid pickling	Raw absorber vent scrubber HCl pickling
pH (units)	<1.0-2.0	<1.0-5.0	<1.0-2.3	<1.0-6.4	1.1-5.0	<1.0-8.2	<1.0-3.7	1.7-7.6
Dissolved iron	49,300	73,230	20,390	3,820	1,658	169	191	970
Oil and grease	25	3.9	2.1	20.3	7.9	2.8	1.9	0.8
Suspended solids	2,140	395	145	96	24	83	3.3	79
Fluoride	NA	NA	6,120	NA	NA	276	1,802	NA
Nitrates	NA	NA	NA	NA	NA	59	NA	NA
Toxic organics	0.425	0.559	NA	0.103	0.744	0.337	0.216	0.874
Chlorinated solvents	0.048	0.118	NA	0.021	0.733	0.197	a	0.81
Phthalates	0.377	0.441	NA	0.068	0.011	0.14	0.195	0.064
Phenols	NA	a	NA	0.014	a	NA	0.021	NA
Other	NA	a	NA	NA	a	a	NA	NA
Antimony	NA	2.19	NA	NA	0.19	0.05	0.20	0.21
Arsenic	0.1	0.21	NA	0.35	0.24	a	0.08	0.01
Cadmium	0.33	0.22	0.14	0.028	a	0.002	a	a
Chromium	100	16.1	3525	6.22	0.18	36.6	0.89	0.50
Copper	3.43	19.9	167	2.42	0.58	3.15	0.09	0.95
Lead	1.07	388	2.0	0.43	0.32	0.05	a	a
Nickel	23.7	12.4	4600	2.35	0.52	36.0	1.18	0.70
Silver	0.44	NA	NA	0.016	NA	NA	NA	NA
Zinc	50	17.6	10.8	12.7	36.2	0.62	0.15	1.08
Cyanide	NA	NA	NA	NA	NA	0.011	a	NA

a - Concentration less than 0.010 mg/L.
NA - Not Analyzed.

Source: Adapted from Reference 21.

nickel sulfate along with copper (2 percent), lead (1 percent), iron, zinc and nearly all the arsenic, antimony, and bismuth that comes in with the anode copper.[8,18]

Primary lead smelters generate sulfur dioxide off-gases from sintering machines which are treated by sulfuric acid production, as described previously. Combined with copper and zinc smelters, these plants generated 2.8 million tons of sulfuric acid in 1982.[1] Secondary lead smelters generate sulfuric acid wastes from battery breaking and leaching operations. One source estimated that the industry generates over 1 million gallons of 1 percent sulfuric acid waste annually from battery breaking.[8] These effluents contain lead, antimony, lead compounds and metal alloys. Zinc leaching wastewater consists of the spent leaching liquor, dilute sulfuric acid, zinc, antimony, lead, copper, metal sulfides and metal chlorides.[8]

Another nonferrous metals industry which has been identified as producing corrosive acidic wastes is the secondary aluminum industry. Scrubber wastewater samples from chlorine demagging at three plants showed the following range in waste characteristics: pH - 1.65 to 3.70, TSS - 44 to 934 mg/L, Al - 474 to 16,600 mg/L, Zn - 2.4 to 38.7 mg/L and smaller quantities of Cd, Cr, Pb, Cu, and Ni.[5] Waste characteristics from other nonferrous metal forming industries are summarized in Table 3.1.4. These data demonstrate the considerable variation in raw waste characteristics which can be found in a given industry.

3.1.4 Fabricated Metals, Machinery, Electrical Supplies and Transportation Equipment

Metal cleaning, coating, fabrication and plating operations (SIC 34 through 37) are another significant source of corrosive wastes. Metal stripping, cleaning and plating result in both acidic and alkaline spent baths which can be heavily contaminated with heavy metal salts, oil and grease, phosphates, silicates, carbonates, inhibitors, surfactants, organic emulsifiers, solvents and suspended solids.[3] Acidic rinsewaters may also have low enough pH to be characterized as RCRA corrosive wastes, particularly if water conservation techniques such as counterflow rinsing are used. Sludges and solid wastes can result from bath treatment and recovery

TABLE 3.1.4. CORROSIVE WASTE CHARACTERISTICS IN THE NONFERROUS METAL FORMING INDUSTRIES

Industry subcategory	Surface treatment waste stream	Oil and grease (mg/L)	Total suspended solids (mg/L)	pH	Cyanide (mg/L)	Lead (mg/L)	Zinc (mg/L)	Ammonia (mg/L)	Fluoride (mg/L)	Copper (mg/L)	Other metal(s)	
											Type	(mg/L)
Titanium	Spent bath	NA	1-66	0.53-6.90	0.02	0.050-5.90	0.020-0.660	1.7-52	1.1-215	NA	Titanium	3.55-186
Titanium	Rinse water	NA	480-3,360	1.80-2.20	NA	65.0-214	2.0-166	NA	74,000-98,000	NA	Titanium	27,900-60,300
Beryllium	Spent bath	1	240	0.32	0.04	NA	NA	NA	79,000	0.18	Beryllium	15,000
Precious metals	Rinse water	1-8	1-3,000	1.30-2.50	0.02	NA	NA	NA	NA	1.80-60.6	Cadmium Silver	0.02-11.1 0.01-6.70
Refractory metals	Spent bath	1	140	0.80	NA	NA	NA	NA	0.27	6.30	Nickel	12.4
Refractory metals	Rinse water	1-6	8.0-140	1.50-2.10	NA	NA	NA	NA	1.1-3,000	0.035-0.400	Nickel	0.016-10.2
Zirconium-Hafnium	Spent bath	1.87-83.7	8.70-12.6	1-3.9	0.118-0.356	NA	0.17-7.5	104-681	6,500-17,100	NA	Chromium	12-24
Magnesium	Spent bath	1- 100,000	70-270	0.80-12.60	NA	NA	8.00-138	0.3-97	1.6-126	NA	Chromium Magnesium	0.350-83,600 <1.0-12,700

NA - Data not available.

Source: Adapted from Reference No. 19.

operations such as continuous filtering in spray applications. Table 3.1.5 lists surface treatment unit operations which may generate corrosive wastes. Table 3.1.6 provides examples for general types of discharges.

Alkali cleaning is associated with lead, nickel, precious metals, titanium, refractory metals, and zirconium forming operations and cleaning lead/tin/bismuth surfaces.[19] Alkaline cleaning is performed prior to acid pickling to loosen scale (potassium permanganate and sodium hydroxide or soda ash) and for removing oil, grease, waxes, soap, or dirt from metal surfaces (sodium hydroxide, sodium carbonate, sodium metasilicate, sodium phosphates, sodium silicate and wetting agents). These wastes have cleaner concentrations of 30 to 90 g/L[25]. They have lower concentrations of basis metal relative to acid baths since alkaline wastes are not generally corrosive. However, they will tend to be higher in oil and grease as well as chlorinated organic solvents since they are used after degreasing operations to remove hydrophobic residue which interferes with subsequent surface treatment.

Sodium hydroxide is one of the most widely used paint strippers for cleaning painting equipment and salvaging ferrous metal parts with defective finishes.[25] It is used in solutions of approximately 0.4 kg/L with additives such as surfactants, sequestering agents and/or solvent blends to increase stripping effectiveness.[25] Alkalis are also used in electrocleaning (60 to 105 g/L of cleaner) and in the chemical milling of aluminum (32 to 60 g/L NaOH). The latter are discarded when sodium aluminate concentration reaches 111 to 146 g/L.[25]

Other milling baths, etchants and pickling solutions are generally comprised of acids. Traditional acid pickling solutions are based upon inhibited sulfuric, hydrochloric or phosphoric acids with certain applications for nitric, hydrofluoric and chromic acids (Table 3.1.7). Almost all metals other than aluminum are etched in acid baths which contain chlorides such as hydrochloric acid or ferric chloride. Highly concentrated solutions of sulfuric acid and/or phosphoric and/or chromic acids are frequently used for electropolishing. Conversely, brightening involves the use of more dilute solutions of oxidants such as chromic acid, nitric acid or hydrogen peroxide.[25]

TABLE 3.1.5. SURFACE TREATMENT UNIT OPERATIONS

<u>Preparative</u>
Acid or alkaline pickling and etching (of metals)
Etching (of glass)
De-rusting (of metals)
Oil and grease removal (from metals, glass and plastics)
Machining and grinding (of metals and glass)
Bonding preparation (of metals)
Chemical and electrochemical machining (of metals)
<u>Protective and Decorative</u>
Electroplating (on metals and plastics)
Anodizing (of aluminum and magnesium)
Galvanizing
Rustproofing
Enameling
Immersion coating (of metals)
Metallization (of glass)
Chemical and electrochemical polishing
Passivation (of treated metal surfaces)
Painting and lacquering
Surface hardening (of metals)
Electroforming (on metals)
Coloring and bronzing (of metals)

Source: Reference 24.

TABLE 3.1.6. CLASSIFICATION OF TYPES OF DISCHARGE FROM SURFACE TREATMENT PROCESSES

Type	Examples
Acids	HCl, H_2SO_4, H_3PO_4 (and acid phosphates), HF, H_3BO_3 (often discharged with dissolved heavy metals present).
Alkalis	NaOH, Na_2CO_3 (frequently with phosphates, silicates and detergents, and often containing oils and oil emulsions).
Heavy metals in solution	Cd, Co, Cr, Cu, Fe, Mo, Mn, Ni, Pb, Sb, Sn, Zn
Complex-forming agents	CN^-, amines, NH_3, EDTA, NTA, citrate, tartrate, oxalate, gluconate.
Organic additives	Aldehydes, ketones, alcohols, fatty and aromatic carboxylic acids, carbohydrates, sulphonic acids, dyes, phenols.
Solvents	Trichloroethylene, toluene, xylene, alcohols.
Oils, waxes and greases	Sometimes discharged with detergents in emulsified form.
Inert solids	Grinding materials (oxides, carborundum, etc.).

Source: Adapted from Reference 24.

TABLE 3.1.7. ACIDS USED IN METAL PICKLING SOLUTIONS

Basis metal	Application	Acid type (% by volume)
Ferrous metals excluding stainless steel	Removing heavy scale	Sulfuric (5 to 10%)
	Rapid removal of light scale or rust	Hydrochloric (25 to 50%)
	Removal of light scale or rust, prior to painting	Phosphoric (15% or more)
Stainless steel and high nickel alloys	Heavy scale removal	Hot sulfuric (10% with sodium thiosulfate or hydrosulfite (1 to 2%)
		Hydrofluoric (2%) with ferric chloride (6 to 8%)
	Light scale	Nitric (20%) with hydrofluoric (2 to 4%)
Cuprous metals	Rapid scale removal	Hydrochloric (25 to 100%)
	Slow scale removal	Sulfuric (5 to 10%)
	Brightening	Nitric acid modified with chromic acid
Aluminum and its alloys	Deoxidation	Sulfuric inhibited with chromic acid
		Phosphoric acid
Magnesium	Scale removal	Chromic acid Sulfuric or nitric acid with chromic acid
	Scale removal when alloy contains silica	Hydrofluoric acid
Zinc	Deoxidation	Sulfuric, hydrochloric or phosphoric acid
Galvanized steel	Deoxidation	Phosphoric acid

Source: Adapted from Reference 25.

Other metal containing acidic corrosives are generated from the manufacturing and coating (e.g., plating, electroplating, chromating, phosphating, metal coloring) of metal products. Acidic paint strippers include inhibited solutions of nitric, sulfuric and chromic acid which can be combined with alcohols and glycol ethers. Phosphoric acid is used in hot strippers for aluminum and zinc substrates. Nitric acids find use in removing very tough hard coatings while sodium hydroxide has been applied for removing polyurethane coatings.[25]

As with other highly contaminated baths, these can be recycled to recover metal, acid or alkaline constituents. The resulting treatment residuals (e.g., filters, spent adsorbents, dewatered sludges) may constitute RCRA corrosive wastes.

There is no data available which summarizes waste characteristics for the diverse industries which utilize metal finishing and cleaning processes. However, Table 3.1.8 gives an indication of the general range in corrosive waste constituent concentrations which are generated by industries involved in the manufacturing and coating of metal products. Detailed waste characteristics for these operations can be found in the literature devoted to specific industries.

3.1.5 Electric Utilities

Electric utilities (SIC 49) generate corrosive wastes primarily during boiler and preheater cleaning and water treatment operations. Alkaline boiler cleaning is used to remove oil-based compounds from fireside tubes and metallic copper deposits. Organic and inorganic acids are used to remove boiler scale (e.g., iron and copper oxides) resulting from corrosion. Of these, only spent inorganic acids tend to meet pH requirements for RCRA corrosive wastes. Hydrochloric acid is the most widely used boiler cleaning compound since sulfuric acid produces highly insoluble calcium sulfate.[27]

Typical cleaning solutions contain 5 to 10 percent HCl with inhibitors to reduce attack on boiler surfaces.[27] Copper complexers (e.g., thiorea) may also be used to prevent copper chloride from reacting with iron to form copper deposits. Noncomplexed waste stream sampling data showed an average pH of 1.1 with high chemical oxygen demand (2,867 mg/L), low suspended solids (45 mg/L) and oil and grease (15 mg/L), and high concentrations of iron (2,625 mg/L),

TABLE 3.1.8. RAW WASTE CHARACTERISTICS - VARIATIONS AMONG PLANTS INVOLVED IN THE MANUFACTURING AND COATING OF METAL PRODUCTS

Parameter[a]	Material coating		Chemical and electrochemical operations		Assembly operations		Metal forming (excluding plastics)	
	Minimum	Maximum	Minimum	Maximum	Minimum	Maximum	Minimum	Maximum
pH	1.5	11.3	0.9	7.4	1.8	11.5	1.5	12.0
Turbidity (JTU)	0.300	3800.0	1.2	500.0	2.2	3800.0	2.2	3800.0
Temperature	9.0	63.0	10.0	65.0	2.0	63.0	12.0	65.0
Dissolved oxygen	1.0	12.0	4.0	12.0	1.0	12.0	1.0	12.0
Sulfide	0.010	24.0	0.100	1.2	0.010	4.8	0.010	6.4
Cyanide	0.010	1.6	0.019	0.20	0.048	0.192	0.020	1.8
Total solids	35.0	63090.0	151.0	54210.0	165.0	28770.0	155.0	63090.0
Total suspended solids	0.200	28390.0	1.6	16560.0	6.3	28390.0	2.0	11990.0
Settleable solids	0.200	40.0	0.150	18.0	0.002	60.9	0.20	40.0
Cadmium	0.002	60.9	0.002	0.29	0.002	60.9	0.002	0.43
Chromium, total	0.005	400.0	0.005	119.1	0.005	0.026	0.005	0.417
Chromium, hexavalent	0.005	36.4	0.005	89.6	0.005	0.007	0.005	0.030
Copper	0.011	1060.0	0.017	155.4	0.013	184.6	0.016	145.0
Fluoride	0.130	110.0	0.150	7.4	0.110	325.0	0.120	1.8
Iron, total	0.103	422.2	0.023	600.0	0.070	95.4	0.110	600.0
Iron, dissolved	0.003	367.7	0.138	26.6	0.030	77.5	0.030	250.0
Lead	0.006	102.8	0.018	2.0	0.007	102.8	0.010	103.0
Oil, grease	0.500	13510.0	0.40	1730.0	0.400	13510.0	0.500	8056.0
COD	3.7	40000.0	1.0	6040.0	11.6	40000.0	1.0	19170.0
Total phosphates	0.200	62.4	0.200	17.5	0.250	62.4	0.20	45.4
Zinc	0.020	86.5	0.153	164.3	0.020	33.9	0.020	146.4
Boron	0.050	21.3	0.120	1.8	0.030	17.0	0.030	16.3
Mercury	0.002	0.055	0.008	0.008	0.002	0.055	0.002	0.002
Nickel	0.007	0.950	0.006	84.5	0.004	93.5	0.004	165.2
Silver	0.002	0.100	0.002	0.010	0.002	0.052	0.002	0.44

[a]All parameters measured in mg/L except pH, turbidity (Jackson Turbidity Units) and temperature (°C).

Source: Adapted from Reference 26.

nickel (174 mg/L), calcium (53 mg/L), zinc (43 mg/L) and copper (19 mg/L).[27] Toxic metals which were sometimes present in high enough concentration to qualify as RCRA wastes included lead (up to 5.2 mg/L) and chromium (up to 8.8 mg/L).[27] Complexed wastes showed higher levels of suspended (2,375 mg/L) and dissolved solids (30,980 mg/L), iron (4,078 mg/L), zinc (415 mg/L), copper (392 mg/L), nickel (240 mg/L), and chromium (16.8 mg/L).[27] Similar waste stream constituents are found in corrosive air-preheater cleaning wastes.[28]

Other potentially corrosive wastes from electric power generating facilities include feed water demineralizer/ion exchange regeneration waste and, in coal-fired facilities, fly ash and coal cleaning waste leachate. Ion exchange wastes are very low (H_2SO_4) or very high (NaOH) in pH, have high dissolved solids, and have copper, iron and zinc levels of 20 to 200 mg/L.[28] One sample of fly ash transport water showed a high pH with high levels of calcium (99 mg/L), but low levels of toxic metals.[28] Leachate from coal cleaning wastes can have pH of 2.0 or less and contains high concentrations of iron (3,310 mg/L), aluminum (370 mg/L), calcium (350 mg/L), magnesium (54 mg/L) and zinc (16 /mg/L), but does not exhibit the characteristics of EP toxicity for heavy metals.[29]

3.1.6 Other Industries Which Generate Corrosive Wastes

The textile industry consumes significant quantities of corrosives (e.g., sodium hydroxide with smaller amounts of acetic acid) in textile treating and finishing operations.[1] Sodium hydroxide is used in cotton mercerizing (15 to 30 percent solution) and scouring (1 to 5 percent) and is frequently recovered for reuse.[20]

The paper industry consumes large quantities of sodium hydroxide and lesser amounts of sulfuric acid and sodium carbonate.[1] Sodium hydroxide is used in chemical pulping (e.g., kraft sulfate process) in which wood chips are cooked in digesters in highly alkaline solutions. It is also used to recycle waste paper and for pulp bleaching.[1] Some of these uses are non consumptive and result in the generation of corrosive wastes.

Other miscellaneous industries which generate corrosives include dye works and printing facilities which generate spent process solutions or equipment cleaning wastes.[20] The largest source of waste corrosives from small quantity generators is vehicle maintenance shops which dispose of used batteries.[30] Much of these wastes are recycled by secondary lead smelters.[8]

3.2 RCRA CORROSIVE WASTE GENERATION AND MANAGEMENT

The most recent, comprehensive source of data regarding corrosive hazardous waste generation and management is a study performed by Camp, Dresser and McKee, Inc. (September, 1984) for the USEPA.[3] This study was based on the results of the National Survey of Hazardous Waste Generators and Treatment, Storage and Disposal Facilities (TSDFs) Regulated Under RCRA in 1981. Details on the survey methodology and limitations can be found in the reference.[31] The CDM analysis differed from the original survey results in that an adjustment was made for a large volume (8.4 billion gallons) waste stream which was found to be exempt from regulation under RCRA. Highlights of the survey results are as follows:[3]

- The total quantity of corrosive waste (D002 and K062) generated in the United States in 1981 was 21.8 to 25.6 billion gallons. This represents approximately 40 percent of all hazardous waste generation. However, it includes mixtures of corrosive and non-corrosive wastes which significantly inflates the total estimate.[31]
- The number of generators of D002 corrosive wastes was 4,705 which represents over one-third of all RCRA waste generators. These wastes were handled at 513 TSDFs in 1981. In addition, 64 TSDFs handled K062 corrosive wastes.
- The total quantity of corrosive waste that was reportedly land disposed, and thus affected by the proposed land disposal ban, was 3.6 to 4.2 billion gallons.
- Nearly 95 percent of land disposed corrosives were liquid acidic wastes.
- Disposal by deep-well injection accounted for 87 percent of land disposed corrosives with another 6 percent disposed in surface impoundments and 5 percent in landfills.

3.2.1 Corrosive Waste Generation Estimates

As indicated, CDM presented a range for waste quantity generated and managed at TSDFs. The upper bound of corrosive waste generation was determined by summing the quantity of onsite generation at TSDFs, as reported in the National Survey TSDF Questionnaire, and total generation at facilities without onsite TSDFs, as reported in the Generator Questionnaire.[31] This quantity exceeded total waste quantity which had an identified management practice by a factor of 1.235. Thus, this was used as a scale-up factor to estimate maximum quantities handled by various management practices.

As a result of misinterpretation of the survey questionnaire, this procedure resulted in an overestimation of hazardous waste generation, particularly for corrosives.[31] Most significantly, waste stocks carried over from storage in 1980 and wastes handled in RCRA-exempt processes were inadvertently included in the data. Exempt processes include discharges to POTWs, wastes treated exclusively in tanks and discharged under NPDES permits, and wastes which are 100 percent recycled. A significant, but unknown, fraction of corrosives are handled in these ways,[14] and thus may have inflated the waste quantity estimates.

The lower bound was taken as the summation of the highest quantity reported for specific waste codes at individual facilities in either treatment, storage, or disposal processes. This procedure eliminated possible double counting for waste volumes reported under multiple management methods. However, at the same time, it excluded fractions of some waste streams which were legitimately handled in lower volumes via other management methods. The resulting quantity exceeded total quantity with known management practices by a factor of 1.05. Consequently, this was used as the scale-up factor for generating minimum waste quantity estimates.[3] However, this data was still inflated, but to a lesser extent, due to the inclusion of exempt wastes.

It becomes clear that the majority of corrosive wastes reported in the National Survey must be dilute aqueous streams when waste generation quantities are compared to acid/alkali production figures (Table 3.1.1). Total production is approximately 19 billion gallons[1,2] (100 percent acid/alkali basis) versus waste generation estimates of 21.8 to 25.6 billion gallons.[3] These figures correspond roughly to the following waste profile.

Five percent of acid/alkali (e.g., nonconsumptive applications) results in the generation of concentrated waste (10 percent concentration) at a rate of 0.1 gallons of waste produced per gallon of acid/alkali consumed. The remaining 95 percent (e.g., consumptive applications) results in 0.01 gallons of waste (1 percent concentration) generated per gallon consumed. Acid consumption[1] and waste generation in steel pickling (K062)[3,21] corresponds well to the concentrated waste profile described above. However, there are no available data with which the dilute waste profile can be compared.

Two other surveys which were national in scope support the waste quantity estimates derived in the EPA National Survey. The U.S. Congressional Budget Office collected waste quantity data through a survey of TSDF and industrial facility manifests. These data were extrapolated to national totals for each SIC code based on percentage of employees in the surveyed sample. The CBO estimated that 27 billion gallons of acidic and basic waste are generated.[32] This accounted for 38 percent of all hazardous waste generation reported in the survey (versus 40 percent reported in the National Survey).

Another source, based on a survey of waste generation in 21 states, concluded that acid/alkali wastes (liquids only) accounted for 32 percent of all hazardous waste generated.[33] Spent pickle liquor generation was estimated to be nearly 1.4 billion gallons (36 percent H_2SO_4, 58 percent HCl, 6 percent combination acid). These data also compare favorably with the National Survey results.

These waste quantity estimates must be interpreted in terms of the types of wastes included in the analyses. These surveys were designed to assess waste quantity handled by the facility as opposed to waste quantity initially produced at the source; i.e., prior to subsequent mixing or treatment.[31] Thus, waste quantity estimates were significantly inflated as a result of the mixture rule (40 CFR 261.3(a)(2)(iii)); i.e., the largest volume of diluted corrosive wastes which still met the definition of corrosivity (40 CFR 261.22) could have been reported. As a basis of comparison, the ISDB survey of the Chemicals and Allied Products Industry (SIC 28) projected a corrosive waste generation of 9.8 billion gallons/year[4] versus 15.5 to 18.3 billion gallons as estimated by the National Survey.[3] The ISDB survey was specifically designed to assess waste quantities and characteristics at the point of

TABLE 3.2.1. MANAGEMENT PRACTICE SUMMARY FOR CORROSIVE WASTES NATIONAL ESTIMATES (Million Gallons)

	High quantity estimate[a]			Low quantity estimate[b]		
	Corrosive waste D002	Spent pickle liquor K062	Total	Corrosive waste D002	Spent pickle liquor K062	Total
Handled[c]	24,596	1,048	25,644	20,912	891	21,803
Disposed[c]	3,970	236	4,206	3,375	200	3,576
Injection well	3,635	28	3,663	3,090	24	3,114
Landfill	85	131	217	73	112	184
Land treatment	18	-	18	15	-	15
Surface impoundment	206	56	262	175	48	223
Other	26	20	46	22	17	39
Treated[c]	16,127	220	16,347	13,711	187	13,898
Tanks	7,040	139	7,180	5,986	118	6,104
Surface impoundment	5,614	39	5,653	4,773	33	4,806
Incineration	6	-	6	5	-	5
Other	3,252	40	3,292	2,764	34	2,799
Stored[c]	10,122	320	10,441	8,605	272	8,877
Tanks	1,542	211	1,754	1,311	180	1,491
Containers	9	-	9	8	-	8
Surface impoundment	6,530	47	6,577	5,552	40	5,592
Waste piles	6	7	13	5	6	11
Other	2,007	-	2,007	1,706	-	1,706
Recycled[d]	373	354	727	317	301	618
Onsite:	330	34	364	280	30	310
Generator	42	6	48	35	6	41
TSDF	288	28	316	245	24	269
Offsite:	43	320	363	36	272	308
Generator	14	170	184	11	143	154
TSDF	29	150	179	25	129	154

[a]1.235 x base data.

[b]1.05 x base data.

[c]Source of base data: Reference 3.

[d]Source of base data: Reference 35.

production, thereby excluding the effects of subsequent mixing or treatment.[32] As a result, the waste generation estimate is 37 to 46 percent lower than that given in the National Survey.

3.2.2 Corrosive Waste Management Practices

High and low waste quantity estimates handled in different management practices are summarized in Table 3.2.1, as extrapolated from the National Survey.[3,35] However, since the data includes mixtures of wastes, actual quantities may be less, particularly for wastewater handling processes in which mixing was most likely to occur. For example, surface impoundment and tank quantities are expected to be more inflated than land disposed quantities, since waste segregation is more likely to be practiced in the latter.

This assertion is supported by comparison of the data with the ISDB.[14] Overall, the Chemical and Allied Products Industry accounts for 71 percent of corrosive waste generation. The ISDB, when extrapolated to national totals (factor of 2.1), shows that the chemical industry accounts for virtually all of the deep well injected corrosives. Thus, these wastes are probably infrequently mixed with other (e.g., nonhazardous) waste streams prior to disposal. The data also suggest that the National Survey underestimated recycled quantities of D002 by nearly a factor of three (6 percent of waste generation versus 2 percent in the National Survey). This is not surprising since the National Survey was not designed to assess recycling, particularly methods which may be exempt from RCRA regulations.[31] Finally, the ISDB also identified higher quantities of waste being incinerated although this management practice still remains infrequently used.

The National Survey understated actual recycling (particularly onsite recycling) since many of these activities are exempt from regulation. Total quantity recycled was estimated to be 618 to 727 million gallons in 1981 (Table 3.2.1). Of this, 48.7 percent was spent pickle liquor.[20] Although pickle liquor (K062) accounts for only 4.1 percent of total corrosive waste generation,[3] it is frequently recovered since it is concentrated (5 to 15 percent) and used in high volumes.[21] Roughly ten times as high a percentage of pickle liquor is recycled as compared to D002 wastes.[3]

Wastes accepted for offsite recycling tend to be highly concentrated (i.e., greater than 10 percent) and are generally not accepted if they contain high levels of contaminants such as toxic metals.[36] Conversely, wastes recycled onsite are not restricted by contaminant concentration unless it interferes with the recovery process. Onsite recycling has been shown to be economical for acid/alkali concentrations of lower concentrations (Section 5.0).

One source estimated offsite shipments of sulfuric acid destined for recycling was 527 million gallons in 1981 (743 MG in 1982), 60 percent of which was generated by the petroleum industry.[2] The remainder originated from steel picklers, soap and detergent producers, industrial chemicals, and phosphate fertilizer manufacturers. In addition, large quantities of hydrochloric and sulfuric acid pickle liquors, corrosive textile finishing solutions, pesticide wastes, and metal plating baths are currently recycled in onsite facilities (Sections 3.1 and 5.0). The most frequently recovered waste constituents in the chemical industry are sulfuric acid, hydrochloric acid, caustic soda and sodium carbonate.[14] Out of 29 recovered liquid wastes reporting concentration data, all reported acid/alkali concentrations in excess of 10 percent.[14]

3.2.3 Corrosive Waste Generation by Industrial Classification (SIC)

Table 3.2.2 summarizes National Survey waste quantities handled and recycled by various industrial classification codes. Table 3.2.3 shows the number of facilities handling and disposing corrosives. As shown, the Chemical and Allied Products Industry accounts for 71.5 percent of the waste quantity. This industry tends to generate a high fraction of its waste in the form of large-volume aqueous streams relative to other industries. Electric, Gas, and Sanitary Services is the second largest waste generator (9 percent), followed by the Primary Metal Industries, Petroleum Refining, and Paper and Allied Products, each with approximately 4.5 percent of the total. The remaining corrosive waste is primarily generated in other metal-related industries.

The data shown previously in Tables 3.2.2, 3.2.3, and 3.1.1 permit comparisons to be made between corrosive raw material consumption, waste generation, and waste management practices. The chemical industry accounts

TABLE 3.2.2. CORROSIVE WASTE QUANTITY HANDLED AND RECYCLED BY INDUSTRIAL CLASSIFICATION (Million gallons/year)

SIC code	Industry description	Waste quantity handled[a] (Million gallons/year) High	Low	Percent	Waste recycled[b] (Million gallons/year) High	Low	Percent	Percent onsite
28	Chemicals and allied products	18,337	15,590	71.5	377	320	45.4	95
49	Electric, gas and sanitary services	2,305	1,960	9.0	0.2	0.2	-	0
29	Petroleum refining	1,150	978	4.5	11	9	1.3	6
33	Primary metals	1,143	972	4.5	350	297	42.2	87
26	Paper and allied products	1,126	957	4.4	26	22	3.1	97
36	Electric and electronic machinery, equipment and supplies	581	495	2.3	14	12	1.6	0
35	Machinery, except electrical	417	354	1.6	25	21	3.0	98
32	Stone, clay, glass, concrete	190	162	0.7	2	2	0.2	100
34	Fabricated metals	183	156	0.7	9	7	1.0	59
37	Transportation equipment	136	115	0.5	14	12	1.6	0
20	Food and kindred products	27	23	0.1	0.1	0.1	-	0
42	Motor freight, transportation, warehousing	10	8	-	0.0	0.0	-	NA
11	Agricultural production - crops	7	6	-	0.0	0.0	-	NA
50	Wholesale trade - durable goods	6	5	-	0.1	0.1	-	100
95	Administration of environmental quality programs	2	2	-	0.0	0.0	-	NA
97	National security and international affairs	2	1	-	0.9	0.7	0.1	75
-	Other industries	23	20	0.1	1.3	1.2	0.2	30
	Total:	25,645	21,803	100.0	829	705	100.0	56

[a]Includes D002 and K062 only.
Source: National Survey, Reference 3.

[b]Includes all corrosive wastes of which D002 and K062 represented 87.7 percent.
Source: National Survey, Reference 20.

TABLE 3.2.3. NUMBER OF FACILITIES HANDLING AND DISPOSING CORROSIVE WASTE BY INDUSTRIAL CLASSIFICATION (million gallons/year)

SIC code	Industry description	No. of facilities handling corrosive waste[a]			No. of facilities disposing corrosive waste		
		High	Low	Percent	High	Low	Percent
28	Chemicals and allied products	503	427	26.7	58	49	34.1
49	Electric, gas and sanitary	141	120	7.5	35	29	20.4
29	Petroleum refining	80	68	4.3	15	13	8.7
33	Primary metals	136	116	7.2	24	20	13.8
26	Paper and allied products	17	15	0.9	-	-	-
36	Electric and electronic machinery, equipment and supplies	290	247	15.4	2	2	1.4
35	Machinery, except electrical	98	83	5.2	-	-	-
32	Stone, clay, glass, concrete	17	15	0.9	1	1	0.7
34	Fabricated metals	178	151	9.5	15	13	8.7
37	Transportation equipment	122	104	6.5	1	1	0.7
20	Food and kindred products	5	4	0.3	-	-	-
42	Motor freight, transportation, warehousing	25	21	1.3	1	1	0.7
11	Agricultural production - crops	1	1	0.1	-	-	-
50	Wholesale trade - durable goods	6	5	0.3	-	-	-
95	Administration of environmental quality programs	2	2	0.1	-	-	-
97	National security & international affairs	52	44	2.7	2	2	01.4
-	Other industries	207	173	11.0	16	14	9.4
	Total:	1,880	1,596	100.0	170	145	100.0

[a]Includes D002 and K062 only.

Source: Adapted from Reference 3.

for nearly 80 percent of known acid/alkali consumption[1,2] and generates nearly 72 percent of the total corrosive waste volume.[3] However, it only accounts for 34 percent of waste disposal[3] and 45 percent of total recycling.[20] Thus, most corrosives in SIC 28 are large volume, dilute wastewater streams which are treated onsite. Of that which is eventually disposed, over 99 percent is deep-well injected and the remainder is landfilled.[14] Organic chemicals industries generate the majority of SIC 28 wastes (86.6 percent) followed by plastics and resins (5.8 percent), pesticides (5.3 percent), dyes and pigments (2.1 percent), and inorganic chemicals industries (0.2 percent).[4,34]

Primary metal and metal-related industries (SIC 33 through 37) account for less than 3 percent of strong acid/alkali use,[1,2] but these are primarily applied in concentrated form in nonconsumptive applications such as metal cleaning. As a result, these industries account for a disproportionate amount of waste generation (9.6 percent),[3] waste disposal (24.6 percent),[3] and waste recycling (49.5 percent).[20] The metals industry also accounts for the largest percentage of waste generators (44 percent).[3] Thus, these facilities tend to generate significantly smaller volume waste streams than the chemical industry. These wastes also tend to be more highly concentrated and are more frequently disposed in landfills. In addition to neutralization, many of these will require treatment to remove or immobilize heavy metals when the land disposal restrictions become effective.

The electric and gas industries (SIC 49) consume less than 3 percent of all corrosive reagents for water and stack gas treatment and boiler cleanout.[1,2] The latter results in a concentrated, metals contaminated waste stream which is infrequently recycled. As a result, these industries account for a disproportionately high percentage of waste volume generation (9.0 percent)[3] and facilities practicing waste disposal (20.4 percent)[3]. In addition, they account for very little corrosive waste recycling (less than 0.1 percent).[20] Many of these disposed wastes will require treatment for both neutralization and metals removal to comply with proposed land disposal restrictions (see Section 3.1.5).

The petroleum refinery industry accounts for 5.1 percent of known corrosive raw material consumption,[1,2] and generates 4.5 percent of corrosive wastes.[3] The National Survey suggests that this industry performs little onsite recycling. However, other data indicates a large quantity of spent sulfuric acid (330 million gallons in 1980, 440 million gallons in 1982) is shipped offsite for reclaiming.[1] This is spent sulfuric alkylation acids which tend to be highly concentrated (95 percent acid or more).[36]

Finally, the pulp and paper industry (SIC 26) consumes 3.8 percent of known acid/alkali consumption,[1,2] and generates 4.4 percent of the total waste.[3] However, almost none of this waste is land disposed.[3] Instead, it is handled in onsite wastewater treatment facilities or recycled. Other industries which use large quantities of corrosives, but generate little waste, are the glass manufacturing industry and food products. Together, these industries consume 7.2 percent of known end uses,[1,2] but generate less than 1 percent of total corrosive wastes.[3] Acids and alkalis in these industries are largely used in consumptive applications.

3.2.4 Recent Changes in Corrosive Waste Generation

Total waste generation estimates may have changed from the 1981 estimates provided by the National Survey. Since that time, corrosive raw material consumption has been essentially stagnant,[1] however, small quantity generators (100 to 1,000 kg/month) have been included under RCRA regulations, waste management practices have shifted toward increased compliance, waste minimization and recycling, and wastes with free liquids have been banned from landfills.

A survey of the eight largest commercial hazardous waste disposal firms by ICF[37] showed a 40 percent increase in landfilling between 1981 and 1984. However, much of this increase was due to disposal of bulk liquids in advance of the ban on these wastes and the disposal of remedial action clean-ups. It is unlikely that currently landfilled quantities of corrosives exceed levels reported in the National Survey. Instead, solidification and stabilization requirements and increased transport and disposal costs would have shifted management practices to increased volume reduction prior to disposal.

Deep well injection at the surveyed commercial firms showed a 24 percent decrease while chemical treatment increased 33 percent between 1981 and 1984.[37] Corrosive waste management practices have probably shifted in similar fashion. Additionally, recycling practices have become more widespread in response to increased disposal costs. However, comprehensive data does not exist which would allow the extent of this practice to be quantified.

On August 1, 1985, the EPA proposed lowering the small quantity generator exclusion limit from 1,000 to 100 kg/month.[38] The effect of this regulatory change (effective 22 September, 1986) will be to add 15.4 million gallons/year of corrosive wastes to that already managed under RCRA.[30] Of this, approximately 52 percent is used lead-acid batteries. Another 72.3 million gallons of spent batteries (90 percent of the total) are currently reclaimed (e.g., by secondary lead smelters) and are thus exempt from regulation under RCRA.[30] It is likely that much of the currently disposed batteries will also be reclaimed when these generators become subject to regulation. The remaining corrosives are primarily small volume streams, a high percentage of which will probably be landfilled. Overall, small quantity generators will account for no more than a 5 percent increase in the quantity of corrosive wastes which are landfilled.

In summary, corrosive waste generation subject to RCRA-handling requirements has probably not changed significantly since the 1981 National Survey. It is likely that waste quantity affected by the land disposal ban has decreased somewhat, primarily due to the decrease in waste volume which is deep well injected. Increased waste minimization efforts (e.g., waste segregation) and recycling have probably also contributed to a decrease in the quantity land disposed. For example, offsite sulfuric acid recovery reportedly increased 35 percent from 1980 (2.3 million tons) to 1982 (3.1 million tons). The Congressional Budget Office projected an overall increase in waste reduction of 20 percent from 1985 to 1990 for nonmetallic inorganic liquids and 40 percent for metal-containing liquids.[32] Thus, with stagnant demand for corrosive raw materials, it is likely that corrosive waste quantity will decline in the next few years.

3.2.5 Corrosive Waste Characteristics Summary

Waste characterization data for corrosive wastes is limited. CDM presented physical profile data for disposed corrosives, as summarized below.[3] Other sources are limited to specific waste management practices or industries, and thus cannot be extended to include the universe of corrosive wastes.

Available data indicate that, overall, acidic wastes are generated in significantly higher quantity than alkaline wastes. A survey of waste generation in 21 states estimated that 65 percent of all corrosives generated were acidic.[33] This estimate is consistent with acid/base consumption figures which show 68 percent represented by acids (Table 3.1.1). Other data show that the majority of corrosives (80.6 percent) do not contain heavy metals.[32] In particular, only 3 percent of the waste streams reported in the chemical industry contain heavy metals.[14] However, since the chemical industry generates 72 percent of all corrosives,[3] this suggests that over 40 percent of other industry wastes contain metals.

Waste characterization data for land disposed corrosives which will be affected by the land disposal ban, is also limited. The National Survey data base provided waste characteristic information for 14 percent of land disposed D002 wastes. Of this, 92.5 percent was characterized as liquid, 7.4 percent were sludges, and only 0.1 percent were solids. The majority of wastes were acidic (82 percent), inorganic (82 percent), and characterized as dilute (liquids only, 94.3 percent). Solids, sludges, and concentrated wastes had a significantly higher tendency to be characterized as organic relative to combined wastes (48, 93, and 31 percent, respectively). The most predominant sludges were waste alkylation acid and wastewater treatment sludge, while solids were dominated by spent catalysts, adsorbents, and filter residue. Other waste characterization data generated from the survey was based on a very limited number of responses, and thus cannot be used to identify overall trends in land disposal. However, the data presented above show good agreement with the ISDB.

Other data sources are either not specific to land disposal or do not cover the entire range of waste generating industries. A review of waste manifests by the State of California showed 38 percent of landfilled corrosives containing metals at levels which exceeded proposed land disposal ban treatment standards.[36] The most common toxic metals in decreasing order were Cr, Ar, Ni, Pb, and Cd. The National Survey data showed 60 percent of landfilled waste was spent pickle liquor (K062) which would have characteristics similar to those presented previously in Table 3.1.3. Of the remainder, only wastes from the chemical industries have been characterized. These are described below.

The ISDB summarized characteristics of landfilled wastes from SIC 28.[14] The most common constituents were water, caustic soda, sulfuric acid, calcium chloride, sodium chloride, and phosphoric acid, often present in concentrations ranging up to 75 percent. Sludges and solids were reported as frequently as liquids, but accounted for only 8.0 percent of the total waste volume. These wastes included spent adsorbents (e.g., activated carbon), filters or filter acid (e.g., diatomaceous earth), and alumina. Very few wastes contained oil or metals.

Wastes which are deep well injected originated almost exclusively from the chemical industries, and thus are well characterized by the ISDB.[14] Of these, 75.9 percent of the total volume was characterized as aqueous liquids, 16.6 percent as sludge, and 7.5 percent as organic liquid. Other data describing constituent types and concentrations were not available at the time this document was published.

Other disposed methods for which waste characterization data are available include incineration and use as a fuel. Incineration is infrequently applied to corrosive wastes. The most commonly incinerated corrosives are organic acids or wastes contaminated with solvents.[14] Data also show some incineration of concentrated inorganic acids such as sulfuric (up to 85 percent concentration) and hydrochloric (up to 30 percent). However, these account for only a small fraction of incinerated corrosives. Similarly, few corrosives are disposed through fuel blending in boilers. Those that are managed in this manner contain high organic concentrations (25 percent or more) of constituents such as adipic, acetic, butyric, propionic and formic acids as well as other organics.[14]

REFERENCES

1. Mansville Chemical Products Corporation, Cortland, N.Y. Chemical Products Synopsis. 1983.

2. Chemical Marketing Reporter. Chemical Profiles. Schnell Publishing Company. 1982 through 1986.

3. Camp, Dresser & McKee, Inc. Technical Assessment of Treatment Alternatives for Wastes Containing Corrosives. Prepared for U.S. EPA under Contract No. 68-01-6403. September 1984.

4. Huppert, M. Science Applications International Corporation. Telephone conversation with M. Breton, GCA Technology Division, Inc., regarding the Industry Studies Data Base. September 1986.

5. U.S. EPA Supplement for Pretreatment to the Development Document for the Inorganic Chemicals Manufacturing Point Source Category. U.S. EPA 440/1-77/087A. July 1977.

6. U.S. Environmental Protection Agency. Treatability Studies for the Inorganic Chemicals Manufacturing Point Source Category. U.S. EPA Effluent Guidelines Division. EPA 440/1-80/103. July 1980.

7. Sittig, M. Fertilizer Industry Processes, Pollution Control, and Energy Conservation. Noyes Data Corp., Park Ridge, N.J. 1979.

8. Coleman, R.T., Radian Corp. Sources and Treatment of Wastewater in the Nonferrous Metals Industry. Prepared for USEPA IERL under Contract No. 68-02-2608. EPA-600/2-80-074. April 1980.

9. Mooney, G.A., CH2M Hill. Two-Stage Lime Treatment in Practice. Environmental Progress, Vol. 1, No. 4. November 1982.

10. Ricker, N.L., and C.J. King. Solvent Extraction of Wastewaters from Acetic Acid Manufacture. Prepared for USEPA ORD, EPA-600/2-80-064. April 1980.

11. Federal Register. 40 CFR Part 261- Hazardous Waste Management System; Identification and Listing of Hazardous Waste. Federal Register Vol. 49, No. 90, pg. 19608. May 8, 1984.

12. U.S. EPA Development Document for Expanded Best Practicable Control Technology, Best Conventional Pollutant Control Technology, Best Available Technology, New Source Performance Technology, and Pretreatment Technology in the Pesticide Chemicals Industry. U.S. EPA Effluent Guidelines Division. EPA-440/1-82-079B. November 1982.

13. Shelby, S.E., and R.W. McCollum. A Case Study for the Treatment of Explosives Wastewater from an Army Ammunitions Plant. Presented in the 39th Industrial Waste Conference Proceedings. Purdue University, West Lafayette, IN, Ann Arbor Books. May 8, 9, 10, 1984.

14. Science Applications International Corporation. Industry Studies Data Base. August 1985.

15. PEDCo Environmental, Inc., Arlington, TX. Petroleum Refinery Enforcement Manual. Prepared for U.S. EPA Division of Stationary Source Enforcement under Contract No. 68-01-4147. EPA-340/1-80-008. March 1980.

16. Bider, W.L., and R. G. Hunt. Industrial Resource Recovery Practices: Petroleum Refineries and Related Industries (SIC 29). Prepared for U.S. EPA OSW under Contract No. 68-01-6000. June 1982.

17. Booze, Allen & Hamilton, Inc. Enhanced Utilization of Used Lubricating Oil Recycling Process By-Products. Performed for U.S. DOE/BETC under Contract No. DE-AC19-79BC10059. June 1981.

18. Franklin Associates, Ltd. Industrial Resource Recovery Practices: Metal Smelting and Refining (SIC 33). Prepared for U.S. EPA OSW under Contract No. 68-01-6000. January 1983.

19. U.S. EPA Development Document for Effluent Limitations Guidelines and Standards for the Nonferrous Metals Forming and Iron and Steel, Copper Forming, Aluminum Metal Powder Production and Powder Metallurgy Point Source Category. U.S. EPA Effluent Guidelines Division. EPA-440/1-84/019B. February 1984.

20. Versar, Inc. National Profiles Report for Recycling/A Preliminary Assessment. Prepared for U.S. EPA Waste Treatment Branch under Contract No. 68-01-7053. July 1985.

21. U.S. EPA Development Document for Effluent Limitations Guidelines and Standards for the Iron and Steel Manufacturing Point Source Category. Vol. V. U.S. EPA Effluent Guidelines Division. EPA-440/1-82-024. May 1982.

22. U.S. EPA Development Document for Proposed Effluent Limitations Guidelines and Standards for the Iron and Steel Manufacturing Point Source Category. Volume 1. U.S. EPA Effluent Guidelines Division EPA-440/1-79-024a. October 1979.

23. U.S. EPA Development Document for Effluent Limitations Guidelines and Standards for the Iron and Steel Manufacturing Point Source Category. Vol. VI. U.S. EPA Effluent Guidelines Division. EPA-440/1-82-024. May 1982.

24. Environmental Protection Service, Fisheries and Environment Canada. Proceedings Technology Transfer Seminar on Waste Handling, Disposal and Recovery in the Metal Finishing Industry. Toronto, Ontario, November 12-13, 1975. Report No. EPS 3-WP-77-3. March 1977.

25. Kirk-Othmer Encyclopedia of Chemical Technology. John Wiley & Sons, New York, N.Y. Third Edition. 1978.

26. U.S. EPA. Controlling Pollution from the Manufacturing and Coating of Metal Products. Water Pollution Control. U.S. EPA Environmental Research Information Center. EPA-625/3-77-009. May 1977.

27. U.S. EPA. Development Document for Effluent Limitations Guidelines and Standards, and Pretreatment Standards for the Steam Electric Point Source Category. U.S. EPA Effluent Guidelines Division. EPA-440/1-82-029. November 1982.

28. Benjamin, M.M. et al. Removal of Toxic Metals from Power Generation Waste Streams by Adsorption and Coprecipitation. Journal of Water Pollution Control Federation, Vol. 54, No. 11. November 1982.

29. Wanger, L.E., and J.M. Williams. Control by Alkaline Neutralization of Trace Elements in Acidic Coal Cleaning Waste Leachates. Journal of Water Pollution Control Federation, Vol. 54, No. 9. September 1982.

30. Ruder, E., et al., ABT Associates. National Small Quantity Generator Survey. EPA-530/SW-85/004, U.S. EPA, Washington, D.C. February 1985.

31. Deitz, S., et al., Westat, Inc. National Survey of Hazardous Waste Generators and Treatment, Storage, and Disposal Facilities Regulated Under RCRA in 1981. Rockville, MD. U.S. EPA/OSW. April 1984.

32. U.S. Congressional Budget Office. Hazardous Waste Management - Recent Changes and Policy Alternatives. CBO Congress of the United States. May 1985.

33. Noll, K.E. et al. Recovery, Recycle and Reuse of Industrial Wastes. Lewis Publishers, Inc., Chelsea, MI. 1985.

34. U.S. EPA. Report to Congress on the Discharge of Hazardous Wastes to Publically-Owned Treatment Works. U.S. EPA Office of Water Regulations and Standards. EPA-530/SW-86/004. February 1986.

35. DPRA, Inc. Written Communication to M. Arienti, GCA Technology Division, Inc., regarding analysis of Recycling Data from the National Survey Data Base. Data Request No. M850415W. June 10, 1985.

36. Radimsky, J. et al. Recycling and/or Treatment Capacity for Hazardous Wastes Containing Dissolved Metals and Strong Acids. California Department of Health Services. October 1983.

37. ICF, Inc. Survey of Selected Firms in the Commercial Hazardous Waste Management Industry: 1984 Update. Prepared for U.S. EPA Office of Policy Analysis. September 1985.

38. Federal Register. 40 CFR Part 261 (50 FR 31278). August 1, 1985.

4. Neutralization Treatment Technologies

All neutralization processes operate under the same fundamental chemical principles and utilize similar types of equipment and process configurations. Additionally, pretreatment requirements and residual post-treatment options are comparable, regardless of the specific neutralization method under investigation. Therefore, in an effort to minimize redundancy, similar aspects of neutralization systems are addressed prior to discussion of specific reagent/waste combinations. Section 4.1 serves as introduction to the basic theory of acid-base chemistry and proceeds to identify considerations in reagent selection, pretreatment requirements, neutralization equipment, process configurations, post-treatment and disposal of residuals. The remaining subsections (Sections 4.2 through 4.7) are organized according to specific neutralization reagents. These highlight the unique aspects of each, including compatable waste types, treatment costs, sludge generation and special considerations in equipment design and reagent handling practices. The reagents include:

- Other acid/alkali wastes;
- Limestone;
- Lime slurry (lime, waste carbide lime, cement kiln dust);
- Caustic soda (sodium hydroxide, sodium carbonate);
- Mineral acids (hydrochloric and sulfuric acids); and
- Carbonic acid (carbon dioxide, boiler flue gas, submerged combustion).

Each of the reagent subsections covers the following topics:

- General process description including typical operating characteristics;
- Case study data which identifies the range in potential applications, processing equipment, and system configurations;
- Capital and operating costs; and
- Status of the technology.

4.1 GENERAL CONSIDERATIONS

4.1.1 Acid Base Theory

The principle mechanism of neutralization involves the reaction between an acid and a base. An acid is any substance that dissociates in solution to produce a proton (H+), and a base is any substance that combines with or accepts a proton. Strong acids or bases are characterized by complete disassociation, while weak acids or bases will disassociate only slightly. In general, acid-base reactions form a salt and water as illustrated in the following reaction between hydrochloric acid and potassium hydroxide:

$$\underset{\text{acid}}{HCl} + \underset{\text{base}}{KOH} \longrightarrow \underset{\text{salt}}{K^{+}:Cl^{-}} + \underset{\text{water}}{H_2O} + \text{Heat} \quad (1)$$

Neutralization reactions are exothermic in nature, but with careful reagent addition and adequate mixing, the excess heat can be safely dissipated. The concentrations of H^+ and OH^- ions and the equilibrium constants are usually very small numbers, therefore, it is convenient to express them in logarithmic terms. The conventional pH scale, which represents the negative logarithm of the hydrogen ion concentration, employs numbers ranging from 0 to 14 to indicate relative acidity or alkalinity. The pH of acidic solutions is less than 7, while the pH of alkaline solutions is greater than 7.

Other means of expressing acidity or alkalinity include the basicity factor and concentration. The basicity factor is a measure of the theoretical neutralizing power of alkali reagents. It is calculated the total weight of the reagent's potential hydroxyl ions. A neutralizing value of 1.0 is assigned to pure calcium oxide (CaO) and the values for other alkali reagents are expressed in relative terms. For example, to neutralize 98 lbs of sulfuric acid, it requires either 80 lbs of sodium hydroxide or 56 lbs of quicklime (CaO) based on stoichiometry. Thus, the basicity factor of sodium hydroxide becomes 56/80 or 0.69. The alternate method of expressing pH is to report the concentration of the acid or alkali in milligrams per liter. For example acidity can be expressed as mg/L of H_2SO_4 while alkalinity can be expressed as mg/L of $CaCO_3$ (limestone).

Most neutralization applications consist of adjusting an acidic or alkaline waste stream with the appropriate reagent to a final pH of 6 to 9 which meets surface water discharge requirements established under the Clean Water Act. However, it is sometimes only necessary to adjust the pH to approximately 5 to 6 (i.e., partial neutralization) to achieve certain treatment objectives. In other applications it may be necessary to neutralize an acid to pH 9 or higher to precipitate metallic ions or to completely clarify a waste for acceptable discharge. These techniques are called under- and over-neutralization, respectively[1].

With few exceptions, reactions involving pH go to completion in a fairly short time. However, an understanding of the reactivity of neutralization reagents is necessary for proper design and sizing of equipment, particularly tankage and space. Reagent reactivity and kinetics are discussed individually in later sections to provide a more detailed characterization for specific reagent/waste combinations.

In contrast, one general process variable that affects reaction kinetics and reagent consumption is the buffer capacity (the ability of a solution to resist change in pH) of the waste stream. If a certain amount of acid (or alkali) is added to a specified volume of buffered process solution, the change in pH will be much less than if the same addition were made to a completely ionized, unbuffered solution. All solutions have some buffering capacity, but the presence of organic salts, salts of strong acids and weak bases, or salts of weak acids and strong bases will increase buffering capacity.

4.1.2 Reagent Selection

Table 4.1.1 summarizes several of the more prevalent neutralization reagents and their characteristics. The selection of the appropriate reagent for wastewater neutralization processes is site specific and dependent on the following considerations: wastewater characteristics, reagent costs and availability, speed of reaction, buffering qualities, product solubility, costs associated with reagent handling and residual quantities and characteristics. Typically, the first step in reagent selections is to characterize the wastewater. General parameters of interest include flow (rate, quantity), pH, pollutant loading, physical form of waste, and waste/reagent compatibility. These characteristics narrow the range of reagents and treatment configurations available for consideration.

Following the selection of candidate reagents, the quantity of reagent required to neutralize the waste to the desired end point must be determined. Reagent quantity is usually calculated by developing a titration curve for each candidate reagent using representative wastewater samples. Figure 4.1.1 illustrates a typical titration curve developed for the neutralization of a ferric chloride etchant using high calcium hydrated lime as an alkaline reagent.[4] These data determines the quantity of reagent required to bring the sample volume of wastewater to the desired pH.

The next step in the experimental procedure is the preparation of reaction rate curves and development of kinetic rate equations for each candidate reagent. Reagent reactivity is an important factor in determining retention time and consequently the size of the treatment facility, the final effluent quality, and the ease or difficulty of process system control. These parameters, in turn, will affect both capital and operational costs associated with the wastewater treatment system. Reaction rate curves for various quantities of residual reagent (i.e., excess above stoichiometric requirements) are determined by plotting pH as a function of time. Other variables which should be monitored include temperature rise, agitator speed, density, viscosity, color, sludge volume, and settleability.

TABLE 4.1.1. ACID/ALKALINE NEUTRALIZATION AGENT CHARACTERIZATION

Reagant	Molecular formula	Chemical name	Molecular weight	Common form and commercial strenght	Bulk density Kg/m^3	Solubility g/100 g water	Typical use	Advantages	Disadvantages	Shipment form	Equivalent basicity factor[a]	Approximate cost/ton ($)	Cost/ton basicity[b] ($)	Equivalent weight
High Calcium Limestone	$CaCO_3$	Calcium Carbonate	100.1	Powder 95% $CaCO_3$ granules	2000-2800	0.0014^{25} (@ 25°C)	Acid Neutralization	Relatively inexpensive	Contains impurities slow reacting	Bulk	0.489	6	12.27	50.5
High Calcium Hydrated Lime	$Ca(OH)_2$	Calcium Hydroxide	74.1	Powder 72-74% CaO	400-640	0.15^{30} (@ 30°C)	Acid Neutralization	Relatively inexpensive	Contains impurities slow reacting	Bulk	0.710	46	64.79	37.05
High Calcium Quicklime	CaO	Calcium Oxide	56.1	Pebble 93-98% CaO	770-1120	Converted to $Ca(OH)_2$	Acid Neutralization	Relatively inexpensive	Requires dry storage	Bulk	0.941	39	41.45	28.05
Carbon Dioxide	CO_2	Carbon Dioxide	54.0	Gas liquified under pressure	--	--	Alkali Neutralization	May be available from flue gas	--	Tank car	---	200	---	27
Dolomitic Hydrated Lime	$Ca(OH)^2$ MgO	Normal Dolomitic	114.4	Powder 46-48% CaO 33-34% MgO	416-666	0.15^{30} (@ 30°C)	Acid Neutralization	Relatively inexpensive	Less reactive than high calcium lime	Bagged	0.912	46	50.44	33.9
Dolomitic Quicklime	CaO-MgO	Calcium Magnesium	96.4	Pebble 55-58% CaO 38-41% MgO	801-1165	Converted to $Ca(OH)_2$ and $Mg(OH)_2$	Acid Neutralization	Relatively inexpensive	Less reactive than high calcium lime	Bulk	1.110	39	35.14	24.9
Soda Ash	Na_2CO_3	Sodium Carbonate	106.0	Powder 58% Na_2O	560-1041	27.6^{30} (@ 30°C)	Acid Neutralization	Highly reactive soluble	Higher costs than calcium reagants	Bulk	0.507	83	163.71	53.0
Caustic Soda	NaOH	Sodium Hydroxide	40.0	Liquid 73% NaOH	----	347^{100} (@ 100°C)	Acid Neutralization	Highly reactive, easy handling	High cost, requires heated storage	Tank car	0.687	205	298.4	40
Magnesia	MgO	Magnesium Oxide	40.31	Powder	1017.9	0.0086^{30} (@ 30°C)	Acid Neutralization	Highly reactive	High Cost	Bagged	0.929	365	392.89	20.16
Sulfuric Acid	H_2SO_4	Sulfuric Acid	98.1	Liquid 77% H_2SO_4 93% H_2SO_4	1704-1834	Complete	Alkali Neutralization	Highly reactive, inexpensive	Forms calcium sulfate sludge with calcium	Tank car	---	59 71	---	49.05
Muriatic Acid	HCL	Hydrochloric acid	36.5	Liquid 20°Be 22°Be	1157-1177	Complete	Alkali Neutralization	Highly reactive	More expensive than sulfuric acid	Tank car	---	66 74	---	36.5

[a]Reference 2

[b]Reference 3

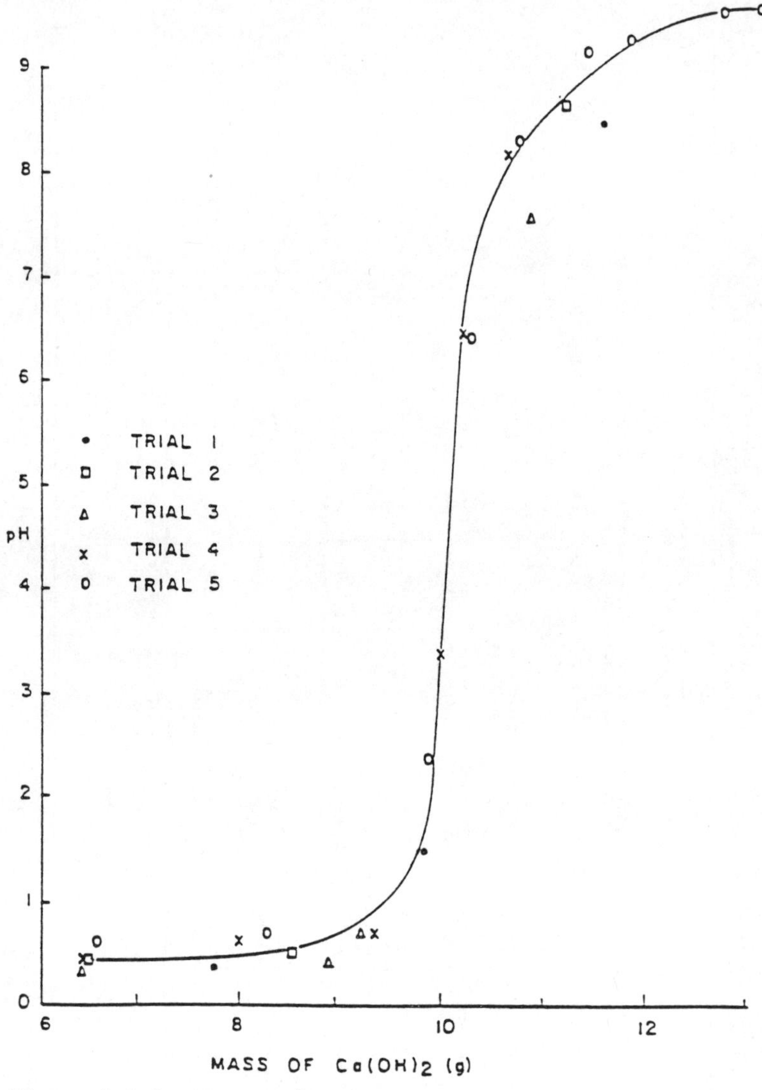

Figure 4.1.1. Neutralization of Ferric chloride etchant waste by calcium hydroxide.

Source: Reference 4

From the titration and reaction rate curves, kinetic rate equations can be developed. One simplified method numerically approximates the relation between the residual reagent (B) and time (t), as follows:

$$r = \frac{dB}{dt} = \frac{B}{t} = \frac{(B_1 - B_2)}{(t_1 - t_2)} = kB^n \qquad (2)$$

Where r is the reaction rate, k is the rate constant and n is a constant which expresses the order of the reaction (typically between 1 and 2). If the differences $(B_1 - B_2)$ and $(t_1 - t_2)$ are small enough, adequate accuracy is obtained provided that the general kinetic model assumed (i.e., $r-kB^n$) fits well with the experimental data results. More difficult, but also more accurate methods for determining the kinetic rate equation (e.g., integral or statistical analysis) are discussed in the literature.[5] Similarly, the reader is referred to standard engineering texts for reactor and costing methodologies based on flow parameters and kinetic rate equations.[6]

In the final selection, the optimal reagents and reagent/waste feed ratio will be those which incur the least overall cost, including not only the cost of the reagent itself, but also the cost of purchasing and maintaining the reagent and neutralization systems, and the costs associated with residual handling. The combination of all such factors may make a slightly more expensive reagent less expensive overall. For example, limestone is the least expensive alkali reagent available on a unit cost basis, but the added expenditure for complex grinding and feeding equipment combined with its slow reactivity and insoluble sludge products, make it the least utilized.

4.1.3 Pretreatment Requirements

Pretreatment of corrosive wastes prior to neutralization typically consists of gross solids removal (e.g., filtration), flow equalization, or treatment of individual waste streams prior to combination with other process wastes. These treatments of segregated wastes result in economic benefits from reduced reagent costs and smaller equipment sizing. Common pretreatment processes include cyanide destruction, chromium reduction, metals precipitation from highly chelated waste, and oil removal

Cyanide wastes cannot be mixed with acid due to formation of toxic hydrogen cyanide gas. Instead, cyanide is first oxidized to carbon dioxide and nitrogen gas through alkaline chlorination. In two-stage chlorination, pH is typically maintained around 11.0 in the first reaction vessel and 8.0 to 8.5 in the second vessel through addition of NaOH, as required.[7]

Chromic acid wastes contain hexavalent chromium which must be reduced to the trivalent form prior to precipitation. Reduction typically occurs at pH 2.0 to 3.0 through addition of acid (e.g., sulfuric) and a reducing agent (e.g., sulfur dioxide, ferrous sulfate, sodium meta-bisulfate, and sodium bisulfite).[8] However, alkaline reduction (pH 7 to 10) using ferrous iron has also been demonstrated. It has proven to be cost effective for highly buffered alkaline waste and the treatment of mixed metal wastes containing less than 10 mg/L of hexavalent chrome.[9]

Chelated metal-containing wastes typically require pH adjustment into the highly alkaline range in order to effectively precipitate metals. Commonly used flocculants/coagulants include aluminum sulfate, aluminum chloride, dithiocarbamate, sodium hydrosulfite, ferric chloride, ferrous sulfate, and various polyelectrolytes and anionic polymers.[7] Selection of flocculant/ coagulant and use of sodium hydroxide versus lime is dependent on the types and quantities of chelators in the waste. Common chelators include ammonia and its derivatives, phosphates, EDTA, citric acid, amines and thiourea.[7]

Removal of oil through emulsion breaking, dissolved air flotation skimming or coalescing may also be performed prior to combination of oily corrosives with other wastes. Traditionally, emulsified oils have been treated at low pH (e.g., pH of 2.0) with alum. However, this form of treatment is giving way to the use of more effective emulsion breaking coagulants such as cationic polymers and other specialty chemicals.

While in some cases filtration or segregated waste pretreatment may be utilized prior to neutralization, the most prevalent form of pretreatment is flow equalization. It is generally used in facilities which experience a wide variation in the flow or pollutant concentration of the wastewater. Figure 4.1.2 illustrates a number of ways that flow equalization can be achieved.

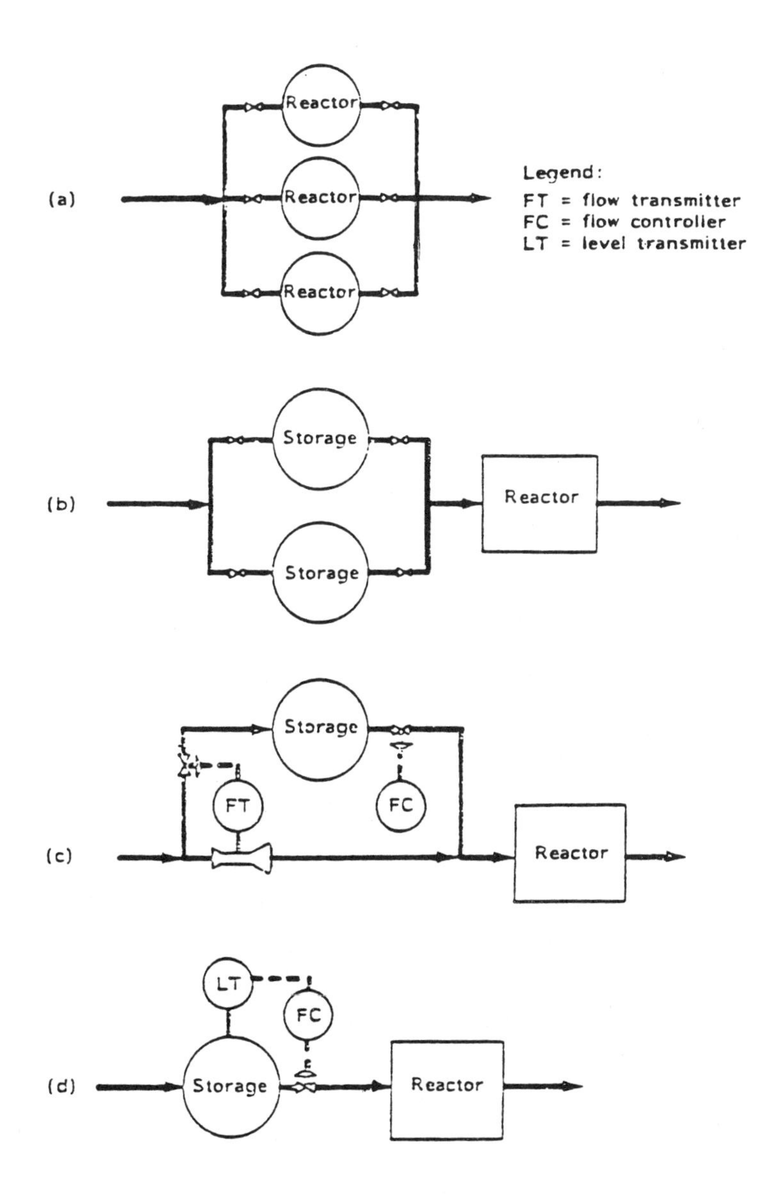

Figure 4.1.2 Alternative concepts for wastewater equalization: (a) batch reactor system, (b) batch equalization continuous processing, (c) side-stream equalization, and (d) flow-through equlization.

Source: Reference 5.

Flow equalization through batch processing involves the storage and batch neutralization of the wastewater in the same vessel. The system is designed with the reactor vessels arranged in parallel. When one vessel is full, the flow is switched to another vessel while neutralization takes place in the first. Equalization through continuous processing from batch storage involves two storage tanks arranged in parallel, operating on a fill-and-draw cycle. While one tank is being filled, the other is discharging its contents to the neutralization system. In this manner, stream segregation may be achieved and the neutralization facility receives a waste stream with uniform characteristics.

In facilities that regularly dump concentrated solutions, sidestream equalization is sometimes used. In this method, the concentrated waste or flow surge is diverted to a storage tank prior to processing. This technique is particularly useful in acid/alkali mixing or in cases where the solution must be slowly bled into the wastewater system. Finally, flow-through equalization attenuates sudden changes in the wastewater characteristics that would adversely affect process control. This can be done through any device that will provide flow resistance such as a sump, valve, weir, or surge tank.

In all methods of flow equalization care must be exercised during the wastewater analysis to completely characterize any peak flows or concentrations that might overload the system. In addition, flexibility in system design should be provided for any future expansion, change in location, or deviation in flow rates.

4.1.4 General Neutralization Processing Equipment

A wide variety of treatment options and configurations are available, however, fully-engineered component neutralization systems generally consist of the following equipment:

- Neutralization System
  - Tanks
  - Mixers
  - pH control instrumentation

- Chemical Feed System
  - Tanks
  - Mixers
  - Level instrumentation
  - Metering equipment

- Miscellaneous
  - Flow monitoring
  - Effluent pH recorder
  - Electrical and mechanical fit-up
  - Incremental engineering requirements

In addition, there is a need for facilities and equipment to collect and segregate the wastewaters, transport the wastewaters to equalization sumps, pump the wastewaters to the treatment system, perform liquid/solid separation, and convey the treated wastewaters to the point of discharge.

4.1.5 Neutralization System

Neutralization tanks are fabricated from a wide range of construction materials such as masonry, metal, plastic, or elastomers. Corrosion resistance can be enhanced with coatings or liners which prevent the premature decomposition of tank walls. For example, concrete reactors susceptible to corrosion can be installed with a two-layer coating of a 6.3 mm base surface (glass-reinforced epoxy polyamide) covered by a 1.0 mm coating of polyurethane elastomer to extend service lifetimes.[5]

Vessel geometrics can be either cubical or cylindrical in nature with agitation provided overhead in line with the vertical axis. While cubical tanks need no baffling, cylindrical vessels are typically constructed with suitable ribs to prevent swirling and maintain adequate contact between the reactants. A general rule of thumb in the design of neutralization reactors is that the depth of the liquid should be roughly equivalent to the tank diameter or width.[10]

Reactors can be arranged in either single- or multi-stage configurations and operate in either batch or continuous mode. Multi-stage continuous configurations are typically required to neutralize concentrated wastes with variable feed rates. In these units most of the reagent is added in the first vessel with only final pH adjustments (polishing) made in the remaining reaction vessels. This is particularly true when using sluggish reagents which require extensive retention time. Single-stage continuous or batch neutralization is suitable for most applications with highly buffered solutions or dilute wastewaters not subject to rapid changes in flow rate or pH.

A holdup period is required to provide time for the neutralization reaction to go to completion. This factor is especially critical where a dry feed (lime or limestone) or slurry is used as the control agent. In these systems, the solids must dissolve before they react, increasing the required holdup time and tank capacity. For example, liquid reagents used in continuous flow operations generally require 3 to 5 minutes of retention time in the first tank. Three minutes corresponds to the absolute minimum size that will not cause considerable splashing or trapping of air. In comparison, solid-based reagent systems such as lime or limestone typically require 30 and 45 minutes retention time, respectively.

Agitation serves the purpose of equalizing the hydrogen or hydroxide concentration profile within the reaction vessel as the influent is dispersed in the tank. Vessels with large stagnant areas provide little mixing between reactants and causes large disturbances when concentrated materials are released into the system. For accurate pH control, Hoyle has suggested that agitator capacity should be measured as a ratio of the system dead time (the interval between the addition of a reagent and the first observable pH change) to the retention time (volume of the vessel divided by the flow through the vessel). A ratio of dead time to retention time of 0.05 approaches an optimum value.

The pH control systems for batch neutralization processes can be quite simple with only on-off control provided via solenoid or air activated valves. Control system designs for continuous flow neutralization systems are more complicated because the wastewater feeds often fluctuate in both flow and concentration. Systems currently available include; proportional, cascade, feedforward, or feedback pH control. Each system has distinct advantages and disadvantages which are discussed in detail in the literature.[10,11,12,13] The pH control equipment usually consists of a pH probe, monitor, and recorder. In addition, there is typically a control panel with an indicator, starters and controls for metering pumps, all relays, high/low pH alarms, switches, and mixer motor starters.

Chemical feed apparatus is similar to that of neutralization systems in that they require storage tanks, agitation, level instrumentation, and metering pumps. Storage tanks should be sized according to maximal feed rate, shipping time required, and quantity of shipment. The total storage capacity should be more than sufficient to guarantee a chemical supply while awaiting delivery. Storage containers must be individually suitable to the reagent being used. For example, hydroscopic reagents such as high calcium quicklime or sulfuric acid must be stored in moisture-proof tanks to prevent atmospheric degradation. Others like sodium hydroxide must be heated or carefully diluted because they will freeze at temperatures slightly below room temperature or when stored at concentrations greater than 40 percent. Agitation and feed equipment specifications are particular to each reagent and is, therefore, covered in detail in the following sections.

In addition to the chemical feed and neutralization systems, both flow monitoring and effluent pH recording equipment are necessary to prevent discharge of insufficiently treated waste resulting from surges or upsets. Also, spare parts such as pH probes, pH controller circuit board, metering pump ball valves, o-rings, and strainers should be kept on hand to prevent any excessive downtime.

4.1.6 Post-Treatment

Most treatment processes for corrosive wastes are typically preceeded by neutralization or pH adjustment in order to enhance processing (e.g., metal precipitation, emulsion breaking), prevent interference with downstream treatment (e.g., biological treatment, carbon adsorption), or to minimize corrosion of subsequent processing equipment. These processes are required because the effluent from neutralization frequently fails to meet NPDES or POTW discharge specifications as a result of the presence of toxic metals, priority organics or high levels of oil, grease or solids. As a result, physical, chemical, and biological treatment processes are commonly employed to improve overall quality.

As discussed in Section 3.0, the vast majority of land disposed corrosive wastes are aqueous liquids. These typically contain low concentrations of organics and/or heavy metals which will frequently require post-treatment to meet discharge requirements. A smaller percentage of land disposed RCRA corrosives consists of concentrated acid/alkali solutions and only 7.5 percent consists of solids or sludges. However, these wastes cover the entire spectrum in organic constituent concentrations and will be most effectively handled by a correspondingly wide range in post-treatment processes.

Post-treatment options are, therefore, most conveniently discussed in terms of the physical and chemical characteristics of the raw waste. Previous investigators[14] have categorized corrosive waste treatment options as follows:

- Aqueous treatment of inorganic acids and bases which do not contain toxic organics or heavy metals at levels which require treatment (Waste C, Figure 4.1.3);
- Aqueous treatment of corrosive liquids with trace organics (Waste A, Figure 4.1.3);
- Aqueous treatment of dilute organic corrosive wastes (Waste D, Figure 4.1.3);
- Treatment of heavy metal sludges (Waste B, Figure 4.1.3);
- Incineration of combustible sludges (Waste E), solids (Waste F), and liquids (Waste G) with optional neutralization (Figure 4.1.3); and
- Recovery of concentrated liquid organics (e.g., oils, solvents) which contain acids or bases (Waste H) which is corrosive (Figure 4.1.4).

Treatment trains for these waste categories are summarized below. This is followed by a more in-depth discussion of clarification and sludge consolidation, since these unit process operations will be applied in the majority of corrosive waste neutralization treatment trains.

4.1.7 Treatment Trains

The generic process for neutralization of corrosives without metals or organics (Process C) consists of two-stage neutralization followed by solid/liquid separation (e.g., in a clarifier) to remove insoluble salts. The sludge would then be dewatered (e.g., filter press) and disposed in a secure

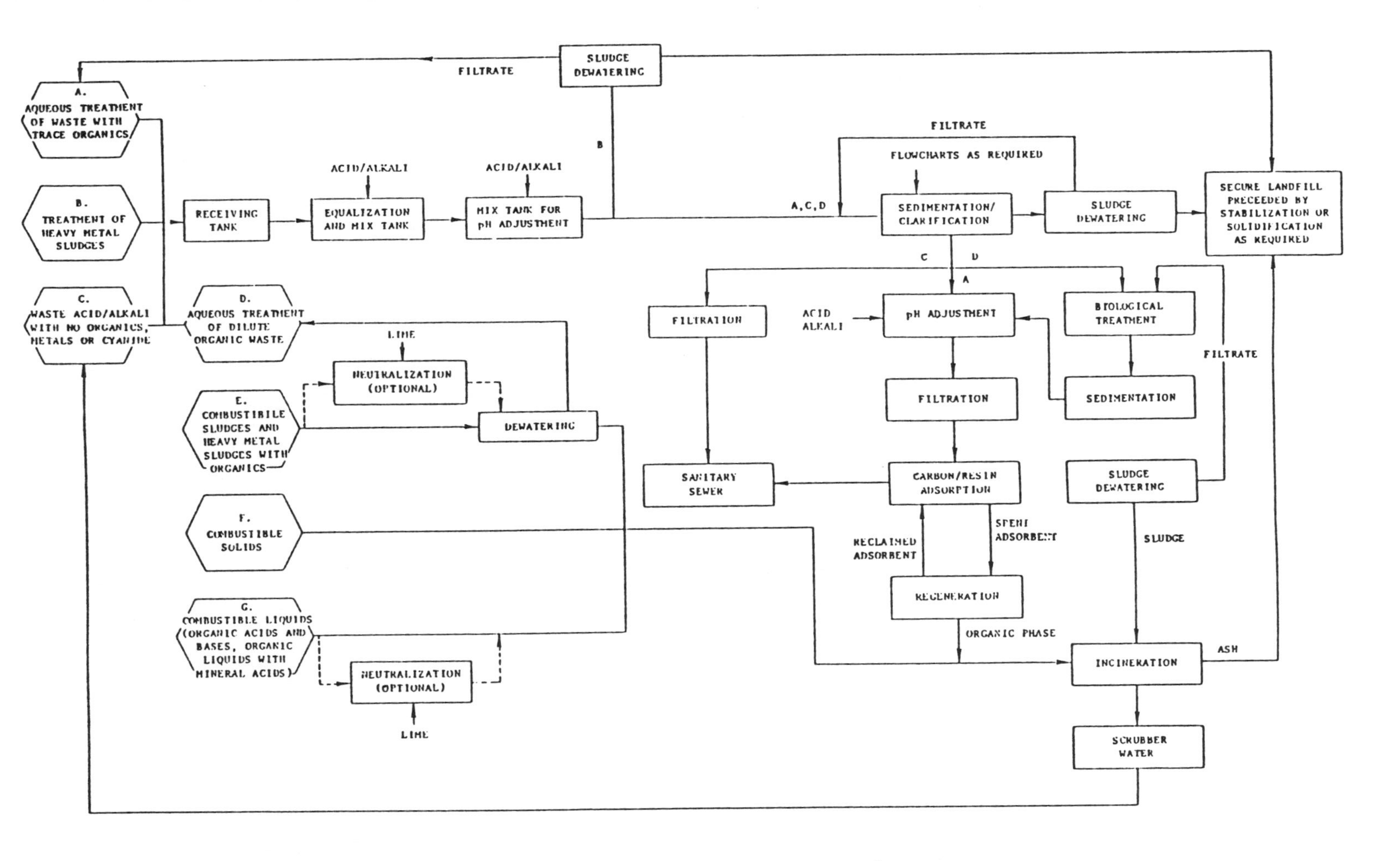

Figure 4.1.3. Treatment trains for corrosive wastes.

Source: References 15 and 16.

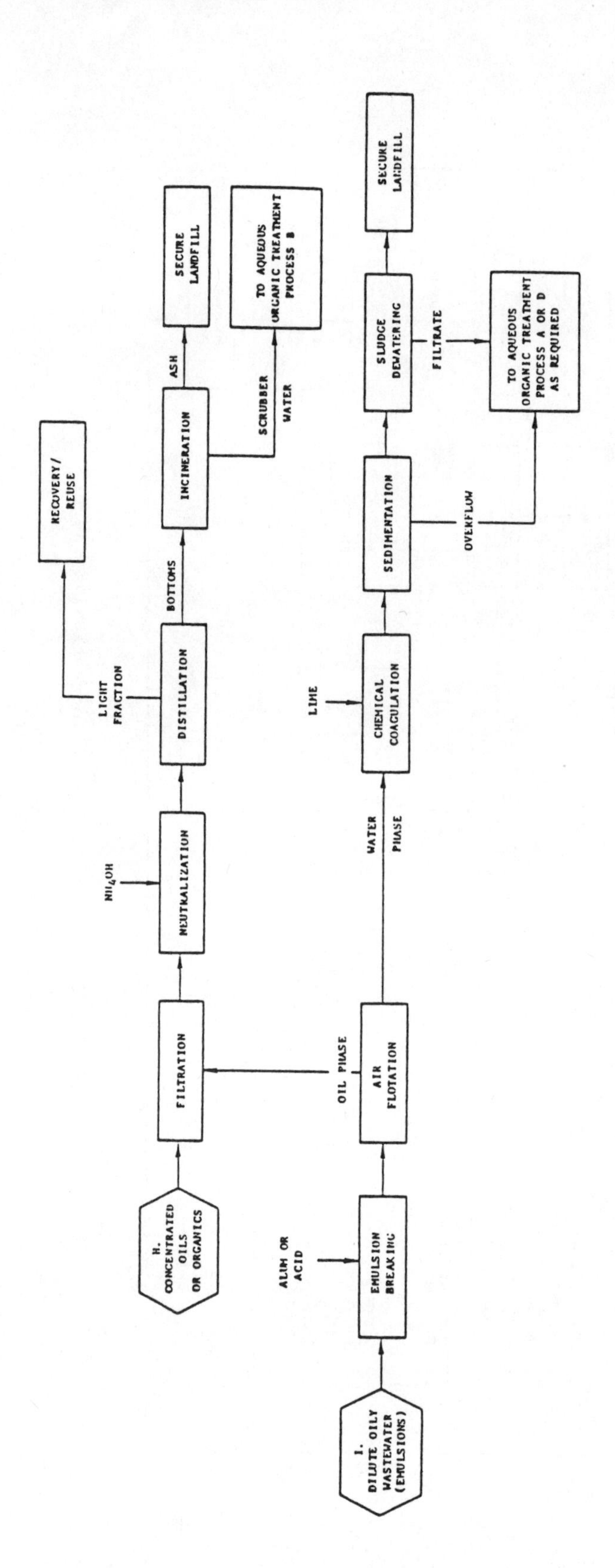

Figure 4.1.4. Treatment of concentrated organics and oily wastewater emulsions.

Source: References 15 and 16

landfill. The clarified aqueous stream would be filtered as required to meet NPDES or POTW discharge requirements. Filter backwash would be recycled to the clarifier inlet.

Aqueous corrosives with trace organics (less than 500 ppm) would require additional treatment of the clarified effluent. Applicable technologies for trace organic removal include adsorption (e.g., activated carbon, resin), air stripping, and ozone oxidation with UV radiation. Dilute organics (up to 10,000 ppm) are often more economically treated via biological degredation as indicated in Figure 4.1.3 (Waste Stream D). Neutralization is required primarily to prevent a microorganism kill in the bioreactor. The aqueous effluent concentration from biological treatment will contain organic levels of approximately 10 to 50 ppm which can be filtered and polished (e.g., activated carbon) prior to discharge. Sludges would contain toxic organics and require disposal such as incineration.

Metal-containing wastes must undergo precipitation at elevated pH (e.g., pH of 9.0) in the clarifier and the supernatant will require pH adjustment prior to additional treatment or discharge. In addition, clarifier sludge will require stabilization/encapsulation prior to disposal in a secure landfill if the Toxicity-Characteristic Leaching Procedure indicates that the waste exceeds maximum permissible metal concentrations in the leachate (see Section 1, Table 1.1).

Concentrated organic corrosive wastes, typically containing organic acids/bases, solvents or oils, will frequently be ammenable to recovery processes following neutralization and phase separation (Figure 4.1.4). If emulsions are present, they can be broken through addition of a highly ionized soluble salt of an acid followed by phase separation; e.g., dissolved air flotation.[16] Aqueous effluent will be treated as previously described for wastes with low organic concentrations.

The concentrated organic phase can be recovered using distillation, steam stripping, solvent extraction or thin film evaporation, following solids removal and neutralization. Alternatively, if recovery is not judged to be cost-effective, the organic waste can be destroyed via incineration or some other form of oxidation; e.g., use as a fuel, chemical oxidation, wet air oxidation, or supercritical fluid processes.

Selection of an appropriate recovery or disposal technology for organic residuals is dependent on the waste characteristics and volume, since these factors will generally determine overall economic feasibility. Approximate ranges of applicability of various treatment/disposal techniques as a function of organic concentration are shown in Figure 4.1.5. Table 4.1.2 summarizes applicable waste characteristics, development status of the technology, performance capability and residuals generation for organic waste treatment processes. Additional details on these processes can be found in the literature.

Neutralization of organic streams prior to incineration is optional, provided the incineration combustion chambers are lined with an acid resistant refractory or fire brick.[16] In particular, combustible solids would probably not be neutralized since this would require addition of water which would reduce the waste's heating value. Incineration flue gas would require scrubbing, followed by aqueous treatment of the resulting liquid waste (Waste C).

Neutralization of concentrated organic corrosives would generally be required prior to handling in other treatment or destruction processes. Lime is not recommended as a neutralization agent due to scale formation. Similarly, sodium-based alkalis are not recommended when treating in thermal processes, due to formation of eutectic solids which create ash and clinkering problems. Instead, ammonium hydroxide has been recommended as a suitable reagent.[16]

4.1.8 Clarification and Sludge Consolidation

As discussed previously, clarification and sludge consolidation unit operations will be applied to the majority of corrosive wastes which are affected by the land disposal ban.

Typically, wastewaters undergo chemical treatment and enter a clarifier where the flow is decreased to a point at which solids with a specific gravity greater than that of the liquid settle to the bottom. For liquid/solid mixtures with a slight density difference, an organic polymer (flocculant) can be added to allow the solids to agglomerate and improve the settling characteristics. The supernatant in the overflow is drawn off and residual

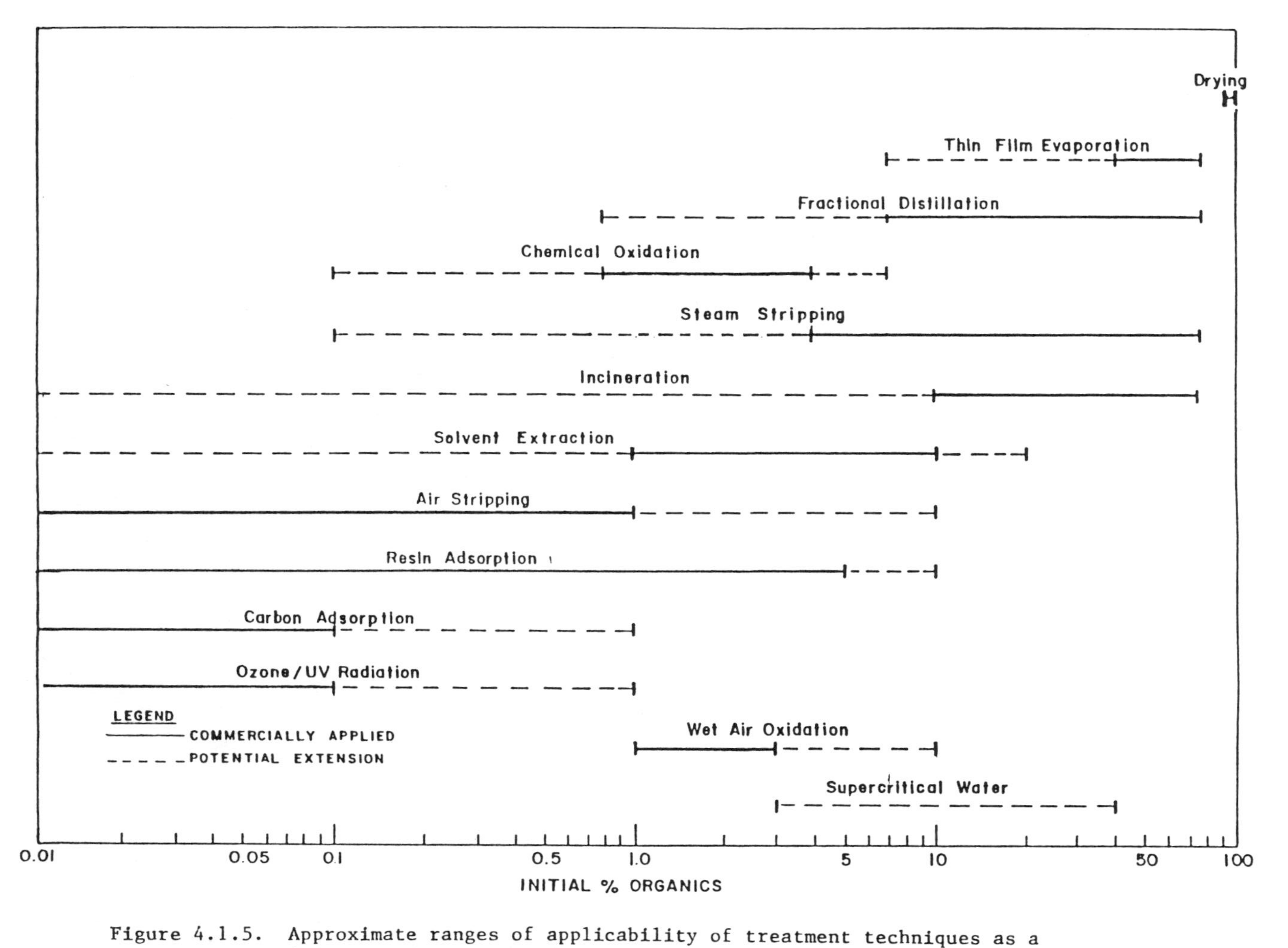

Figure 4.1.5. Approximate ranges of applicability of treatment techniques as a function of organic concentration in liquid waste streams.

Source: Reference 15

TABLE 4.1.2. SUMMARY OF ORGANIC RESIDUAL TREATMENT PROCESS

Process	Applicable waste streams	Stage of development	Performance	Residuals generated
Incineration				
Liquid injection incineration	All pumpable liquids provided wastes can be blended to Btu level of 8500 Btu/lb. Some solids removal may be necessary to avoid plugging nozzles.	Estimated that over 219 units are in use. Most widely used incineration technology.	Excellent destruction efficiency (>99.99%). Blending can avoid problems associated with residuals, e.g., HCl.	TSP, possibly some PICs and HCl. Only minor ash if solids removed in pretreatment processes. Scrubber water must be neutralized.
Rotary kiln incineration	All wastes provided Btu level is maintained.	Over 40 units in service; most versatile for waste destruction.	Excellent destruction efficiency (>99.99%).	Requires APCDs. Residuals should be acceptable if charged properly and neutralized.
Fluidized bed incineration	Liquids or nonbulky solids.	Nine units reportedly in operation-circulating bed units under development.	Excellent destruction efficiency (>99.99%).	As above.
Fixed/multiple hearths	Can handle a wide variety of wastes.	Approximately 70 units in use. Old technology for municipal waste combustion.	Performance may be marginal for hazardous wastes, particularly halogenated organic wastes.	As above.
Use As A Fuel				
Industrial kilns	Generally all wastes, but Btu level, chlorine content, and other impurity content may require blending to control charge characteristics and product quality.	Only a few units now burning hazardous waste.	Usually excellent destruction efficiency (>99.99%) because of long residence times and high temperatures.	Requires APCDs. Residuals should be acceptable.
High temperature industrial boilers	All pumpable fluids, but should blend HCl and halogenated organics. Solids removal particularly important to ensure stable burner operation.	Several units in use for hazardous wastes.	Most units tested have demonstrated high DRE (>99.99%). unless boilers equipped	Wastes must be blended to meet emission standards for TSP and HCl with APCDs.

(continued)

TABLE 4.1.2 (continued)

Process	Applicable waste streams	Stage of development	Performance	Residuals generated
Physical Treatment Methods				
Distillation	This is a process used to recover and separate concentrated organics. Fractional distillation will require solids removal to avoid plugging columns.	Technology well developed and equipment available from many suppliers; widely practiced technology.	Separation depends upon reflux (99+ percent achievable). This is a recovery process.	Bottoms will usually contain levels of organic in excess of 1,000 ppm; condensate may require further treatment.
Evaporation	Agitated thin film units can tolerate higher levels of solids and higher viscosities than other types of stills.	Technology is well developed and equipment is available from several suppliers; widely practiced technology.	This is an organic recovery process. Typical recovery of 60 to 70 percent depending on initial volatile content.	Bottoms will contain appreciable organics. Generally suitable for incineration.
Steam Stripping	A simple distillation process to remove volatile organics from aqueous solutions. Preferred for low concentrations and organics with low solubilities.	Technology well developed and available.	Not generally considered a final treatment, but can achieve low residual organic levels.	Aqueous treated stream will probably require polishing. Further concentration of overhead steam generally required.
Air Stripping	Generally used to treat low concentration aqueous streams.	Technology well developed and available.	Not generally considered a final treatment, but may be effective for highly volatile wastes.	Air emissions may require treatment.
Liquid-Liquid Extraction	Generally suitable only for liquids of low solid content.	Technology well developed for industrial processing.	Can achieve high efficiency separations for certain waste combinations.	Organic or metal solubility in aqueous phase should be monitored.
Carbon Adsorption	Suitable for low solid, low concentration aqueous waste streams.	Technology well developed; used as polishing treatment.	Can achieve low levels of residual organics in effluent.	Adsorbate must be processed during regeneration. Spent carbon and wastewater may also need treatment.
Resin Adsorption	Suitable for low solid waste streams. Consider for recovery of valuable organics.	Technology well developed in industry for special resin/solvent combinations. Applicability to waste streams not demonstrated.	Can achieve low levels of residual organics in effluent.	Adsorbate must be processed during regeneration.

(continued)

TABLE 4.1.2 (continued)

Process	Applicable waste streams	Stage of development	Performance	Residuals generated
Other Thermal Technologies				
Circulating bed combustor	Liquids or nonbulky solids.	Only one U.S. manufacturer. No units treating hazardous waste.	Manufacturer reports high efficiencies (~99.99%).	Bed material additives can reduce HCl emissions. Residuals should be acceptable if neutralized.
Molten glass incineration	Almost all wastes, provided moisture and metal impurity levels are within limitations.	Technology developed for glass manufacturing Not available yet as a hazardous waste unit.	No performance data available, but DREs should be high (>99.99%).	Will need APC device for HCl and possibly PICs; solids retained (encapsulated) in molten glass.
Molten salt destruction	Not suitable for high (20%) ash content wastes.	Technology under development since 1969, but further development on hold.	Very high destruction efficiencies for organics (six nines for PCBs).	Needs some APC devices to collect material not retained in salt. Ash disposal may be a problem.
Furnace pyrolysis units	Most designs suitable for all wastes.	One pyrolysis unit RCRA permitted. Certain designs available commercially.	Very high destruction efficiencies possible (>99.99%). Possibility of PIC formation.	TSP emissions lower than those from conventional combustion will need APC devices for HCl. Certain wastes may produce an unacceptable tarry residual.
Plasma arc pyrolysis	Present design suitable only for liquids.	Commercial design appears imminent, with future modifications planned for treatment of sludges and solids.	Efficiencies exceeded six nines in tests with solvents.	Requires APC devices for HCP and TSP, needs flare for H_2 and CO destruction.
Fluid wall advanced electric reactor	Suitable for all wastes if solids pretreated to ensure free flow.	Ready for commercial development. Test unit permitted under RCRA.	Efficiencies have exceeded six nines.	Requires APC devices for TSP and HCl; Chlorine removal may be required.
In situ vitrification	Technique for treating contaminated soils, could possibly be extended to slurries. Also use as solidification process.	Not commercial, further work planned.	No data available, but DREs of over six nines reported.	Off gas system needed to control emissions to air. Ash contained in vitrified soil.

(continued)

TABLE 4.1.2 (continued)

Process	Applicable waste streams	Stage of development	Performance	Residuals generated
Chemical Treatment Processes				
Wet air oxidation	Suitable for aqueous liquids, also possible for slurries. Organic concentrations up to 15%.	High temperature/ pressure technology, widely used as pretreatment for municipal sludges, only one manufacturer.	Generally applicable as pretreatment for biological treatment. Some compounds resist oxidation.	Some residues likely which need further treatment.
Supercritical water oxidation	For liquids and slurries containing optimal concentrations of about 10% organic.	Supercritical conditions may impose demands on system reliability. Commercially available in 1986.	Supercritical conditions achieve high destruction efficiencies (~99.99%) for all organic constituents.	Residuals not likely to be a problem. Neutralization can be accomplished in process.
Ozonation	Oxidation with ozone (possibly assisted by UV) suitable for low solid, dilute aqueous solutions.	Now used as a polishing step for wastewaters.	Not likely to achieve residual organic levels in the low ppm range for most wastes.	Residual contamination likely; will require additional processing of off gases.
Other chemical oxidation processes	Oxidizing agents may be highly reactive for specific constituents in aqueous solution.	Oxidation technology well developed for cyanides and other species (phenols), not yet established for general utility.	Not likely to achieve residual levels in the low ppm range for most wastes.	Residual contamination likely; will require additional processing.
Biological Treatment Methods				
	Aerobic technology suitable for dilute wastes although some constituents will be resistant.	Conventional treatments have been used for years.	May be used as final treatment for specific wastes, may be pretreatment for resistant species.	Residual contamination likely; will usually require additional processing such as absorption.

Source: Reference 15

solids are removed in a final polishing step such as carbon filtration or ion exchange. The solids in the underflow can then be discharged to a holding tank for subsequent dewatering.

Figure 4.1.6 shows investment costs for flocculation/clarification units as a function of flow rate. The unit is assumed to have a separate flocculation tank, a polymer feed system, a slant-tube separator, and a zone in which sludge collects prior to discharge.

The sludge generation and solids settling rates are usually determined by laboratory analyses conducted by equipment vendors. Table 4.1.3 is an example of a laboratory analysis of the neutralization characteristics of 3 alkali reagents applied to a 1.75 pH spent acid plating waste.[18] While sodium hydroxide yielded the lowest sludge volume (due to the extreme solubility of sodium salts) it was almost seven times as expensive ($15,170/million gallons) to utilize as hydrated lime ($2,200/million gallons). A limestone/lime dual alkali reagent system was ultimately selected for this application since it yielded the lowest effective sludge volume (5,480 mg/L) at the least overall cost ($3,170/million gallons). Since the application was a one-time neutralization of a contaminated surface impoundment, the reagent selection criteria were limited to reagent cost, sludge dewatering characteristics and volume requiring disposal and the added cost of grinding and slurrying equipment.

In addition to different degrees of quantity sludge generation, each reagent imparts to the sludge variable settling characteristics, thereby affecting the sizing parameters of downstream equipment. For example, lime neutralized sludge exhibits a granular nature that settles fairly rapidly and dewaters effectively (4 to 20 lb of dry solids/hr/ft^2 yielding a 3/16 to 3/8 in. cake). Conversely, sodium hydroxide sludge results in a fluffy gelatinous precipitate with low settling rates.[19] Figure 4.1.7 shows the results of three settling tests conducted on power plant effluents with both lime and sodium hydroxide. In all three cases sodium hydroxide settled more slowly and in subsequent filtration tests, dewatered about half as effectively.[9] However, the use of lime or limestone generates greater sludge weight and volume. This is primarily due to insoluble acid salts and calcium sulfates formed when neutralizating sulfate containing wastes such as

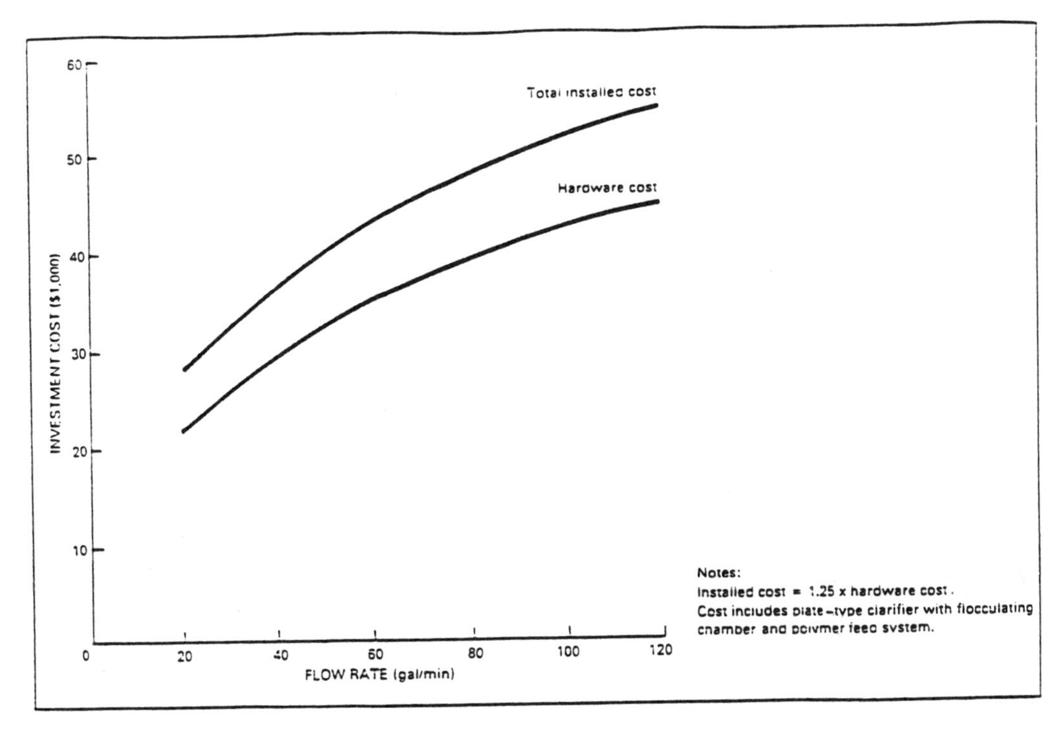

Figure 4.1.6 Investment cost for flocculation/clarification units.

Source: Reference 17

TABLE 4.1.3 LABORATORY ANALYSIS OF THE NEUTRALIZATION OF SPENT ACID PLATING WASTE

Pretreatment	Treatment	Final PH (S.U.)	Supernatant TSS (mg/L)	Supernatant Zn (mg/L)	Supernatant Pb (mg/L)	Sludge[a] TSS (mg/L)	Sludge[a] Volume %
Raw waste	None	1.75	41	191	0.4	-	-
	10 g/L $CaCO_3$	3.0	1400	72	0.3	3100	10
	8.0 g/L $Ca(OH)_2$	9.5	162	0.15	0.1	6200	27
	6.5 g/L NaOH	11.5	-	0.25	0.1	-	-
10 g/L $CaCO_3$	3 g/L $Ca(OH)_2$	10.5	130	0.20	0.1	5480	25
10 g/L $CaCO_3$	2 g/L NaOH	10.3	90	0.03	0.1	5700	40

[a]Sludge volume measured after 15 hours.

Source: Reference 18

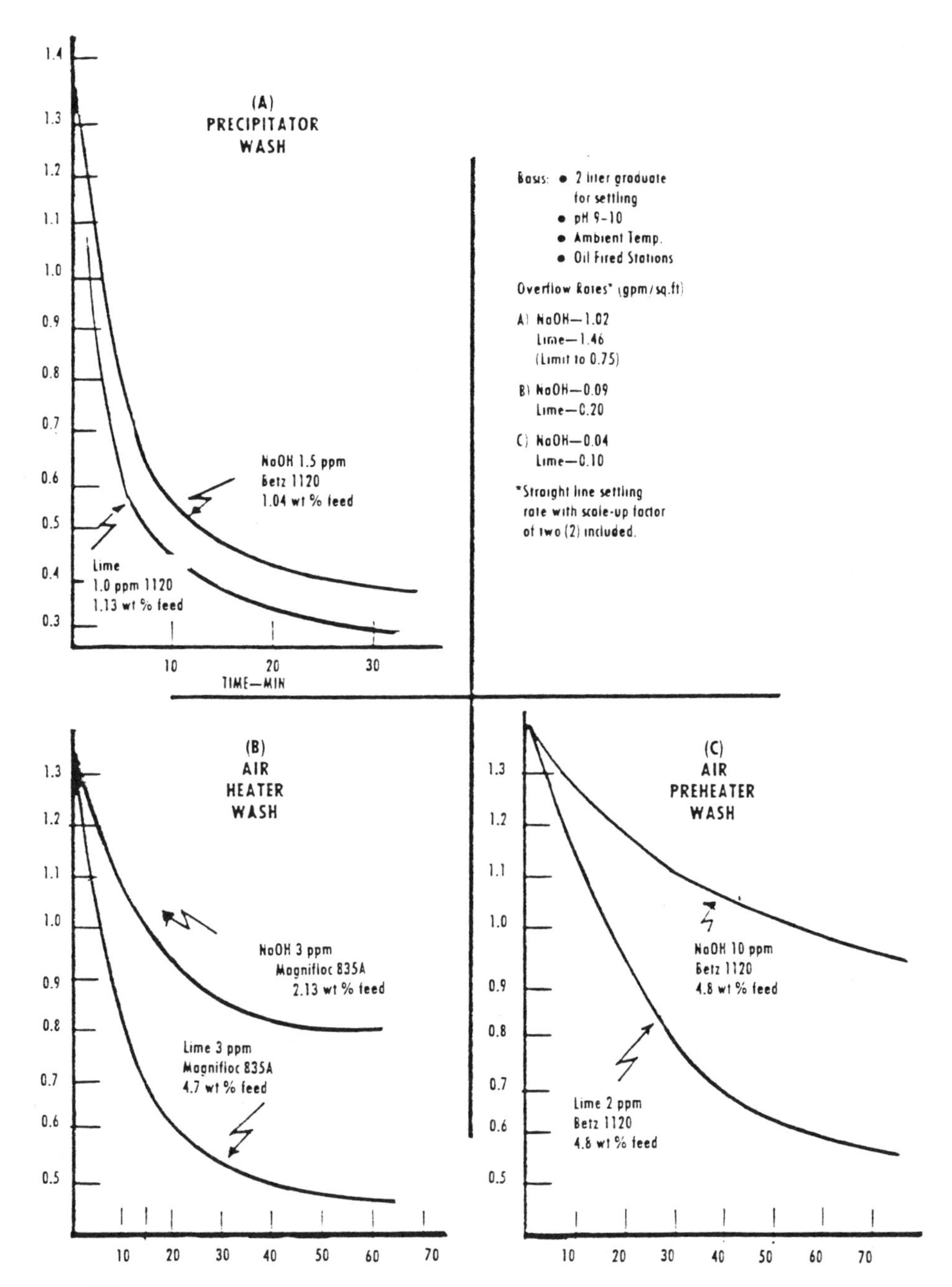

Figure 4.1.7 Settling rate curves.

Source: Reference 19

sulfuric acid. Therefore, as landfill and hauling costs become more significant, sodium hydroxide becomes more competitive with lime and limestone as a neutralization agent.

Few, if any, sludges settle at a rate sufficient to utilize only clarifiers or thickeners to accumulate sludge for disposal on land. Therefore, the underflow from the clarifier is typically concentrated through the use of mechanical dewatering equipment such as centrifuges, rotary vacuum filters, belt filters, drying ovens, and recessed-plate filter presses. The obtainable degree of cake dryness can be determined by bench-scale tests by the equipment vendor to identify the suitability of a particular dewatering device (see Table 4.1.4). The low solids content of sodium hydroxide after sedimentation (3 to 10 percent) requires the use of a filter press.[20] Conversely, supspended solids removal from lime neutralized sludges can be accomplished through use of a wider range of equipment including rotary vacuum or continuous belt filters. Figures 4.1.8 and 4.1.9 present the unit costs for both sludge storage/thickening units and recessed plate filter presses, respectively. Items that were not included in these figures, but will add to the cost of installation include; high pressure feed pumps, filtrate return lines (to clarifier) and cake solids handling equipment. The feed volume capacity of the unit is based on a feed solids concentration of 2 percent, a cake solids concentration of 20 percent and a press cycle of 8 hours.[17]

4.1.9 Land Disposal of Residuals

Installation of wastewater treatment systems inevitably results in the problem of sludge disposal. The cost of hauling the sludge to a licensed hazardous waste landfill will depend on the volume of sludge, the distance hauled, and the sludge composition. Figure 4.1.10 illustrates the annual disposal costs for 100 lbs of sludge generated over a range of sludge concentrations and unit disposal costs.[17] Sometimes it is possible to dispose of calcium based reagent sludges through agricultural or acid pond liming. In one neutralization application, over 200,000 lbs/acre of lime neutralized waste pickle liquor sludge was applied onto Miami silt loam to improve overall crop yields.[21]

TABLE 4.1.4. SUMMARY OF SLUDGE DEWATERING DEVICE CHARACTERISTICS

Parameter	Gravity (low pressure)	Basket centrifuge	Solid Bowl centrifuge	Vacuum filter	Belt filter press	Recessed filter press
Cake Solids %	16 - 24	20 - 30	30 - 42	30 - 40	36 - 46	50 - 60
Operational Variables	-Rate of sludge feed -Polymer concentration -Belt speed -Depth of sludge in cylinder	-Bowl speed -Time at full speed -Depth of skimming -Sludge feed rate	-Bowl/conveyor differential speed -Pool depth -Sludge feed rate	-Quantity of H_2O -Drum speed -Vacuum level -Conditioning chemicals -Filter media	-Belt speed -Belt tension -Washwater flow and pressure -Belt type -Polymer conditioner	-Feed pressure -Filtration time -Use of precoat -Cloth washing frequency -Filter cloth used
Advantages	-Low energy & capital cost -Low space requirements -Requires little operator skill	-Same machine for thickening & dewatering -Very flexible -Little operator attention	-Easy to install -Low space requirement -Either thickening or dewatering -High rate of feed -Can operate on highly variable feeds	-Continuous operation -Long media life -Low maintenance -Easy operation	-Only filter press produces drier cake -Low power -Low noise & vibration -Continuous operation	-High solids filter cake -High solids capture -Only mechanical device capable of meeting some landfill requirements
Disadvantages	-Limited capacity -Low solids concentration -Requires large quantity of conditioning chemicals	-Unit is not continuous -High ratio of capital cost to capacity -Requires complex controls -Requires noise control	-Requires prescreening -Very noisy with high vibration -High power consumption -Requires high maintenance skills	-High power requirement -Vacuum pumps are noisy -Requires at least 3% feed solids for operation	-Very sensitive to incoming feed -Short media life -Greater operational attention and polymer dosage	-High capital cost -Batch discharge -High polymer usage -Media replacement costs are high

Source: Reference 20

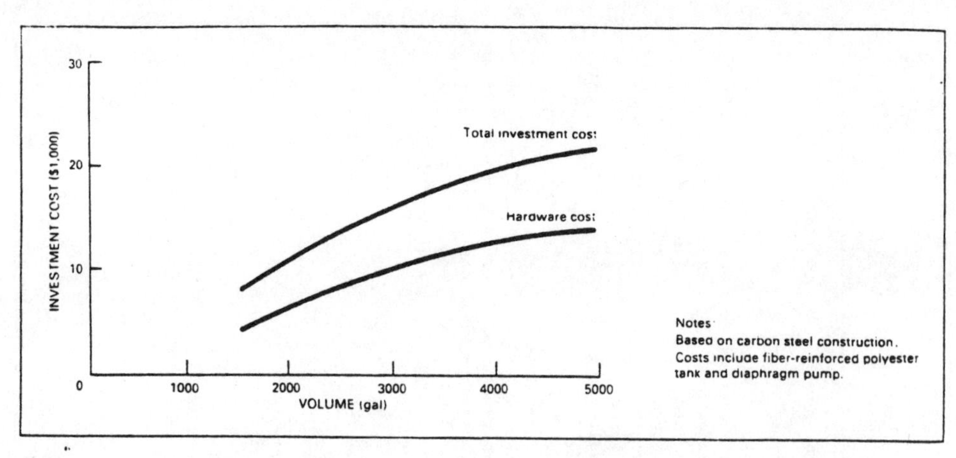

Figure 4.1.8 Investment cost for sludge storage/thickening units.

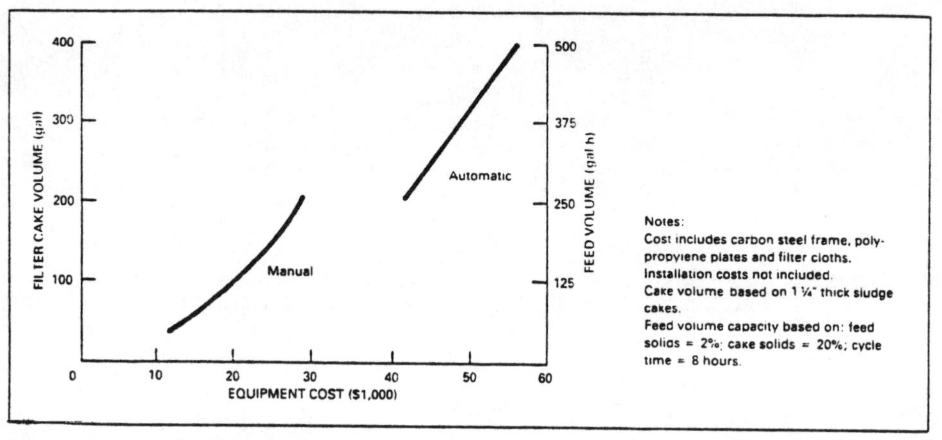

Figure 4.1.9 Hardware cost for recessed plate filter presses.

Source: Reference 17

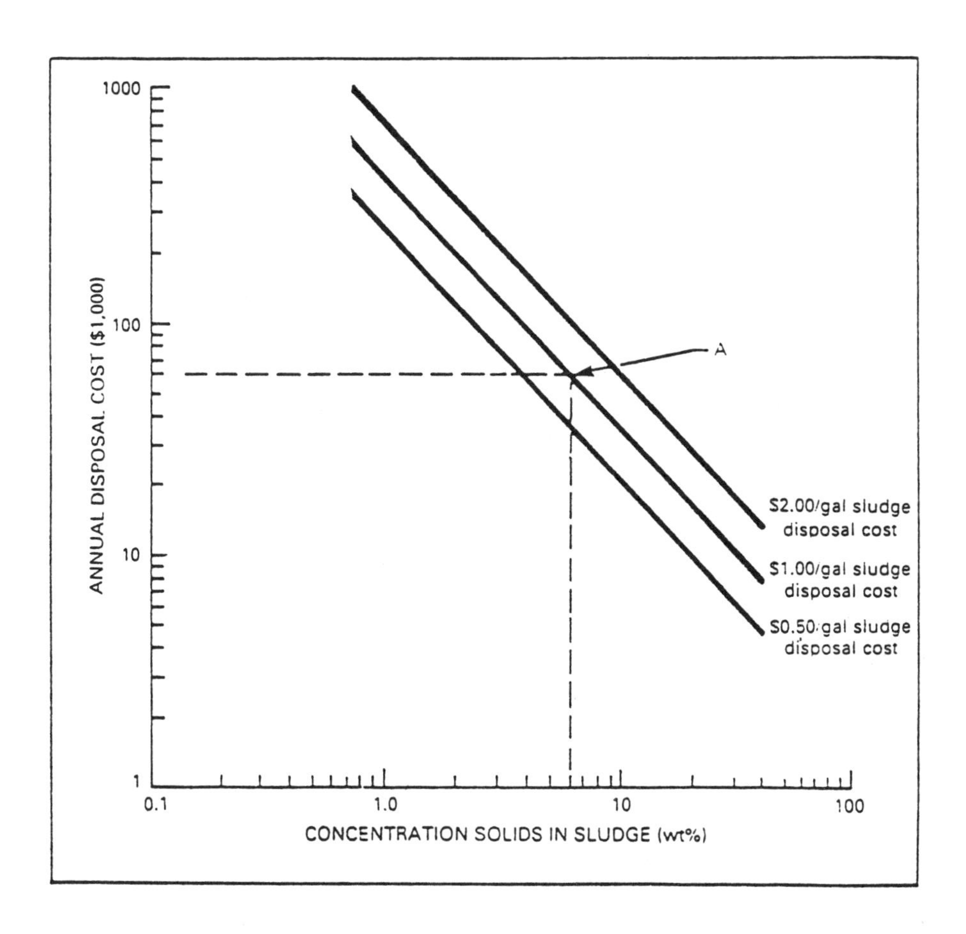

Figure 4.1.10 Annual cost for disposal of industrial sludge (per 100 lb dry solids generated per day).

Source: Reference 17

In the absence of the possibility of land disposal of waste in a location that will be demonstrably supportive of human health and the environment, another option is to treat the waste to immobilize the waste constituents for as long as they remain hazardous. This method of treatment, based on fixation or encapsulation processes, is a possibility for some corrosive wastes; however, it is more likely a treatment that will be undertaken to ensure that residuals from other treatment processes can be safely disposed. Certain of these residuals could be found hazardous for reasons other than pH characteristics; e.g., their heavy metal content may lead to positive tests for EP toxicity. In such cases, encapsulation may be needed to eliminate this characteristic.

The following discussions will summarize available information concerning immobilization techniques, namely solidification/fixation or encapsulation. Chemical fixation involves the chemical interaction of the waste with a binder; encapsulation is a process in which the waste is physically entrapped within a stable, solid matrix.

Solidification can be used to chemically fix or structurally isolate metallic species or trace organics which may be present in neutralization sludges to a solid, crystalline, or polymeric matrix. The resultant monolithic solid mass can then be safely handled, transported, and disposed of using established methods of landfilling or burial. Solidification technologies are usually categorized on the basis of the principal binding media, and include such additives as: cement-based compounds, lime-based pozzolanic materials, thermoplasts, and organic polymers (thermosets).[22] The resulting stable matrix produces a material than contains the waste in a nonleachable form, is nondegradable, cost-effective, and does not render the land it is disposed in unusable for other purposes. A brief summary of the compatibility and cost data for selected waste solidification/stabilization systems is presented in Tables 4.1.5 and 4.1.6.

CEMENT BASED SYSTEMS

These systems utilize type I Portland cement, water, proprietary additives, possibly fly ash, and waste sludges to form a monolithic, rock-like mass. In an EPA publication,[23] several vendors of cement based systems

TABLE 4.1.5 COMPATIBILITY OF SELECTED WASTE CATEGORIES WITH DIFFERENT WASTE SOLIDIFICATION/STABILIZATION TECHNIQUES

	Treatment Type						
Waste component	Cement based	Lime based	Thermoplastic solidification	Organic polymer (UF)*	Surface encapsulation	Self-cementing techniques	Glassification and synthetic mineral formation
Organics:							
1. Organic solvents and oils	May impede setting, may escape as vapor	Many impede setting, may escape as vapor	Organics may vaporize on heating	May retard set of polymers	Must first be absorbed on solid matrix	Fire danger on heating	Wastes decompose at high temperatures
2. Solid organics (e.g., plastics, resins, tars)	Good--often increases durability	Good--often increases durability	Possible use as binding agent	May retard set of polymers	Compatible--many encapsulation materials are plastic	Fire danger on heating	Wastes decompose at high temperatures
Inorganics:							
1. Acid wastes	Cement will neutralize acids	Compatible	Can be neutralized before incorporation	Compatible	Can be neutralized before incorporation	May be neutralized to form sulfate salts	Can be neutralized and incorporated
2. Oxidizers	Compatible	Compatible	May cause matrix break down, fire	May cause matrix break down	May cause deterioration of encapsulating materials	Compatible if sulfates are present	High temperatures may cause undeable reactions
3. Sulfates	May retard setting and cause spalling unless special cement is used	Compatible	May dehydrate and rehydrate causing splitting	Compatible	Compatible	Compatible	Compatible in many cases
4. Halides	Easily leached from cement, may retard setting	May retard set, most are easily leached	May dehydrate	Compatible	Compatible	Compatible if sulfates are also present	Compatible in many cases
5. Heavy metals	Compatible	Compatible	Compatible	Acid pH solubilizes metal hydroxides	Compatible	Compatible if sulfates are present	Compatible in many cases
6. Radioactive materials	Compatible	Compatible	Compatible	Compatible	Compatible	Compatible if sulfates are present	Compatible

* Urea-Formaldehyde resin.

Source: Reference 23

TABLE 4.1.6 PRESENT AND PROJECTED ECONOMIC CONSIDERATIONS FOR WASTE SOLIDIFICATION/STABILIZATION SYSTEMS

Type of treatment system	Major materials required	Unit cost of material	Amount of material required to treat 100 lbs of raw waste	Cost of material required to treat 100 lbs of raw waste	Trends in price	Equipment costs	Energy use
Cement-based	Portland Cement	\$0.03/lb	100 lb	\$ 3.00	Stable	Low	Low
Pozzolanic	Lime Flyash	\$0.03/lb	100 lb	\$ 3.00	Stable	Low	Low
Thermoplastic (bitumen-based)	Bitumen Drums	\$0.05/lb \$27/drum	100 lb 0.8 drum	\$18.60	Keyed to oil prices	Very high	High
Organic polymer (polyester system)	Polyester Catalyst Drums	\$0.45/lb \$1.11/lb \$17/drum	43 lb of polyester-catalyst mix	\$27.70	Keyed to oil prices	Very high	High
Surface encapsulation (polyethylene)	Polyethylene	Varies	Varies	\$ 4.50*	Keyed to oil prices	Very high	High
Self-cementing	Gypsum (from waste)	**	10 lb	**	Stable	Moderate	Moderate
Classification/mineral synthesis	Feldspar	\$0.03/lb	Varies	--	Stable	High	Very high

* Based on the full cost of \$91/ton.
** Negligible but energy cost for calcining are appreciable.

Source: Reference 23

reported problems with organic wastes containing oils, solvents, and greases not miscible with an aqueous phase. For although the unreactive organic wastes become encased in the solids matrix, their presence can retard setting, cause swelling, and reduce final strength.[24] These systems are most commonly used to treat inorganic wastes such as incinerator generated wastes and heavy metal sludges from neutralization/precipitation processes.

LIME BASED (POZZOLANIC) TECHNIQUES

Pozzolanic concrete is the reaction product of fine-grained aluminous siliceous (pozzolanic) material, calcium (lime), and water. The pozzolanic materials are wastes themselves and typically consist of fly ash, ground blast furnace slag, and cement kiln dust. The cementicious product is a bulky and heavy solid waste used primarily in inorganic waste treatment such as the solidification of heavy metal and flue gas desulfurization sludges.[25]

THERMOPLASTIC MATERIAL

In a thermoplastic stabilization process, the waste is dried, heated (260-450°F), and dispersed through a heated plastic matrix. Principal binding media include asphalt, bitumen, polypropylene, polyethylene, or sulfur. The resultant matrix is somewhat resistant to leaching and biodegradation, however the rates of loss to aqueous contacting fluids are significantly lower than those of cement or lime based systems. However this process is not suited to wastes that act as solvents for the thermoplastic material. Also there is a risk of fire or secondary air pollution with wastes that thermally decompose at high temperature.[22]

ORGANIC POLYMERS (THERMOSETS)

Thermosets are polymeric materials that crosslink to form an insoluble mass as a result of chemical reaction between reagents, with catalysts sometimes used to initiate reaction. Waste constituents could conceivably enter into the reaction, but most likely will be merely physically entrapped, within the crosslinked matrix. The crosslinked polymer or thermoset will not soften when heated after undergoing the initial set. Principal binding agents or reactants for stabilization include ureas, phenolics, epoxides, and polyesters. Although the thermosetting polymer process has been used most frequently in the radioactive waste management industry, there are formulations that may be applicable to certain neutralization sludges. It is important to note that the concept of thermoset stabilization, like thermoplastic stabilization, does not require that chemical reaction take place during the solidification process. The waste materials are physically trapped in an organic resin matrix that, like thermoplastics, may biodegrade over extended periods of time and release much of the waste as a leachate.[26] It is also an organic material that will thermally decompose if exposed to a fire.

Encapsulation is often used to describe any stabilization process in which the waste particles are enclosed in a coating or jacket of inert material. A number of systems are currently available utilizing polybutadiene, inorganic polymers (potassium silicates), portland concrete, polyethylene, and other resins as macroencapsulation agents for wastes that have or have not been subjected to prior stabilization processes. Several different encapsulation schemes have been described in Reference[27,28]. The resulting products are generally strong encapsulated solids, quite resistant to chemical and mechanical stress, and to reaction with water. Wastes successfully treated by these methods and their costs are summarized in Tables 4.1.7 and 4.1.8. The technologies could be considered for stabilizing neutralization sludges but are dependent on the compatibility of the neutralized waste and the encapsulating material. EPA is now in the process of developing criteria which stabilized/solidified wastes must meet in order to make them acceptable for land disposal.[29]

TABLE 4.1.7. ENCAPSULATED WASTE EVALUATED AT THE U.S. ARMY WATERWAYS EXPERIMENT STATION

Code No.	Source of Waste	Major Contaminants
100	SO_x scrubber sludge, lime process, eastern coal	Ca, $SO_4^{=}/SO_3^{=}$
200	Electroplating sludge	Cu, Cr, Zn
300	Nickel - cadmium battery production sludge	Ni, Cd
400	SO_x scrubber sludge, limestone process eastern coal	Cu, $SO_4^{=}/SO_3^{=}$
500	SO_x scrubber sludge, double alkali process eastern coal	Na, Ca, $SO_4^{=}/SO_3^{=}$
600	SO_x scrubber sludge, limestone process, western coal	Ca, $SO_4^{=}/SO_3^{=}$
700	Pigment production sludge	Cr, Fe, CN
800	Chlorine production brine sludge	Na, Cl^-, Hg
900	Calcium fluoride sludge	Ca, F^-
1000	SO_x scrubber sludge, double alkali process, western coal	Cu, Na, $SO_4^{=}/SO_3^{=}$

Source: Reference 28

TABLE 4.1.8. ESTIMATED COSTS OF ENCAPSULATION

Process Option	Estimated Cost
Resin Fusion:	
Unconfined waste	$110/dry ton
55-Gallon drums	$0.45/gal
Resin spray-on	Not determined
Plastic Welding	$253/ton = $63.40/drum (80,000 55-gal drums/year)

Source: Reference 28

REFERENCES

1. Lewis, C.J., and R.S. Boynton. Acid Neutralization with Lime for Environmental Control and Manufacturing Processes. National Lime Association, Bulletin No. 216. 1976.

2. Kirk-Othmer Encyclopedia of Chemical Technology. Vol. 14, 3rd, Edition. John Wiley & Sons, New York, NY. 1981.

3. Chemical Marketing Reporter. Week ending July 18, 1986.

4. Oberkrom, S.L., and T.R. Marrero. Detoxification Process for a Ferric Chloride Etching Waste. Hazardous Waste and Hazardous Materials, Vol. 2, No. 1. 1985.

5. MITRE Corporation. Manual of Practice for Wastewater Neutralization and Precipitation. EPA-600/2-81-148. August 1981.

6. Levenspiel, O. Chemical Reaction Engineering. 2nd Edition, John Wiley & Sons, New York, NY. 1972.

7. GCA. Case Studies of Existing Treatment Applied to Hazardous Waste Banned from Landfill, Phase II Case Study for Facility D. July 1986.

8. Kiang, Y.H., and A.A. Metry. Hazardous Waste Processing Technology. Ann Arbor Science Publishers, Inc. 1982.

9. Higgins, T.E., and B.R. Marshall. CH2MHILL, Reston, VA. Combined Treatment of Hexavalent Chromium with Other Heavy Metals at Alkaline pH. Presented in Toxic and Hazardous Wastes: Proceeding of the 17th Mid-Atlantic Industrial Waste Conference. Lehigh University.

10. Hoyle, D.L. Designing for pH Control. Chemical Engineering, November 8, 1976.

11. Cushnie, G.C. Removal of Metals from Wastewater: Neutralization and Precipitation. Pollution Technology Review No. 107. Noyes Publication, Park Ridge, NJ. 1984.

12. Hoffman, F. How to Select a pH Control System for Neutralizing Waste Acids. Chemical Engineering. October 30, 1972.

13. Jungek, P.R., and E.T. Woytowicz. Practical pH Control. Industrial Water Engineering. February/March 1972.

14. Camp, Dresser, and McKee. Technical Assessment of Treatment Alternatives for Wastes Containing Corrosives. Contract No. 68-01-6403. September 1984.

15. GCA. Technical Resource Document: Treatment Technologies for Solvent Containing Wastes. Contract No. 68-03-3243. August 1986.

16. Warner, P.H., et al. Treatment Technologies for Corrosive Hazardous Wastes. Journal of the Air Pollution Control Federation. April 1986.

17. U.S. EPA. Reducing Water Pollution Control Costs in the Electroplating Industry. EPA-625/5-85-016. September 1985.

18. Hale, F.D., et al. Spent Acid and Plating Waste Surface Impoundment Closure. Management of Uncontrolled Hazardous Waste Site. October 31 - November 2, Washington, D.C. 1983.

19. Mace, G.R., and D. Casaburi. Lime vs. Caustic for Neutralizing Power. Chemical Engineering Progress. August 1977.

20. U.S. EPA. Dewatering Municipal Wastewater Sludges. EPA-625/1-82-014, October 1982.

21. Berger. Land Application of Neutralized Spent Pickle Liquor. 17th Industrial Waste Conference, Purdue University. 1962.

22. GCA Technical Resource Document: Treatment Technologies for Dioxin-Containing Wastes. Contract No. 68-03-3243. August 1986.

23. Guide to the Disposal of Chemically Stabilized and Solidified Waste, EPA SW-872. September 1980.

24. Environmental Laboratory U.S. Army Engineer Waterways Experiment Station, Survey of Solidification/Stabilization Technology for Hazardous Industrial Wastes, EPA-600/2-79-056.

25. McNeese, J.A., Dawson, G.W., and Christensen, D.C., Laboratory Studies of Fixation of Kepone Contaminated Sediments, in "Toxic and Hazardous Waste Disposal", Vol. 2 Pojasek, R.B. Ed., Ann Arbor Science, Ann Arbor, Michigan. 1979.

26. Stabilizing Organic Wastes: How Predictable are the Results? Hazardous Waste Consultant. May 1985 pg. 18.

27. Thompson, D.W. and Malone P.G., Jones, L.W., Survey of Available Stabilization Technology in Toxic and Hazardous Waste Disposal, Vol. 1, Pojasek, R.B., Ed. Ann Arbor Science, Ann Arbor Michigan. 1979.

28. Lubowitz, H.R. "Management of Hazardous Waste by Unique Encapsulation Processes." Proceedings of the Seventh Annual Research Symposium. EPA-600/9-81-002b, March 1981.

29. C. Wiles. Hazardous Waste Engineering Research Laboratory, U.S. EPA, private communication; and Critical Characteristics and Properties of Hazardous Waste Solidification/Stabilization, HWERL, U.S. EPA, Contract No. 68-03-3186 (in publication).

4.2 MIXING OF ACID AND ALKALI WASTES

4.2.1 Process Description

The process of acid/alkali mixing (mutual neutralization) may be the most economical method of neutralization available, particularly in cases where compatible acid/alkaline wastes are present in the same plant. Prior to implementation, data are typically collected on the volume and concentration of each waste stream and their respective flow patterns (batch or continuous). In addition, waste stream mixing characteristics are usually investigated in order to pre-determine possible waste incompatibilities that would prevent or limit the use of the technology. For example, the precipitation of metal hydroxides or other insoluble species (e.g., calcium sulfate) may result in increased sludge generation or plugging of the transport or dewatering equipment. If the sludge generation is considerable, the increased dewatering, disposal, and maintenance costs could possibly outweigh the benefit of any savings realized on reagent costs. Also, if incompatible wastes produce a reaction that is too sluggish or difficult to control, or generate reaction products that are toxic (i.e., hydrogen cyanide) or highly exothemmic, then implementation may not be feasible.

As with most neutralization processes, acid/alkali mixing can be operated in either a batch or continuous mode. Operational type depends primarily on the variation in flow rate or concentration of the divergent influent streams. Batch operations are typically utilized in treating concentrated bath dumps or intermittent flow applications. Reactor configurations can be either single- or multi-stage. However, unit-processing of the wastes considered in this document (pH less than 2 or greater than 12.5) generally require multi-stage continuous operation.[1]

Mutual neutralization finds its widest application in waste treatment systems where reagent consumption can be reduced prior to primary neutralization through the averaging of flow rates and pollutant loadings. In cases where the intermixing of the acid/alkali wastestreams result in an effluent suitable for discharge, mutual neutralization can be utilized as a primary neutralization process. General equipment for pretreatment systems typically consists of: storage tanks, metering and pH control equipment, attenuation vessel, pumps, and segregation and collection equipment. Continuous, acid/alkali mixing systems operating as primary neutralization processes usually consist of: flow equalization basin, neutralization vessel(s), emergency reagent storage and feed system, effluent pH recorder and controller, associated pumps and piping.

Figure 4.2.1 illustrates a mutual neutralization pretreatment system which is commonly employed in the metal finishing industry. The influent to the system consists of three primary streams; effluent from cyanide oxidation (pH 9 to 11) chromium reduction (pH 2 to 3), and various metal cleaning and plating operation waste streams (variable pH).[2] The three streams are combined into a collection vessel where hydrogen ion and flow rate averaging occurs through mechanical agitation. The resultant aggregate (pH 5) is adjusted with sodium hydroxide or lime to pH 9 or 9.5 in the main neutralization/precipitation reactor to precipitate the heavy metals as hydroxides.

The method of acid/alkali mixing is a highly site-specific wastewater treatment process. It requires a detailed collection of process data (flow rate, batch dumping frequency, etc.), as well as comprehensive laboratory analysis to determine waste stream characteristics and variability. Once this is completed and the reaction kinetics, products, and processing requirements are determined to be favorable, the two types of waste may be combined approximately in stoichiometric equivalents, to neutralize the component which is above an acceptable level for discharge. For example, a 20,000 gpd concentrated (8 percent sulfuric acid) spent pickling liquor waste stream would require approximately 13,050 gpd of a general purpose alkaline cleaning solution (10 percent sodium hydroxide) to achieve hypothetical neutralization. In actual practice, a greater quantity of sodium hydroxide solution will be required due to excess hydroxide demand for metallic species such as iron present in the pickle liquor.

Typically, in continuous mutual neutralization operations where acid/alkali mixing is the primary treatment, provisions are made for storage tanks if either waste stream is produced in excess of required quantities. In addition, an emergency reagent feed system is sometimes required if either stream may be generated in less than the specified quantity or concentration to ensure a uniform discharge. Retention time is also usually greater than for other neutralization systems due to the dilute nature of the reagent streams.

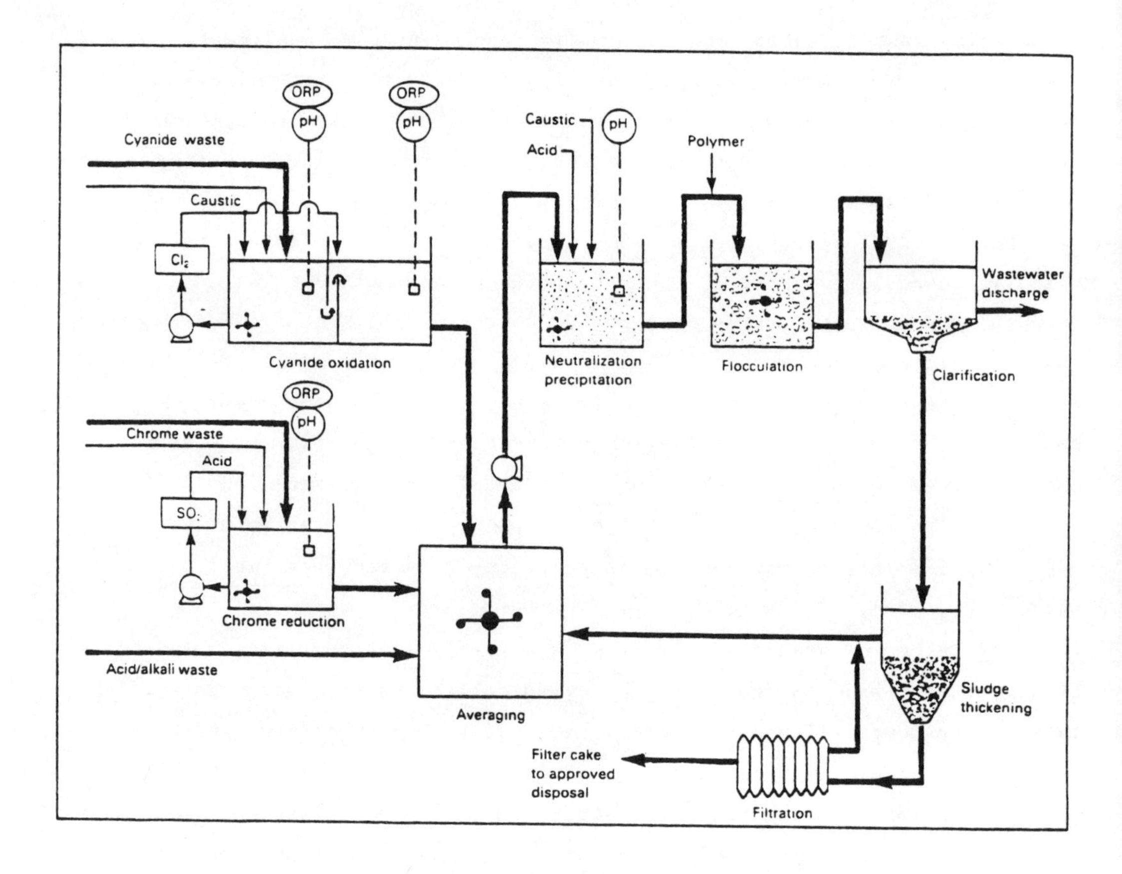

Figure 4.2.1 Conventional wastewater treatment system for electroplating.
Source: Reference 2

Mutual neutralization systems are not limited to liquid/liquid application. They can be operated with either liquid/gaseous (e.g., boiler flue gas, Section 4.7) or liquid/solid (e.g., waste carbide limes, Section 4.3) reagents. Semi-solids such as battery paste have also been utilized in recent mutual neutralization applications.[3]

When onsite waste streams are not available, they might possibly be obtained from a nearby plant that is producing compatible acid/alkaline wastes; e.g., waste pickling liquor from a metal finishing operation and waste carbide lime from an acetylene production plant. Such waste exchanges can be conducted either privately or through the use of a commercial waste exchange firm. A more detailed discussion of waste exchanges can be found in Section 5.9.

4.2.2 Process Performance

The following case studies contain performance data on both simple (two streams) and complex (three streams or more) acid/alkali mixing applications. In addition, the data have been gathered from a variety of industries and manufacturing operations and is intended to illustrate the versatility of applying mutual neutralization to corrosive waste stream treatment.

A battery paste consisting of lead oxide, free lead, and lead sulfate is currently used in the treatment of sulfuric acid wastewater produced in a lead-acid battery manufacturing plant.[3] The paste (a battery manufacturing byproduct) is fed into the system in the form of a dilute slurry (1 to 5 percent solids) at typical strengths of 4 to 10 lbs. of paste/pound of sulfuric acid, depending on the lead oxide content of the paste.

The reaction takes place continuously (16,500 to 22,000 gpd) in multi-stage reactors provided with mechanical agitation. With an influent stream pH of 1, the effluent pH averaged about 6.9 with 6.0 as the lowest recorded pH and 8.4 as the highest in a one month period. In addition, the lead oxide treatment had the added benefit of lowering the sulfate ion content in the effluent to 323 to 450 mg/L. By comparison, a similar plant using sodium hydroxide to regulate the acidity of the wastewater produced sulfate ion concentrations of 1,000 to 2,000 mg/L or more.

A second example of mutual neutralization involves a high-volume, two stream integration at a major automobile manufacturing plant. The wastes to be neutralized include a 200,000 gpd alkaline waste stream and a 60,000 gpd acid waste stream.[4] An engineering study indicated that the combined flow would be of sufficient nature to correct the pH to a range of 6.0 to 10.5 prior to discharge to the sewer system. To prevent any high alkaline conditions that could raise the pH above discharge limits, an auxillary neutralization system was installed. The system consisted of a pH control system, 2-Milton Roy acid pumps, and a 10,000 gallon sulfuric acid storage tank. The sulfuric acid tank was located immediately adjacent to a private railroad spur, so that acid could be purchased in tank car lots for price advantage.

A third case of simple mutual neutralization examined the feasibility of combining selected acid and alkaline wastes from coal-fired power plants.[5] The goals of this research program were to minimize hazardous waste generation and effectively isolating any hazardous constituents, through mutual neutralization and subsequent precipitation. The two waste streams of interest consisted of the supernatant from an aqueous fly ash suspension (pH 12.4) and an acid-iron boiler cleaning solution (pH 0.7). When the two wastes were combined, the neutralization reaction was practically instantaneous (pH 9.0) with both major and minor trace metallic contaminants being effectively adsorbed (see Table 4.2.1). The authors suggest that co-neutralization of acidic wastes with fly ash is readily applicable to other industrial waste streams since, with a small dosage of polymeric coagulant, liquid/solid separation is easily achieved.

The fourth case study involves a slightly more complex mutual neutralization application at a metal finishing facility.[6] In this application, the waste treatment system combined a metals-laden waste acid stream with two highly alkaline waste cleaner streams in a continuous acid/alkali pretreatment operation. In this manner the facility was able to economically reduce neutralization reagent costs prior to final pH adjustment and effectively remove metals (primarily iron) through hydroxide precipitation. The acid waste stream consisted primarily of a spent pickle liquor solution which averaged 14,000 gpd and pH 2.1 or less. The second stream consisted of an industrial detergent with pH 12.9, average flow rate of

TABLE 4.2.1 CHARACTERIZATION OF MUTUAL NEUTRALIZATION WASTE STREAMS FROM COAL-FIRED PLANT APPLICATION

Constituent	Fly ash solution	Boiler cleaning waste	Combined[a] system
pH	12.4	0.7	9
Alkalinity (mg/L)	878	--	--
Cl mg/L	2.5	40,000	1.2×10^{-2}
SO_4 mg/L	---	130	1.5×10^{-5}
Na mg/L	5.3	43	2.4×10^{-4}
K mg/L	1.4	0.31	3.4×10^{-5}
Mg mg/L	0.6	3.7	2.4×10^{-5}
Ca mg/L	99.4	3.3	2.4×10^{-3}
Fe mg/L	---	5,130	1.0×10^{-3}
Cu μg/L	0.4	159,000	2.6×10^{-5}
Zn μg/L	2.0	15,890	2.7×10^{-6}
Cd μg/L	---	22	1.5×10^{-9}
Cr μg/L	119	1,620	2.5×10^{-6}
Se μg/L	---	5	---
As μg/L	0.1	7	1.3×10^{-8}
Pb μg/L	---	20.5	---
Ni μg/L	5	2,910	1.5×10^{-6}
NH_3 mg/L	---	---	Not measured

[a] Concentrations express in M/L.
Source: Reference 5

12,000 gpd and contained 74 mg/L of calcium. The third stream, a highly alkaline (pH 13.1) scrap cleaner had an average flow rate of 19,000 gpd. The resultant flow stream had a pH of 12.3 which was subsequently adjusted by the addition of sulfuric acid to bring the effluent within discharge limits. A summary of waste stream characteristics is provided in Table 4.2.2.

The fifth case study involved a high volume, complex acid/alkali mixing application at a major manufacturer of household appliances in Allentown, PA.[7] The process operates as a primary neutralization system and consisted of the integration of an acid, alkaline, and mixed waste stream. The acid waste stream (60,000 gpd) contained wastes from pickling solutions, acid rinses, and chromium stripping solutions. The alkaline waste stream (43,200 gpd) contained primarily spent cleaning solutions. The mixed waste stream (63,000 gpd) consisted of process spills, demineralizer regeneration waters and miscellaneous acid/alkali rinses. Following sodium bisulfite-hexavalent chrome reduction, the three streams were combined into one of two 12 foot reactor-clarifiers. After clarification, solids were removed through the use of a belt filter and then disposed in a offsite landfill. The characteristics of each of the three waste streams and the final effluent are summarized in Table 4.2.3.

4.2.3 Process Costs

The costs associated with two stream acid/alkali mixing can be described in terms of capital expenditures, operation, and maintenance costs. Specific requirements will depend on the specific nature of the waste streams involved. An example cost analysis is provided herein for a mixed waste system which requires solids removal and sludge consolidation. Operation and maintenance costs for this system include; depreciation, interest, taxes, insurance, maintenance, labor, and sludge disposal. Capital costs include; collection sumps and associative piping, a two stage continuous reactor system, emergency pH control system, a flocculation/clarification unit, sludge storage, and a high pressure filter press. Equipment sizing is based on combined acid/alkali flows of 400, 2,500, and 3,500 gallons per hour, 8 hours a day, 300 days per year. The acid stream is assumed to be 60 percent of the

TABLE 4.2.2 SUMMARY OF APPLICATION OF MUTUAL NEUTRALIZATION TO METAL FINISHING WASTES

Parameter	Stream 1	Stream 2	Stream 3	Combined stream
Flow rate (gpd)	14,000	12,000	19,000	45,000
Percent of total	31	27	42	100
Alkalinity (mg/L $CaCO_3$	--	7,600	10,950	6,400
Acidity (mg/L)	1,030	--	--	--
pH	2.1	12.7	13.1	12.3
Calcium concentration (mg/L)	--	--	74	31

Source: Reference 6

TABLE 4.2.3 AVERAGE QUALITY OF ACID/ALKALI WASTE STREAMS

Parameter[a]	Acid Stream	Alkali Stream	Mixed Stream	Combined Effluent Discharge
Suspended Solids	26	1,400	67	5
Total Alkalinity[b]	--	--	900	125
Total Acidity	7,400	1,730	--	25
Total Cr	307	9.7	2.9	0.75
Cr^{+6}	270	2.8	--	--
Fe	45	--	0.2	0.1
Cl	1,200	130	200	56
SO_4	660	16	12	160
PO_4	575	176	191	70
Ni	35	2	51	0.2
Cu	1.7	0.1	--	0.1
Al	39.6	7.6	3.3	0.05
Ph	1.2	12.4	2.4	7.4

[a]All concentrations expressed in mg/L.

[b]Expressed in mg/L $CaCO_3$.

Source: Reference 7

total flow and contain 8 percent sulfuric acid and 24 mg/L of ferrous iron. The alkaline stream is assumed to be 40 percent of the total flow and contains 10 percent by weight of sodium hydroxide.

Prior to acid/alkali mixing, each waste type must be segregated in separate pipe systems and drained into the appropriate collection sump. Each system consists of two sumps sized for a maximum of 30 minutes of detention time. Each sump is a preformed PVC inground tank with sump pump, level control, and steel grating cover. In addition, provision has been made for two wastewater collection systems, each consisting of; 60 feet of 6 inch diameter pipe with two bends and six connections. The following costs are based on updated (July, 1986) costing information contained in "Reducing Water Pollution Costs in the Electroplating Industry":[8]

- Collection Conduit

Linear Runs	4" - $1.55/ft 6" - $2.40/ft
Each Connection	4" - $210.00 6" - $360.00
Each Bend	4" - $210.00 6" - $360.00

- Collection Sumps

Collection Sumps	100 gallons - $2,100 300 gallons - $2,600 500 gallons - $3,100 1,000 gallons - $3,600

The costs for the two stage continuous reactor system are based on a first stage flow concentration/equalization tank and a second stage final pH adjustment tank equipped with an emergency pH control system. Figure 4.2.2 presents mixed reactor construction costs for the first stage reinforced concrete tank applicable to vessel sizes of 14.2 m^3 to 70.8 m^3. Figure 4.2.3 presents mixed reactor construction and installation costs for the second stage continuous neutralizer complete with pH control and emergency reagent storage and feed systems. The reagent feed and storage system is sized for a 1-week supply and uses caustic soda and sulfuric acid as the neutralizing agents.

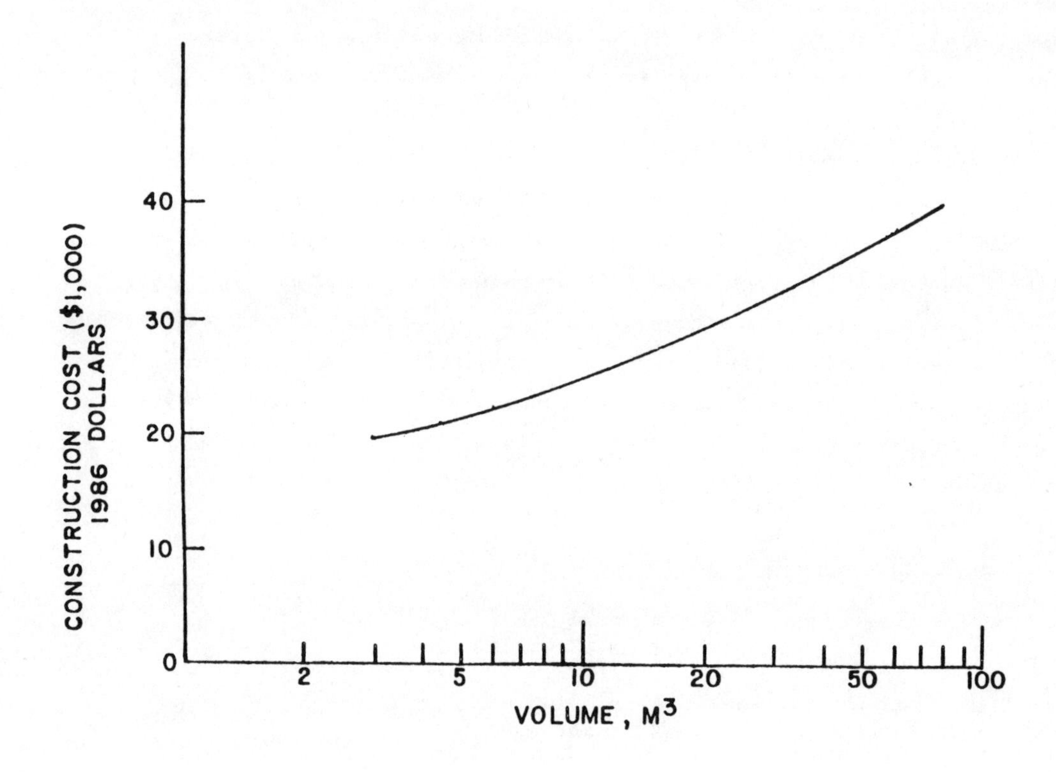

Figure 4.2.2 Construction costs for reinforced concrete reactor.

Source: Reference 1

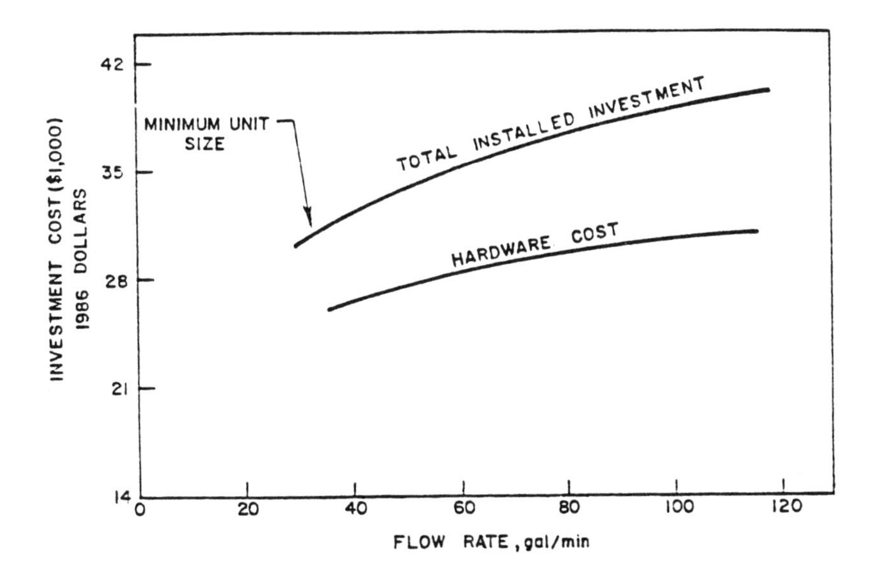

Figure 4.2.3 Investment cost for continuous neutralization unit.

Source: Reference 8

The flocculater/clarifier, sludge storage, and filter press unit costs are based on information contained in Figures 4.1.6, 4.1.8, and 4.1.9, respectively. Sludge generation is assumed to be 0.68 gallons of sludge/gallon of clarified underflow processed. The density of the sludge is 8.34 lbs/gallon of sludge based on a 2 percent solution which is subsequently dewatered to 20 percent solids in the filter press.

Yearly operation and maintenance costs are based on percentages of the total annualized capital. Taxes and insurance are assumed to be 7 percent, while general maintenance and overhead is 5 percent.[8] Labor is assumed to be 4 hours/day, 300 days/year, at a rate of $20/hour. Sludge disposal costs are based on the operation of the filter press for the 400, 2,500, and 3,500 gallon/hour systems at rates of 5, 28, and 40 gallons/hour of clarifier underflow. Approximately 16.7 lbs of dry solids are assumed to be generated per 100 gallons processed. At 20 percent solids and assuming a disposal rate of $2.0/gallon of sludge, disposal costs can be estimated using Figure 4.1.10. A summary of mutual neutralization costs is presented in Table 4.2.4 for each of the three flow rates. As shown, significant economies of scale are acheive, particularly between the 400 and 2,500 gpm flow rates. Labor requirements are essentially the same regardless of system size. As capacity increases, disposal costs and sludge handling equipment show the most significant increase relative to toal treatment cost. Thus, it is more critical for high flow systems to select wastes which result in minimal sludge generation.

4.2.4 Process Status

The mixing of acid/alkali wastes is both technically feasible and widely applied. Its primary advantage is reduced cost since neutralizing reagent requirements are minimized. The main disadvantage is that mixing two waste streams, each with its own variability in composition and flow, may require more conservative system design; i.e., larger equilization and neutralization tanks and back-up reagent grade neutralization systems. Additionally, care must be exercised when combining waste streams or accepting wastes from another firm to prevent any hazardous by products or releases in the neutralization reaction.

TABLE 4.2.4. MUTUAL NEUTRALIZATION COSTS

	Flow rate (gph)		
	400	2,500	3,500
Capital investment ($)			
Collection sumps (2)	6,200	7,200	7,200
Piping systems (2)	5,676	5,676	5,676
1st stage reactor	17,000	19,000	21,000
2nd stage reactor with emergency pH control system	30,000	32,000	35,000
Flocculation/clarification unit	22,000	32,000	40,000
Sludge storage unit	2,000	10,000	12,000
Filter press	11,000	13,500	18,000
Total capital costs	93,876	119,376	138,876
Annualized capital[a]	16,616	21,129	24,581
Operating costs ($/yr)[b]			
Taxes and insurance (7%)	1,163	1,479	1,721
Maintenance and overhead (5%)	830	1,056	1,229
Labor and overhead ($20/hr)	24,000	24,000	24,000
Sludge disposal ($2.0/gal)	2,000	11,190	16,008
Total cost/yr ($)	44,609	58,854	67,539
Cost/1,000 gallons ($)	47	10	8

[a]annualized capitol costs derived by using a capital factor:

$$CRF = \frac{i\ (1+i)^n}{(1+i)^n - 1}$$

where: i = interest rate and n = life of the investment. A CRF of 0.177 was used to prepare cost estimates in this document. This corresponds to an annual interest rate of 12 percent and an equipment life of 10 years.

[b]Does not include utilities.

Source: Adapted from References 1, 2, and 8 using July, 1986 CPI index.

The environmental impact of this technology is similar to that of other neutralization technologies provided that incompatible waste constituents do not generate toxic by-products. However, as with most neutralization technologies, a sludge product is generated through the formation of insoluble by-products and the precipitation of metals.[9] Any technical developments which reduce the amount of sludge and enhance its filtering and settling characteristics will improve the acceptance of this type of neutralization.

REFERENCES

1. MITRE Corporation/Manual of Practice for Wastewater Neutralization and Precipitation. EPA-600/2-81-148. August 1981.

2. U.S. Environmental Protection Agency. Economics of Wastewater Treatment Alternatives for the Electroplating Industry. EPA-625/5-79-016. June 1979.

3. Yehaskel, A. Industrial Wastewater Cleanup, Recent Development. Noyes Data Corporation, Park Ridge, NJ. USA 1979, pp. 250-251.

4. Besselievre, E.B. The Treatment of Industrial Wastes. McGraw Hill Book Company, New York, NY. 1967.

5. Benjamin, M.M. Removal of Toxic Metals from Power Generation Waste Streams by Absorption and Co-precipitation. Journal of the Water Pollution Control Federation. November 1982.

6. Price, F., et al. Report on Business and the Environment. McGraw-Hill Book Company, New York, NY. 1972.

7. Anderson, J.S. Case History of Wastewater Treatment in a General Electric Appliance Plant. Journal of the Water Pollution Control Federation. October 1968.

8. U.S. EPA. Reducing Water Pollution Costs in the Electroplating Industry. EPA-625/5-85-016. September 1985.

9. Arthur D. Little, Inc. Physical, Chemical, and Biological Treatment Techniques for Industrial Wastes. U.S. EPA SW-148. November 1976.

4.3 LIMESTONE TREATMENT

4.3.1 Process Description

Limestone treatment is a well-developed and established technology for the neutralization of acidic waste streams. Limestone is a particularly effective reagent for the neutralization of dilute acid waste streams containing low concentrations of acid salts and suspended solids.[1] With modifications, it may also be a good neutralizing agent for many of the acidic waste streams considered in this document, either as a primary treatment for weak acids or as a pretreatment for other processes i.e., partial neutralization. However, in most applications, limestone has been replaced by more cost-effective reagents such as lime slurry and caustic soda which eliminate solids handling problems. In addition, caustic soda results in reduced sludge generation.

Limestone is available in either high calcium ($CaCO_3$) or dolomitic ($CaCO_3$ $MgCO_3$) form.[2] A summary of physical and chemical properties is provided in Table 4.3.1. Both types of limestone are available as either a powder or crushed stone. Crushed stone diameters are typically 0.074 mm (200 mesh) or less since both the reactivity and completeness of the reaction increase proportionately to the available surface area.[3] High calcium is most commonly used because of its greater reaction rate and its more widespread availability. Dolomitic limestone reactivity will increase if finely ground and sludge production will be minimal due to the formation of soluble magnesium sulfate. However, its reactivity is generally too slow even with grinding, and hence not suitable for most applications.

Various configurations are available for limestone treatment applications as illustrated in Figure 4.3.1. Influent characteristics, effluent criteria, operational and economic constraints will provide the basis for design selection. Waste streams are introduced to the neutralization media in an upflow direction with the application rate dependent on the stone size. For example, at rates of 50 $gal/ft^2/min.$, any stone smaller than 30 mesh will be swept from the bed. Therefore, when utilizing limestone powder (200 mesh), flow rates are lowered to 1 to 5 $gal/ft^2/min$, thereby limiting waste

TABLE 4.3.1 SUMMARY OF HIGH CALCIUM AND DOLOMITIC LIMESTONE PROPERTIES

Parameter	Unit	High calcium	Dolomitic
Chemical Analyses	wt (%)		
CaO		58.84	30.07
MgO		0.26	20.75
CO_2		43.26	46.02
SiO_2		1.14	0.14
Al_2O_3		0.41	1.90
Other		---	1.12
Bulk Density	kg/m^3	2,000 - 2,800	2,050 - 21,870
Specific Gravity	H_2O = 1	2.65 - 2.75	2.75 - 2.90
Solubility	g/100 g water	0.0014	0.0245
Molecular Weight	g/gmole	100.1	184.39
Specific Heat	Cal/g/ C	0.19 - 0.26	0.19 - 0.294
Stability	°C	898	725
Basicity Factor	CaO = 1	0.489	0.564
pH	H_2O = 7.0	8 - 9	8.5 - 9.2

Source: References 2 and 3

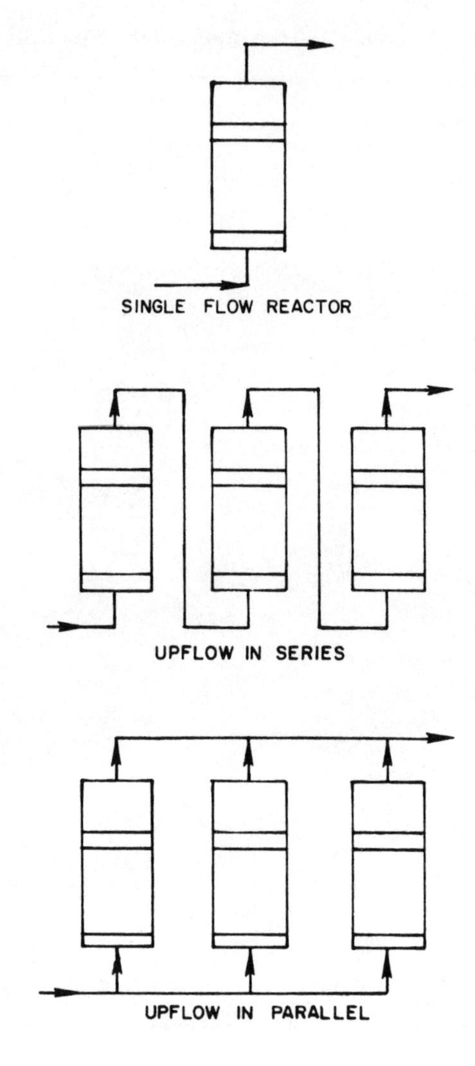

Figure 4.3.1. Upflow limestone bed neutralization process configurations.

throughput.[4] The upflow expanded configuration permits the use of smaller particle size, limits channeling, increases neutralization rates, and decreases required bed size.

There are two basic modes of operation for limestone beds; fixed and moving. In the fixed bed mode, the entire bed is removed from service when the stones become inactive. In moving beds, a high rate of application washes away any insoluble particles which might coat the limestone surface, extending the bed life indefinitely. For arrangements in series, when the first bed becomes saturated, effluent quality is maintained in the second. This method is generally applied to waste streams with high solids content and low application rates. Conversely, beds in parallel are primarily used for high volume applications where suspended solids content in the influent is low.

Figure 4.3.2 is a schematic of a single bed, upflow treatment system utilizing a settling and mixing section prior to neutralization.[5] In this system, a uniform, dilute waste is fed to the bed through a large holding capacity, perforated pipe feed apparatus. Influent dilution with recycled treated product or more neutral process streams is necessary since some strong mineral acids such as sulfuric acid will react with limestone forming insoluble calcium sulfate. Generally, sulfuric acid concentrations greater than 1.3 percent (although greater than 5,000 mg/L is not recommended) will result in the sulfurization of the stone particles.[4] When this occurs, the neutralization capacity of the stone particles is greatly reduced and the bed is eventually rendered inactive.

Aeration of the treated waste stream is frequently desireable since carbon dioxide is evolved during the neutralization process. Stripping of CO_2 through some form of mechanical aeration will prevent the formation of carbonic acid which would otherwise depress the effluent pH.

Process operation data for full-scale applications are incomplete for three primary reasons; either the essential data have not been generated, is considered proprietary, or is outdated due to the increasing preference of alternate alkali reagents in recent years. The available information does, however, contain material gathered from a wide variety of data sources; including manufacturers, trade associations, industrial users, and available literature. Table 4.3.2 gives a composite summary of several important parameters in a "typical" upflow limestone treatment bed operation. These parameters will be covered in more detail in later sections.

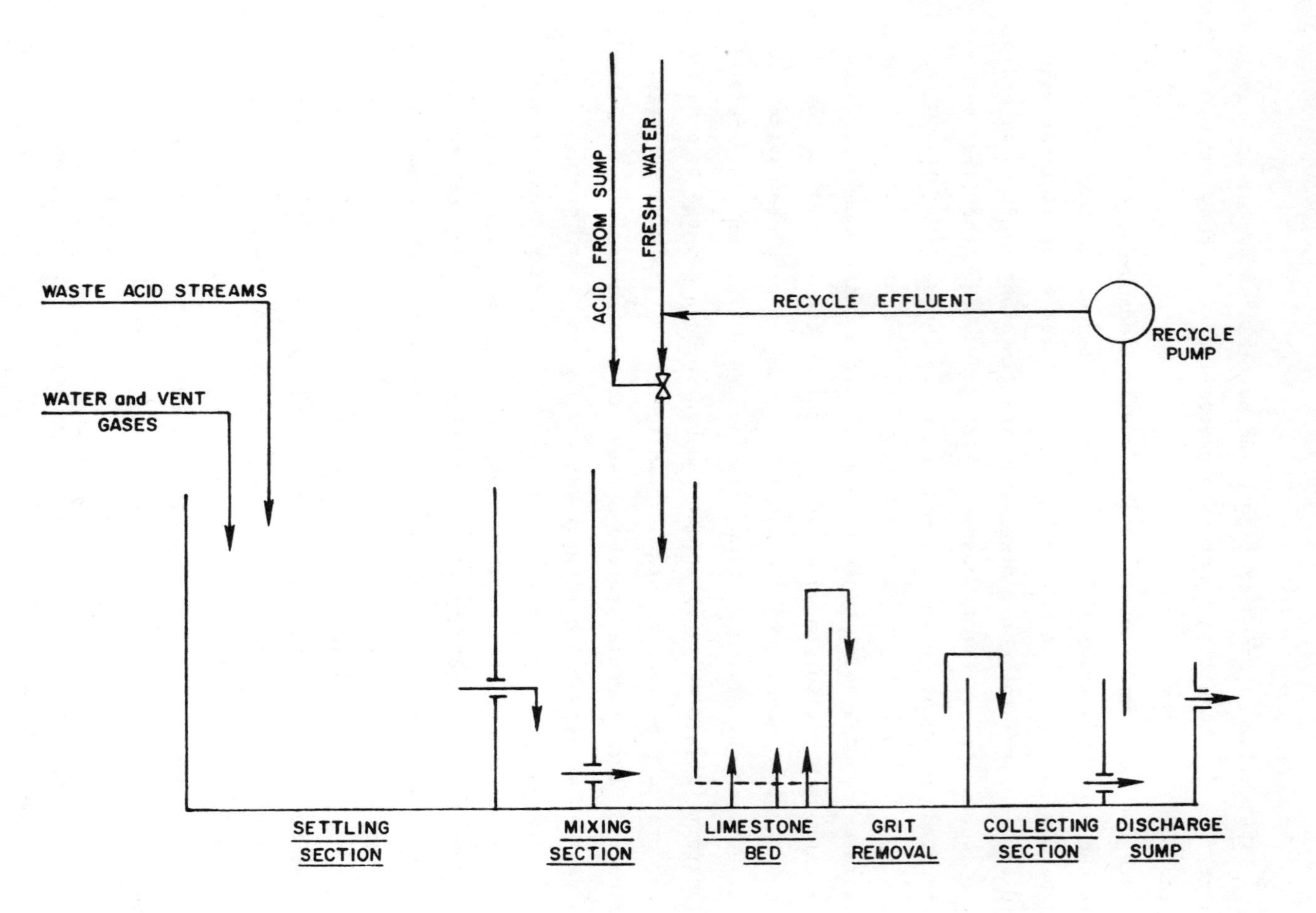

Figure 4.3.2. Single bed, upflow limestone treatment system schematic.
Source: Reference 5.

TABLE 4.3.2 SUMMARY OF TYPICAL OPERATING PARAMETERS

Parameter	Unit(s)	Operating range	Ideal range
Application Rate	$gal/ft^2/min$	1 -80	1 - 5
Type of Stone	% Magnesium Oxide	High Calcium - Dolomitic 5% MgO - 40% MgO	5
Temperature	°C	15 - 30	15
Incoming Mineral Acidity	mg/L	5,000 - 7,000	5,000
Bed Depth	ft	2 - 4	3
Stone Size	Mesh	80 - 200	200
Aeration Time	Minutes	0 - 10	2

Source: References 3, 4, 5, 6

4.3.2 Process Performance

In evaluating the effectiveness of limestone as a neutralization agent, parameters of interest are; type of bed, type of limestone, flow rate, influent mineral acidity, bed depth, effluent pH, rate of reaction and pre- and post-treatment requirements. This section contains several case studies in which limestone treatment was evaluated for neutralizing dilute acidic waste streams. A summary of performance data and process parameters for limestone treatment beds is presented in Table 4.3.3. Data for the individual waste streams are, with few exceptions, for influent concentrations of 5,000 to 10,000 mg/L of mineral acidity.

One case study involves a bench-scale evaluation of the neutralization of a dilute nitric and sulfuric acid waste stream from the manufacture of nitro-cellulose (Table 4.3.4).[4] The parameters evaluated were rate of application, influent mineral acidity, type of stone, effect of aeration, and sludge product formed. The experimental unit employed a bed of fine stone (approximately 8 to 30 mesh) in which the waste was applied at high upflow rates. The rates of application ranged from 20 to 80 $gal/ft^2/min$. The waste acidity was varied from 2,950 ppm to 12,400 ppm, but 5,000 ppm of sulfuric acid acidity was judged optimal due to calcium sulfate formation. Bed depths were also varied, but sufficient depth was allowed for decomposition of the stone by the acid. Depths of 2 to 4 feet were recommended so that expansion was not excessive,

Results of the study indicated that both amorphous and crystalline high-calcium limestone were satisfactory reagents while dolomitic limestone did not achieve satisfactory performance. Wastes containing appreciable concentrations of acid salts were not neutralized at high rates, possibly due to bed clogging or fouling. Finally, acid wastes relatively free of solids can be continuously neutralized by an upflow rather than downflow application to beds utilizing limestone gravel. Since limestone is limited in its ability to neutralize acidic solutions above pH 7.0 (over-neutralization), it generally cannot be used as the sole reagent in applications where metal precipitation is required.

TABLE 4.3.3 SUMMARY OF LIMESTONE BED TREATMENT PERFORMANCE DATA

Application	Type of bed	Type of limestone	Flow rate (gal/ft^2/m)	Concentration (mg/L)	Bed depth (ft)	Bed area (ft^2)	Effluent pH
Resin Manufacturing Waste[a]	Upflow	Unspecified	13.6 g/ft^2/m	1% HCl	3 ft	113.5 ft^2	Unspecified
Silicon Products Wastes[b]	Upflow	94% Soluble High calcium	20 - 30	< 1% HCl	3	27	4.0 - 6.0
Acid Mine Drainage[c]	Upflow (pilot)	Dense High Calcium (England)	6.1	5,000 mg/L H_2SO_4	10	.785	5.8 - 6.5
Nitro-Cellulose Waste[d]	Upflow (lab)	Calcite	21.6	5,660 mg/L 34% NHO_3 66% H_2SO_4	2	.083	5.4
Nitro-Cellulose Waste	Upflow (lab)	Calcite	16.9	2,950 mg/L	1	.083	4.7
Nitro-Cellulose Waste	Upflow (lab)	Amorphous	15.8	2,950 mg/L	1	.083	4.8

[a]Source: Reference 7

[b]Source: Reference 4

[c]Source: Reference 8

[d]Source: Reference 5

TABLE 4.3.4 CASE STUDY 1. BENCH SCALE LIMESTONE TREATMENT DATA SUMMARY

Rate of application ($gal/ft^2/min$)	Influent Mineral acidity (ppm)	Effluent pH	Type of stone	Aeration time (min)	$CaSO_4$ formed (ppm)
a	12,400	--	High calcium	--	11,084
a	6,200	--		--	5,542
a	3,100	--		--	2,271
21.6	5,660	5.4	High calcium	--	--
52.8	5,660	4.9		--	--
60.0	5,660	4.0		--	--
16.9	2,950	4.7	Calcite	--	--
21.2	2,950	4.2		--	--
34.0	2,950	2.7		--	--
15.8	2,950	4.8	Amorphous	--	--
21.6	2,950	4.2		--	--
44.0	2,950	2.6		--	--
b	b	4.3	High calcium	0	--
b	b	7.4		3	--
b	b	8.0		10	--

[a]Liter portions of waste were stirred rapidly after addition of limestone.

[b]Aeration of effluent having a pH of 4.3 with diffused air.

Source: Reference 4

Another case study describes a treatment train in which limestone is used in conjunction with biochemical oxidation for the removal of acid salts. The evaluation consisted of several bench-scale studies followed by a pilot plant application (Table 4.3.5).[8] The waste to be treated was acid mine drainage which contained less than 5,000 mg/L of sulfuric acid (pH 2.8) and a high concentration of acid salts (primarily ferrous sulfate). The pilot plant was designed to convert ferrous-to-ferric salts through biochemical oxidation, followed by neutralization with limestone (see Figure 4.3.3). The neutralization system consisted of two 10 foot (0.785 ft^2 diameter) limestone grit reactors arranged in series. The system was operated for 28 months at an average temperature of 20°C with effluent pH ranging from 5.8 to 6.5.

The lower limit of application for the system was expected to be 10 to 20 mg/L of dissolved iron and a total acidity of approximately 25 mg/L of sulfuric acid. The upper limit was assumed to be 5,000 mg/L H_2SO_4, but higher limits may be achieved at lower operating temperatures (15°C) and through the use of aeration. Peak flow rate for an operational system was estimated at approximately 832,000 gpd with an average flow rate of approximately 240,000 gpd.

The inherent problem with this type of neutralization/precipitation system is that it is only effective for reducing metallic species such as trivalent chrome and iron in its operational (pH) range. In addition, the high solubility of Cr^{+3} (approximately 8 mg/L) in comparison to Fe^{+3} (0.0075 mg/L) at pH 6.0, limits this treatment in potential applications to primarily the removal of acid iron salts.

While it has been demonstrated in the previous two examples that limestone neutralization is possible, the inhibition of the stone particles in the presence of high quantities of sulfate and/or metallic ions make it less attractive than other reagents. Limestone is a solid-based reagent that liberates CaO for neutralization through surface dissolution. The inihibition of the particle surface through calcium sulfate precipitation increases retention times, reagent purchases, equipment sizing, and lowers waste throughput. Improved reaction kinetics can be achieved by increasing the available solid surface area through greater limestone loading. However, both reagent purchase and sludge disposal costs will increase proportionately with the excess limestone applied.

TABLE 4.3.5 SUMMARY OF LIMESTONE TREATMENT PILOT PLANT DATA

Test Series	Limestone particle size	Weight of limestone (kg)	Acid mine drainage feed	Rate of neutralization of strong acid salts	Flow Configuration and mixing	Application rate	Loss of activity or blocking of bed
Preliminary Experiments							
Powder	<60 B.S.S.	0.05 - 1	Aliquot	Fast	Agitated	--	--
First Grit	3/16 x 1/4 in	35	Continuous	Fast initially	Down	20 gpm	Blocked in 48 hours
	1/2 x 3/4 in	300	Continuous	Fast initially	Horizontal	20 gpm	Blocked in few days
Second Grit	12 B.S.S. to 1/2 in	1.3	Aliquot	Fast	Aerated	--	Not determined
Third Grit	12 to 6 B.S.S. and 1/8 - 1/4 in	1.7	Continuous	Fast	Down or up	25 mL/min	Blocked in 70 hours, restored by upflow expansion
Fourth Grit	1/16 to 1/4 in	105 - 150	Continuous	Fast	Down	750 mL/min	Blocked 50 - 150 hours, restored by upflow expansion
Fifth Grit	60 - 22 B.S.S.	0.5	Aliquot	1-4 min	Tumbling	--	None
Sixth Grit	60 - 16 B.S.S.	7	Continuous	4 min	Up	3.3 ft/min	25 hours to loss of activity
Secondary Experiments							
Pilot Plant	60 - approximately 6 B.S.S.	40	Continuous	4 min	Up	3.3 ft/min	No loss of activity with attributors

Source: Reference 8

B.S.S. = British sieve size.

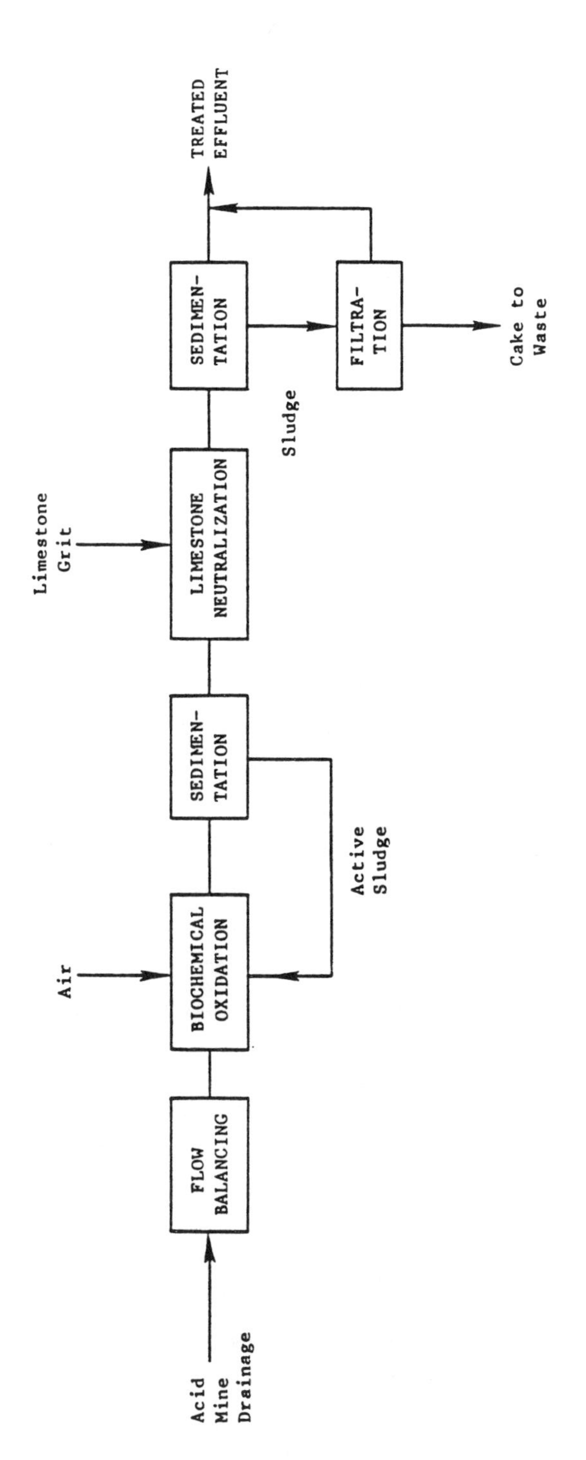

Figure 4.3.3 Flow diagram of complete biochemical oxidation and limestone neutralization process.

Source: Reference 8

In 1981, Volpicelli et al. sought to increase limestone dissolution rates by decreasing the mean particle size of the stone.[9] Crushed stones of various diameters were applied in stoichiometric amounts in a controlled stirred reactor. The waste stream that was neutralized was a sugar plant effluent (pH 0.9) containing 17,500 mg/L of sulfuric acid. The experimental results, which are presented in Table 4.3.6, demonstrated that stone sizes greater than 400 mesh (38 microns) were only effective when applied in excesses of 220 to 550 percent. However reaction rates were slow (35 to 240 minutes) and an excessive quantity of reagent was wasted (54 to 82 percent). When the mean particle size was reduced to 400 mesh or less, complete neutralization (pH 7.0) was achieved in 15 minutes due to the increased dissolution rate. More significantly, only a 10 percent excess of limestone powder was required, resulting in only 9 percent of the available surface area being rendered inactive.

Figure 4.3.4 illustrates a two-stage, backmix flow reactor system proposed by Volpicelli for the continuous application of limestone powder. With optimum agitation, the reagent loading should approach the solid surface area to liquid volume ratio determined in the laboratory. Provision is made in the system design for an aeration vessel prior to sludge thickening to raise the final pH from 5.75 to 7.0. Aeration is required since carbon dioxide is evolved as a by-product in the limestone neutralization process. The carbon dioxide will often combine with free hydrogen to form carbonic acid which will depress the pH endpoint. Therefore some sort of desorption apparatus such as an aerator or a drop from a weir is employed to strip the CO_2 prior to final discharge.

4.3.3 Process Costs

Table 4.3.7 summarizes the process costs for the construction and installation of a continuous limestone powder neutralization system. Equipment sizing and operating costs were based on the neutralization of a 2 percent sulfuric acid waste stream with various flow rates. The flow rates are assumed to be 400, 2,500, and 3,500 gallons/hour, operating 2,400 hours/year. The capital costs include; a collection sump and associative piping, a two-stage continuous reactor, a dry powder feed system,

TABLE 4.3.6 SUMMARY OF LIMESTONE NEUTRALIZATION EXPERIMENTS

Run	Liquid volume (L)	Particle size (mm)	Limestone weight (g)	Limestone Excess[a]	Reaction time (min)	Final pH	Surface area (cm^2/L)	Limestone conversion (%)
1	.75	1-1.25	16	1.18	90	1.12	420	16.4
2	.50	1-1.25	50	5.49	130	1.60	1,960	9.4
3	.50	.42-.6	50	5.49	240	7.00	4,300	18.0
4	.50	.2-.42	50	5.49	35	6.25	7,200	18.0
5	.50	.2-.42	25	2.76	180	2.00	3,600	25.0
6	.50	.09-.125	20	2.20	35	6.90	8,200	46.0
7	.50	Powder	10	1.10	15	6.55	12,000	91.0
8	.25	Powder	5	1.10	15	6.55	12,000	91.0
9	.25	Powder	2.5	.55	15	1.30	6,000	98.0
10	.25	Powder	3.5	.77	15	1.70	8,400	96.0
11	.25	Powder	4	.88	15	2.70	9,600	98.0

[a]Ratio between loaded and stoichiometric weight of limestone.

Source: Reference 9

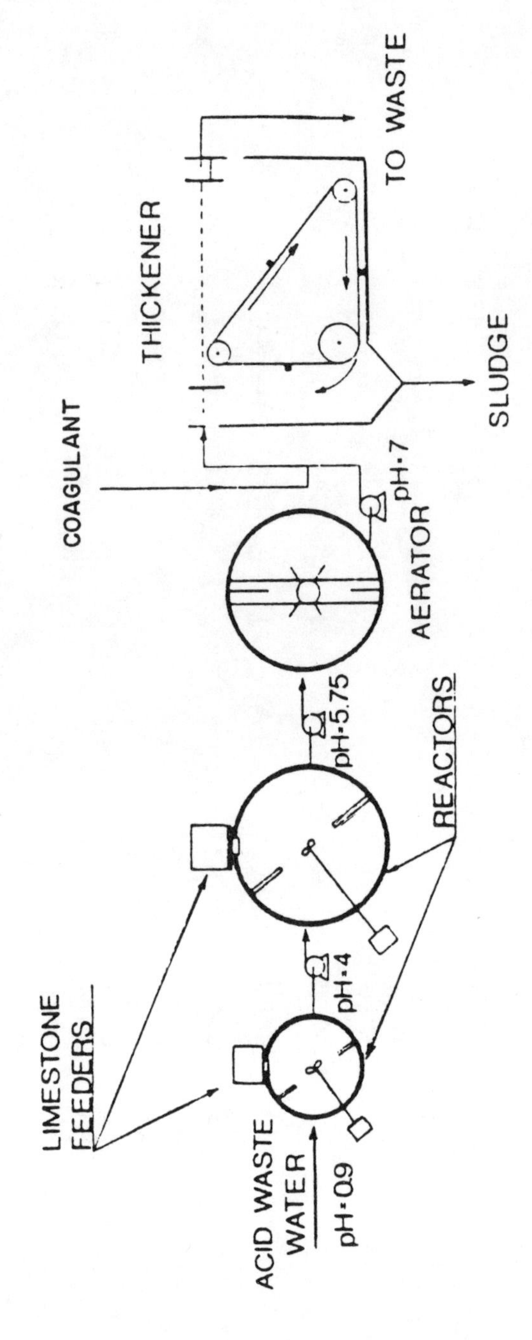

Figure 4.3.4 Continuous limestone powder neutralization process.

Source: Reference 9

TABLE 4.3.7 CONTINUOUS LIMESTONE POWDER TREATEMENT COSTS

	Flowrate (gph)		
	400	2500	3500
Captial Investment($)			
Collection sump	3100	3600	3600
Piping system	2838	2838	2838
2-Stage reactor	34,000	38,000	42,000
Feed system	37,078	118,648	133,479
Aeration vessel[a]	18,419	20,419	22,419
Flocculation/Clarification	22,000	32,000	40,000
Sludge storage	8,000	8,000	10,000
Filter press	11,000	16,500	21,000
Total Capital Cost	136,435	240,000	275,336
Annulized Capital	24,149	42,480	48,735
Operating Costs[b](%)			
Taxes and Insurance (7%)	1,690	2,974	3,412
Maintenance (5%)	1,208	2,124	2,437
Labor ($20/hr)	24,000	24,000	24,000
Sludge Disposal ($200/ton)[c]	64,514	456,809	565,499
Reagent Cost ($84/ton)	37,596	98,864	107,795
Total Cost/Year ($)	153,157	627,251	751,878
Cost/1,000 gallons	159	104	90

[a]Does not include cost of sparger pumps.
[b]Does not include utilities.
[c]Does not include gypsum refund.
Source: Adapted from References 10, 11, 12 using July, 1986 CPI index.

and sludge separation and handling equipment. Operational expenses were assumed to include taxes, insurance, maintenance, labor, sludge disposal, and reagent costs.

The wastewater collection system is comprised of one PVC in-ground tank, a sump pump, a level controller, and 30 feet of 6-inch pipe with one bend and three connections (for cost estimate see acid/alkali mixing). The costs for the reactor system are based on Figure 4.2.2 and consist of two reinforced concrete reactors sized for 30 minutes of retention time. The cost data for the feed system is based on Figure 4.3.5 and adapted construction costs for dry feed systems.[10] A 5 minute detention time is required in the dissolving tank and water is used at a rate of 2 gallons/pound of limestone. Conveyance from the solution tanks to the point of application is by dual head diaphragm retaining pumps.[10] The limestone is stored in mild steel storage hoppers located indoors and directly above the storage tanks (hopper facilities also include dust collectors).

The flocculation/clarification, sludge storage, and filter press unit costs are based on information contained in Figures 4.1.6, 4.1.8, and 4.1.9 respectively. Sludge generation is assumed to be 1.322 lbs of calcium sulfate precipitate per pound of sulfuric acid neutralized based on a calcium sulfate solubility of 0.241 grams/100 milliliters of solution. Final sludge volumes will contain 50 to 60 percent solids, although cost estimates are based on 60 percent as a maximum value attainable by a recessed plate filter press. The aeration vessel is assumed to be a standard reinforced concrete tank provided with 30 feet of sparger system piping. Cost estimates are based on Figure 4.2.2 but does not include purchase and installation of pumps and compressors which will add slightly to the final figure.

Operating costs for: (1) taxes and insurance; and (2) maintenance, are based on percentage of the total annualized capital investment; 7, and 5 percent, respectively.[11,12] Labor costs are based on 1,200 hours of manual operation per year at a basic labor rate of $20/hour, including overhead. Sludge disposal costs are based on dry sludge generation rates of 976, 8,078 and 11,310 lbs/day for the 400, 2,500, and 3,500 gph systems, respectively. Final sludge volumes are assumed to be 60 percent dry solids and disposed of at a cost of $200/ton. Due to the relatively pure state of the neutralization reaction byproducts, the precipated calcium sulfate (gypsum) is suitable for a variety of uses such as agricultural liming or

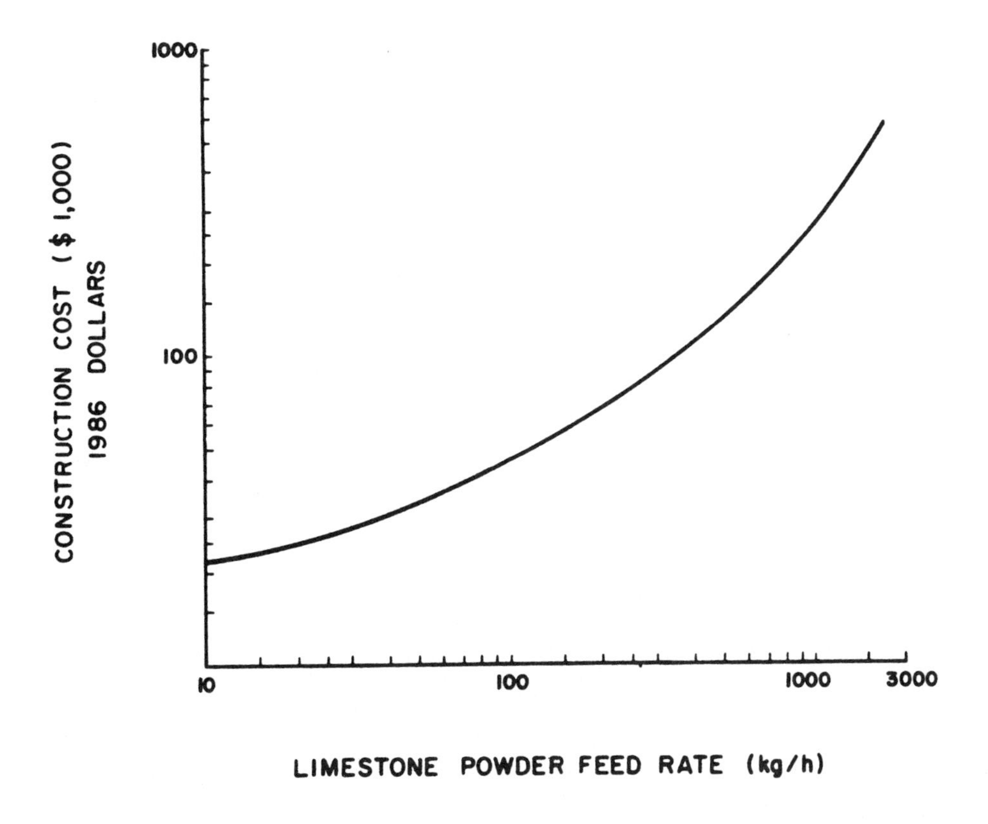

Figure 4.3.5. Construction cost for Limestone Powder Feed System.
Source: Reference 10.

gypsum board manufacture. Therefore, the possibility of waste exchange or reuse can significantly reduce or even eliminate the costs associated with disposal. Reagent costs are based on a quoted price of $84/ton (Lee Lime, Lee Massachusetts, 1986) for extremely fine (200 mesh or less) limestone powder. Reagent demand was assumed to 1.284 lbs of high calcium limestone powder per pound of sulfuric acid neutralized. This figure is based upon a basicity factor of 0.489 and a 10 percent excess due to insoluble limestone feed.[3]

As evidenced by Table 4.3.7 sludge disposal consitutes as much as 75 percent of the total yearly expenditures for this treatment application. While more efficient methods of sludge dewatering such as dyring ovens may reduce the overall costs of disposal, the implementation of a land disposal ban in the near future will make any process generating significant quantities of sludge product, both economically and environmentally unattractive.

4.3.4 Process Status

The primary advantage of limestone neutralization is that limestone is a low cost and widely available reagent. However, limestone is limited in its ability to neutralize over pH 6.0 or acid concentrations greater than 5,000 mg/L. While the process of limestone powder treatment enhances neutralization, it also increases reagent costs nearly eight fold. The added cost of grinding and sifting to mesh sizes of 200-400 to achieve the same neutralizing power as hydrated lime makes it twice as expensive. This, combined with the rising costs and environmental awareness in disposing of solid hazardous waste residues, typically makes limestone treatment unattractive except under special circumstances.

There have been attempts to use limestone in combination with lime in a dilute, dual alkali mode. The limestone is used as a pretreatment to raise the pH to approximately 3.0 or 6.0 with lime completing the process of neutralization. The limestone/lime process is usually more complicated than a simple lime slurry process, resulting in higher projected costs and limited application. However, in large volume applications (see Section 4.4.2) the large savings in reagent (when used in pebble form) may offset any increases in capital expenditures.

REFERENCES

1. Camp, Dresser, and McKee. Technical Assessment of Treatment Alternatives for Wastes Containing Corrosives. Contract No. 68-01-6403. September 1984.

2. Kirk-Othmer Encyclopedia of Chemical Technology. Vol. 14, 3rd. Edition. John Wiley & Sons, New York, NY. 1981.

3. Boynton, R.S. Chemistry and Technology of Lime and Limestone. Interscience Publishers, New York, NY. 1966.

4. Gehm, H.W. Neutralization of Acid Waste Waters with Upflow Expanded Limestone Bed. Sewage Works Journal 16:104-120. 1944.

5. Tully, T.J. Waste Acid Neutralization. Sewage and Industrial Wastes 30:1385. 1985.

6. Levine, R.Y., et al. Sludge Characteristics of Lime Neutralized Pickling Liquor. 7th Industrial Waste Conference, Purdue University. 1952.

7. Arthur D. Little, Inc. Physical, Chemical, and Biological Treatment Techniques for Industrial Wastes. U.S. EPA SW-148. November 1976.

8. Gloves, H.G. The Control of Acid Mine Drainage Pollution by Biochemical Oxidation and Limestone Neutralization Treatment. 22nd Industrial Waste Conference, Purdue University. 1967.

9. Volpicelli, G., et al. Development of a Process for Neutralizing Acid Wastewaters by Powdered Limestone. Environmental Technology Letters, Vol. 3, pp. 97-102. 1982.

10. MITRE Corporation Manual of Practice for Wastewater Neutralization and Precipitation. U.S. EPA-600/2-81-148. August 1981.

11. U.S. EPA. PB-220-302. Processes Research, Inc. Neutralization of Abatement-Derived Sulfuric Acid.

12. Peters, M.S., et al. Plant Design and Economics for Chemical Engineers. 3rd Edition, McGraw-Hill Book Company, New York, NY. 1980.

4.4 LIME SLURRY TREATMENT

4.4.1 Process Description

Lime Slurry--

Lime slurry treatment of corrosive waste streams is analogous to that of limestone neutralization. It is one of the oldest and perhaps most prevalent of all industrial waste treatment processes.[1] It is used extensively as an alkaline reagent in the neutralization of pickling wash waters, plating rinses, acid mine drainage, and process waters from chemical and explosive plants.[2,3,4] Lime slurry has replaced limestone in many applications as a low-cost alkali due to its greater available surface area, pumpable form, continuous application, and greater effectiveness in removing Ca salts from the process.[1] However, similar to the use of limestone, a major disadvantage of the process is the formation of a voluminous sludge product.

Limes are formed by the thermal degradation of limestone (calcination), and are available in either high calcium (CaO) or dolomitic (CaO-MgO) form (Table 4.4.1).[5] The pure, oxidized calcium product is referred to as quicklime. Quicklime varies in physical form and size, but can generally be obtained in lump (63 to 255 mm), pebble (6.3 to 63 mm), ground (1.45 to 2.38 mm) or pulverized (0.84 to 1.49 mm) form.[5] As with limestone, experimental evidence has shown an increase in dissolution as the size of a lime particle diminishes. For example, a 100 percent quicklime of 100 mesh (0.149 mm) will dissolve twice as fast as one of 48 mesh (0.35 mm).[3]

Although lime can be fed dry, for optimal efficiency it is slaked (hydrated) and slurried before use. Slaking is usually carried out at temperatures of 82 to 99°C with reaction times varying from 10 to 30 minutes. Following slaking, a wet plastic paste is formed (lime putty) and then slurried with water to a concentration of 10 to 35 percent.[4]

While most lime is sold as quicklime, small lime consumers often cannot economically justify the additional processing step that slaking entails. Therefore, high calcium and dolomitic lime are also available in hydrated form (either $Ca(OH)_2$ or $Ca(OH)_2$ MgO). This product is made by the lime manufacturer in the form of a fluffy, dry, white powder. It is supplied either in bulk or in 23 kg (50 lb) bags. Hydrated lime is suitable for dry

TABLE 4.4.1 SUMMARY OF HIGH CALCIUM AND DOLOMITE QUICKLIME PROPERTIES

Parameter	Unit	High calcium lime	Dolomite lime
Molecular Formula	--	CaO	CaO - MgO
Molecular Weight	g/gmole	56.1	96.4
Components	% by weight		
CaO		93.25 - 98.0	55.50 - 57.50
MgO		0.30 - 2.50	37.60 - 40.80
SiO_2		0.20 - 1.50	0.10 - 1.50
Fe_2O_3		0.10 - 0.40	0.05 - 0.40
Al_2O_3		0.10 - 0.50	0.05 - 0.50
H_2O		0.10 - 0.90	0.10 - 0.90
CO_2		0.40 - 1.50	0.40 - 1.50
Bulk Density	Kg/m^3	770 - 1120	801 - 1165
Specific Gravity	H_2O = 1.0	3.2 - 3.4	3.4 - 3.6
Solubility	g/100g water	converts to $Ca(OH)_2$	converts to $Ca(OH)_2$ MgO or $Ca(OH)_2$ $Mg(OH)_2$
Specific Heat	Cal/g/°C	0.175 - 0.286	0.19 - 0.294
Stability	°C	Stable at any temperature but will readily hydrolyze	Same
Basicity Factor	CaO=1	0.941	1.110
pH	H_2O = 7.0	11.27 - 12.53	Same

Source: Reference 5

feeding or for slurrying and the storage characteristics, purity, and uniformity are generally superior to slaked lime prepared onsite. High calcium hydrate is far more reactive than dolomitic hydrate, neutralizing acids in minutes instead of hours. It is also more efficient with dilute acids and in cases where over-neutralization is required. Dolomitic hydrate, which possesses greater basicity (approximately 1.2 times), is a much slower reactant, although heat (76.4°C) and agitation can accelerate its inherently slow reactivity.[3] Generally, dolomitic hydrate is most efficient with strong acids where complete neutralization (above a pH of 6.5) is not necessary.

Both quicklime and hydrated lime deteriorate in the presence of carbon dioxide and water (air-slaking), therefore prior to dry feeding, lime is generally stored within moisture-proof containers and consumed within a few weeks after manufacture. The storage characteristics of dry hydrated lime are superior to quicklime, but carbonation may still occur causing physical swelling, marked loss of chemical activity, and clogging of discharge valves and pipes.

Dry chemical feed systems consist of either manual addition of 50 lb bags or, in large operations where lime is stored in bulk, automatic mixing and feeding apparatus. Automatic feeders are often positioned directly at the base of the bulk storage bins to minimize potential clogging due to excessive dry lime transport distance. Two types of automatic feed systems are available. Volumetric feed systems deliver a predetermined volume of lime while gravimetric systems discharge a predetermined weight. Gravimetric feeders require more maintenance, are roughly twice as expensive, but can guarantee a minimum accuracy of 1 percent of set rate versus 30 percent for volumetric feeders.[6]

Figure 4.4.1 is an example of a typical lime slurry system with storage and slaking equipment. A slurry tank with agitator is used followed by a slurry recirculation line.[7] The process flow lines bleed off a portion of the recirculation slurry to reactors arranged in parallel. The process line is as short as possible (to prevent caking) and the control valves are located close to the point of application. Research has indicated that the most successful control valves used have been automatic pinch valves.[7]

Figure 4.4.2 is a schematic of a small volume, high calcium, hydrated lime application for the neutralization/precipitation of electroplating

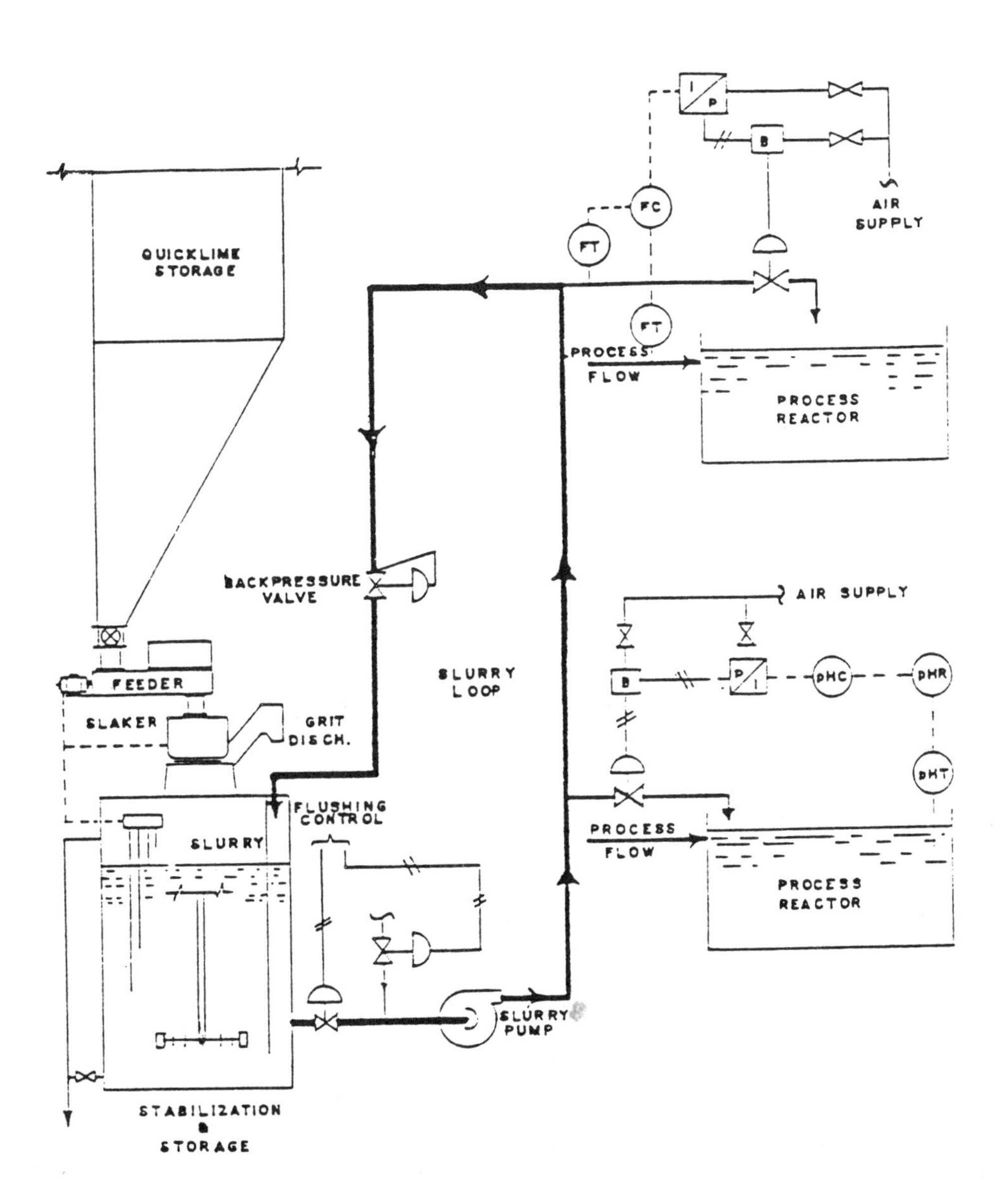

Figure 4.4.1 Flowsheet of a lime slurry system.

Source: Reference 7

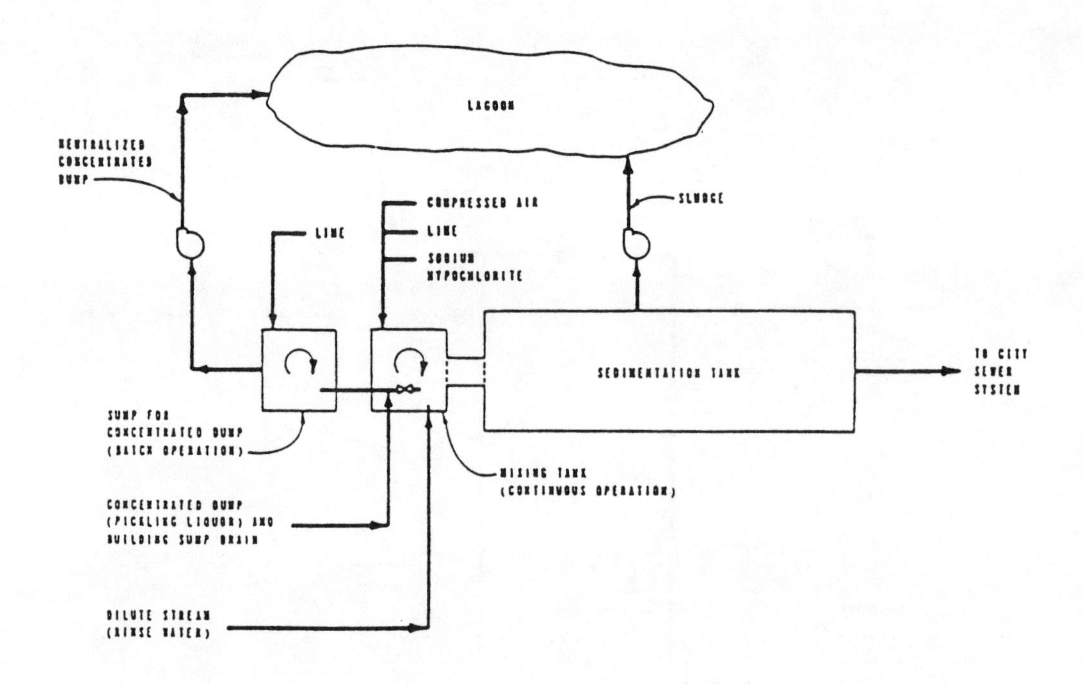

Figure 4.4.2. Small volume, high calcium, hydrated lime slurry treatment system.

Source: Reference 8

wastes.[8] The system operates in either batch or continuous mode and includes a sedimentation tank for liquid/solid separation and a lagoon for sludge disposal. A pH probe controls the supply of a lime slurry to either a sump for the neutralization of concentrated bath dumps (high in acids and metals) or a continuous mixing tank (57 gpm) for the neutralization of less acidic wastes; e.g., electroplating rinses. Mixing is provided by both compressed air and a mechanical agitator. The sedimentation tank is operated in a semi-continuous manner with sludge being pumped to a lagoon once or twice a month and the supernatant being discharged to the sewer.

Lime slurry operations are typically conducted under atmospheric conditions and room temperatures. The neutralizing unit is usually a reinforced tank with acid-proof lining and some sort of agitation to maintain intimate contact between the acid wastes and the lime (slurry) solution. Vertical ribs can be built into the perimeter to keep the contents from swirling instead of mixing.

During operations, adequate venting may have to be provided due to the possible evolution of heat and noxious gases. Calcium hydroxide will evolve 31,140 calories/gram molecular weight of sulfuric acid neutralized, and approximately 27,900 calories with hydrochloric acid.[3] Table 4.4.2 presents a summary of process parameters gathered from various lime slurry neutralization systems. However, while these provide an indication of typical system design, testing under actual or simulated conditions is the only sound basis for the determination of individual waste treatment parameters.

Waste Carbide Lime--

One alternative to the purchase of virgin lime for neutralization applications, is the use of carbide lime.[5] Calcium carbide and water are reacted together in the acetylene manufacturing process. To produce acetylene and by-product carbide lime:

$$CaC_2 + 2H_2O = C_2H_2 + Ca(OH)_2 \qquad (1)$$

TABLE 4.4.2 SUMMARY OF TYPICAL LIME-SLURRY OPERATING PARAMETERS

Parameter	Unit(s)	Operating range	Optimum range
Type of Stone	% MgO	5 - 40	5
Stone Size	mm	0.149 - 255	0.149
Slaking Temperature	°C	82 - 99	Same
Slurry Solids	%	5 - 40	a
Retention Time	Min	5 - 15	5
Sedimentation Time	Min	15 - 60[b]	15 - 30
Mineral Acidity	Mg/L	10,000 - 100,000	20,000

[a]is dependent on site specific factors
[b]High calcium lime will settle in 15 minutes with 1-2% acid wash streams and 30-60 minutes with 3-10 percent acid streams. Dolomite will typically take 15-60 minutes.

Source: References 2, 3, 4

However, since commercial calcium carbide is usually produced by burning coke and limestone in an electric furnance, all of the nonvolatile impurities in the coke and limestone are retained in the calcium carbide. Many of these impurities are in the reduced oxidation state and, therefore, non-reactive. Table 4.4.3 contains an example of two typical carbide lime samples which illustrates the typically high levels of such impurities. In addition, a study by the National Lime Association[1] showed that the physical characteristics of carbide lime, such as degree of porosity and surface area were poorer than commercial lime.

When carbide lime is utilized in dry powder form, feeding, slaking, and neutralization process requirements are equivalent to that of hydrated lime. However, the presence of large quantities of reduced impurities will increase reagent requirements, sludge volumes, and processing equipment dimensions. For example, to neutralize 100 lbs of reagent grade (98 percent pure) sulfuric acid, stoichiometric requirements for a 98 percent pure high calcium hydrated lime are approximately 87 lbs, (10 percent insoluble feed and 2.2 lbs of inerts). In comparison, when an 84 percent $Ca(OH)_2$ - carbide lime is used to neutralize the same acid, 117.7 lbs is required (10 percent excess and 18.8 lbs of inert impurities). Storage, slaking, and feeding equipment dimensions in this case would increase by approximately 35 percent to handle the greater volume of reagent required on an equivalent neutralization basis. Similarly, sludge handling equipment and disposal costs will increase by a minimum of 14 percent based on a greater quantity of insolubles in the feed. Consequently, before choosing between virgin reagent and carbide lime wastes, one should compare the 20 percent price advantage[1] enjoyed by carbide lime with increased capital expenses, slower reactivity, and variability in the quality of the waste materials.

Cement Kiln Dust--

Cement kiln dust has also been proposed as an alternative to lime for the neutralization of acidic waste streams.[10] Cement kiln dust is a cement manufacturing byproduct derived from reacted or partially reacted raw materials (limestone) and fuel used in cement rotary kiln operations. The dust particles (6 microns or less) become entrained in the rotary kiln hot exhaust gases and are subsequently removed through collection devices such as

TABLE 4.4.3 WASTE CARBIDE LIME COMPOSITION

Compound	Weight Percent (%)	
	Sample A	Sample B
$Ca(OH)_2$	84.3	96.30
$CaCO_3$	6.9	-
$CaSO_3$	1.8	-
$CaSO_4$	1.0	0.34
SiO_2	1.7	1.41
Al_2O_3	0.7	1.33
Fe_2O_3	1.1	0.12
CO	-	0.14
CaS	-	0.08
CNS	-	0.01

Source: Reference 9

baghouses. Kiln dust is similar to quicklime in that it contains calcium oxide, is a dry powder, and must be hydrated (i.e., slaked and slurried) prior to use.

The primary advantage of kiln dust is that it costs about 66 to 75 percent less than virgin lime and consistently displays a pH in the range of 11.2 to 12.1.[11] However, physical and chemical characteristics vary widely according to collection method or kiln efficiency. Table 4.4.4 presents elemental and anion variation in U.S. cement kiln dust from 113 different cement kiln operations. As shown the maximum calcium concentration analyzed was 36.7 weight percent. Therefore, depending on the degree of calcination in the kiln exhaust stack, the maximum theoretical CaO concentration would be approximately 51.3 percent with a median value of 42.6 percent. Facilities replacing virgin quicklime with cement kiln dust will require a minimum increase in reagent of 93 to 128 percent. In addition, both feed and sludge handling facilities will have to be expanded in proportion, with attendant capital and sludge disposal cost increases. Similar to waste carbide lime, savings in reagent purchase costs will be balanced against these increases.

4.4.2 Process Performance

Table 4.4.5 presents a summary of bench-scale performance data on the lime neutralization of six acidic manufacturing wastes conducted by Faust.[13] Both high calcium and dolomitic quicklimes were examined in various dosages to evaluate effects on final effluent pH and sludge characteristics. The results showed that while dolomitic formed up to 44 percent less sludge (on a dry weight basis), it was less effective in treating concentrated acids in a timely manner. For example, after treating a paint manufacturing waste stream of 12,910 ppm mineral acidity for 6 minutes, dolomitic lime only achieved a final pH of 4.5. Conversely, when high calcium quicklime was applied to dye manufacturing waste containing 10,410 ppm of mineral acidity, a final pH of 9.2 was reached in the same time frame.

Table 4.4.6 summarizes the operating characteristics and performance of an automatic, continuous two-stage neutralization system used by a manufacturer of automobile bumpers.[14] The wastewater flow to the plant was

TABLE 4.4.4 - Elemental and anion variation in U.S. cement kiln dust, μg/g

Element or anion	Range	Mean[1]	Median
Ag................	<3 - 17	5.4	4.8
Al................	9,900 - 50,200	23,200	23,100
As................	1.3 - 518	24	9.3
Ba................	<55	<55	<55
Be................	<2	<2	<2
Bi................	<50	<50	<50
Ca................	106,000 -367,000	295,000	305,000
Cd................	<1.5 - 352	21	7.3
Co................	<10	<10	<10
Cr................	11 - 172	41	34
Cu................	7 - 206	30	24
Fe................	1,000 - 44,400	14,700	14,100
Hg	<0.13- 1.0	<.13	<.13
K.................	3,400 -232,000	36,600	26,800
Li................	<4 - 76	18	16
Mg................	1,980 - 19,100	7,820	6,820
Mn................	63 - 2,410	383	280
Mo................	<50	<50	<50
Na................	495 - 27,700	4,700	3,190
Ni................	<12 - 91	22	29
Pb................	17 - 1,750	253	148
Sb................	<1.6 - 70	3.2	<1.6
Si................	26,900 -111,000	63,500	65,100
Sn................	<100	<100	<100
Sr................	62 - 8,750	670	430
Ti................	500 - 2,900	3,530	1,100
Tl................	<60 - 185	<60	<60
V	<100	<100	<100
Zn................	32 - 8,660	462	167
Br^-...............	<200	<200	<200
Cl^-...............	<100 -123,000	6,900	4,900
F^-................	100 - 3,600	1,300	1,000
NO_2^-..............	<200	<200	<200
NO_3^-..............	200 - 16,700	<200	<200
PO_4^{3-}..............	200 - 1,600	<200	<200
SO_4^{2-}.............	4,100 -316,000	77,800	68,600

[1]A value of 1/2 the detection limit was arbitrarily used to calculate the mean for those elements having concentrations both above and below the detection limit.

[2]Mercury value based on only 16 samples.

Source: Reference 12

TABLE 4.4.5 SUMMARY OF LIME-SLURRY NEUTRALIZATION DATA

	Type of waste					
Parameter	Dye	Pharma-ceutical	Paint	Paint	Chemical	Dye
pH	0.5	0.7	1.5	1.0	1.0	1.5
Mineral Acidity (ppm)	30,500	19,900	12,910	12,430	12,530	10,410
Type of Lime	Dolomite Quicklime	Dolomite Quicklime	Dolomite Quicklime	High Calcium Quicklime	High Calcium Quicklime	High Calcium Quicklime
Lime Dosage (g/L acid)	17.3	11.9	10.0	8.2	7.8	7.9
pH (6 min)	3.9	4.1	4.5	10.8	10.2	9.2
% Sludge[a] (Volume)	38.0	22.4	33.0	38.4	24.0	17.0
Dry Sludge[b] (g/L)	21.4	12.1	13.7	21.4	15.9	14.7
% Dry Solids (Neutralization Mix)	1.9	1.1	1.25	1.98	1.5	1.38
% Dry Solids (Settled Sludge)	5.7	5.4	4.16	5.6	6.7	8.63

Acid Volume = 500 ml.
Acid Temperature = 25°C.

[a]Expressed as percent of original acid volume after one hour of sedimentation.

[b]From neutralization of one liter of acid.

Source: Reference 13

TABLE 4.4.6 FULL-SCALE AUTOMATIC LIME NEUTRALIZATION SYSTEM CHARACTERISTICS

Parameter	Unit	Value
Lime Type		High calcium quicklime
Stone Size	mm	6.3 - 63
Influent Feed Rate	gph	100,000 - 125,000
Slaker Feed Rate	lbs/hr lime	750
Slaker Temerature	°C	57 - 68
Slurry Make-Up Ratio	H_2O/Lime	3.5
Dry Solids Concentration	%	10
Slurry Holding Capacity	gals	400
Influent pH	S.U.	2.0
Effluent pH	S.U.	6.3 - 6.7

Source: Reference 14

100,000 gal/hour, with a peak capacity of 125,000 gallons. Approximately 11,500 lbs of sulfuric acid (spent pickle liquor) were discharged to the system daily, primarily in the form of intermittent, concentrated (2 lbs acid/gal) bath dumps. To prevent reagent wastage due to acid surges, the reactors were arranged in series and provided with high-speed agitation. A 10 percent lime slurry is added into the first tank (2,500 gallons), regulated by an automatic pH probe. This is followed by a second 2,500 gallon agitated reactor and a 6,000 gallon tank where the reaction mixture is allowed to settle. The supernatant flows over a weir into a receiving sewer. While solids are collected and concentrated by vacuum filtration prior to incineration.

A third case study, conducted by Mooney et al.,[15] focused on two lime neutralization systems for the treatment of acidic cooling waters from the production of phosphoric acid in the fertilizer industry. One system (Plant A) treated highly-acidic process cooling water, and other (Plant B) treated reduced strength seepage waters.

At Plant A, use of a combination of calcium carbonate and lime was selected, due to the high acidity, to reduce chemical costs. First-stage facilities were newly constructed, including lime and calcium carbonate storage and feed systems, reactors, and an earthen-basin thickener clarifier of 48.8-meter diameter. Calcium carbonate was added at a fixed dosage in the first reactor, and lime was added to the second reactor based on pH control to approximately pH 5.5. Sludge was pumped from the collection sump directly to the gypsum-slurry area for disposal in the gypsum stack. Overflow from the thickener clarifier was partially reused for the slurry of calcium carbonate and lime, and the remainder was pumped to the second-stage system.

An original liming station was revamped for use as second-stage facilities, where lime was added based on pH control to about pH 9.5. An existing plate settling device was relocated to provide clarification. However, as predicted from pilot testing, only 30 to 40 percent of the overflow nameplate capacity could be achieved due to the gelatinous nature of the solids. At very high polymer dosages, plugging of the plates occurred. Thus, settling and sludge storage was conducted in an existing 4050-m^2 pond. Pond overflow is monitored for effluent limits, and settled sludge was periodically transferred to the gypsum stack.

At Plant B, lime was added in the first and second stages to about pH 5.5 to 6.0 and pH 9.0 to 9.5, respectively. Settling was conducted in identical 27.5 meter diameter earthen-basin thickener clarifiers, and sludge was separately pumped from each to the gypsum-slurry area for disposal in the gypsum stack. No reuse of effluent was practiced, and the effluent was monitored for permit limits. An existing cooling pond was modified to serve as an equalization/storage basin to provide consistent influent flow and strength. This system can also treat direct cooling pond water at reduced rates due to higher strength. Table 4.4.7 contains a summary of nominal design criteria, while Table 4.4.8 contains typical operating data.

These studies demonstrated that two-stage lime treatment of phosphoric acid process cooling water was capable of achieving EPA guidelines for fluoride (25 mg/L) and phosphorus (35 m/L). Major advantages of the two-stage method are minimum lime requirements, separate handling of readily dewaterable first-stage solids, and effective removal of numerous other constituents.

Addition of calcium carbonate in conjunction with lime for treatment of high-acidity cooling water allows optimization of chemical efficiency, minimizes sludge protection, and ensures the proper first stage pH for optimum fluoride removal.

4.4.3 Process Costs

Table 4.4.9 summarizes process costs developed for the construction and operation of a continuous, high calcium, hydrated lime neutralization system. Equipment sizing and operational cost were based mainly on precepts presented in the two previous costing sections.[16,17,18] However, based on the increased reactivity of lime as compared to mutual neutralization or limestone, the 400 gph reactor system was assumed to be a single-stage reinforced concrete reactor sized for 30 minutes of retention time. In addition, both the 2,500 and 3,500 gph systems were assumed tó be two-stage, reinforced concrete reactors sized for a 15-minute retention time in the first stage and 30 minutes in the second stage.

The cost of the hydrated lime slurry feed system is based on Figure 4.4.3. The equipment includes a slurry tank, pH control, recycle loop, electrode assembly, signal transmitter, pH recorder-controller, controller

TABLE 4.4.7 NOMINAL DESIGN CRITERIA FOR TWO STAGE LIME NEUTRALIZATION SYSTEM

Parameter		Plant A	Plant B
Flow, m^3/min.		3.8	--
Stage I			
$CaCO_3$	Reactor Dentention, Min	30	--
$Ca(OH)_2$	Reactor Detention, Min.	30	30
Thickener-Clarifier Dia., Meters		48.8	27.5
Stage II			
$Ca(OH)_2$	Reactor Dentention, Min.	30	35
Thickener-Clarifier Dia., Meters		(4050-m^2 Pond)	27.5
$CaCO_3$	Storage, Metric Tons	563	--
$CaCO_3$	Feed Capacity, Metric Tons/Hr	13.6	--
CaO	Storage, Metric Tons	454	454
CaO	Feed Capacity, Metric Tons/Hr	7.3	7.3
Operating Power, kW		224	231

Source: Reference 15

TABLE 4.4.8 TYPICAL OPERATING DATA FOR THE TWO-STAGE LIME NEUTRALIZATION OF ACIDIC PROCESS COOLING WATER

Parameter	Plant	
	A	B
Influent Flow, M^3/min	3.8	3.4
Raw Water		
Fluoride, mg/L as F	11,100	1,460
Phosphorous, mg/L as P	6,730	2,230
pH, std. units	1.5	2.6
$CaCO_3$ Kg/m^3	10	--
CaO Kg/m^3	24	6.8
Net Effluent Flow, M^3.min	2.7	2.7
Treated Water		
Fluoride, mg/L as F	1 - 14	6 - 12
Phosphorous, mg/L as P	1 -15	5 - 70
pH, std. units	9.2 - 9.7	9.1 - 9.6

Source: Reference 15

TABLE 4.4.9 CONTINUOUS HYDRATED LIME NEUTRALIZATION TREATMENT COSTS

	Flow rate (gph)		
	400	2,500	3,500
Capital investment			
Collection sump	3,100	3,600	3,600
Piping system	2,838	2,838	2,838
Reactor system	17,000	36,000	38,000
Feed system	74,000	125,000	150,000
Flocculation/clarification	22,000	32,000	40,000
Sludge storage	8,000	8,000	10,000
Filter press	11,000	15,000	18,000
Total capital cost	137,938	222,438	262,438
Annualized capital	24,415	39,371	46,452
Operating cost[a] ($)			
Taxes and insurance (7%)	1,709	2,756	3,252
Maintenance (5%)	1,221	1,969	2,323
Labor ($20/hr)	24,000	24,000	24,000
Sludge disposal ($200/ton)[b]	64,514	435,809	565,499
Reagent cost ($40/ton)	4,615	28,892	40,449
Total cost/year ($)	102,474	532,798	681,975
Cost/1,000 gal ($)	126	89	81

[a]Does not include utilities.
[b]Does not include gypsum refund.

Source: Adapted from References 6, 16, 17, and 18 using July 1986 CPI index.

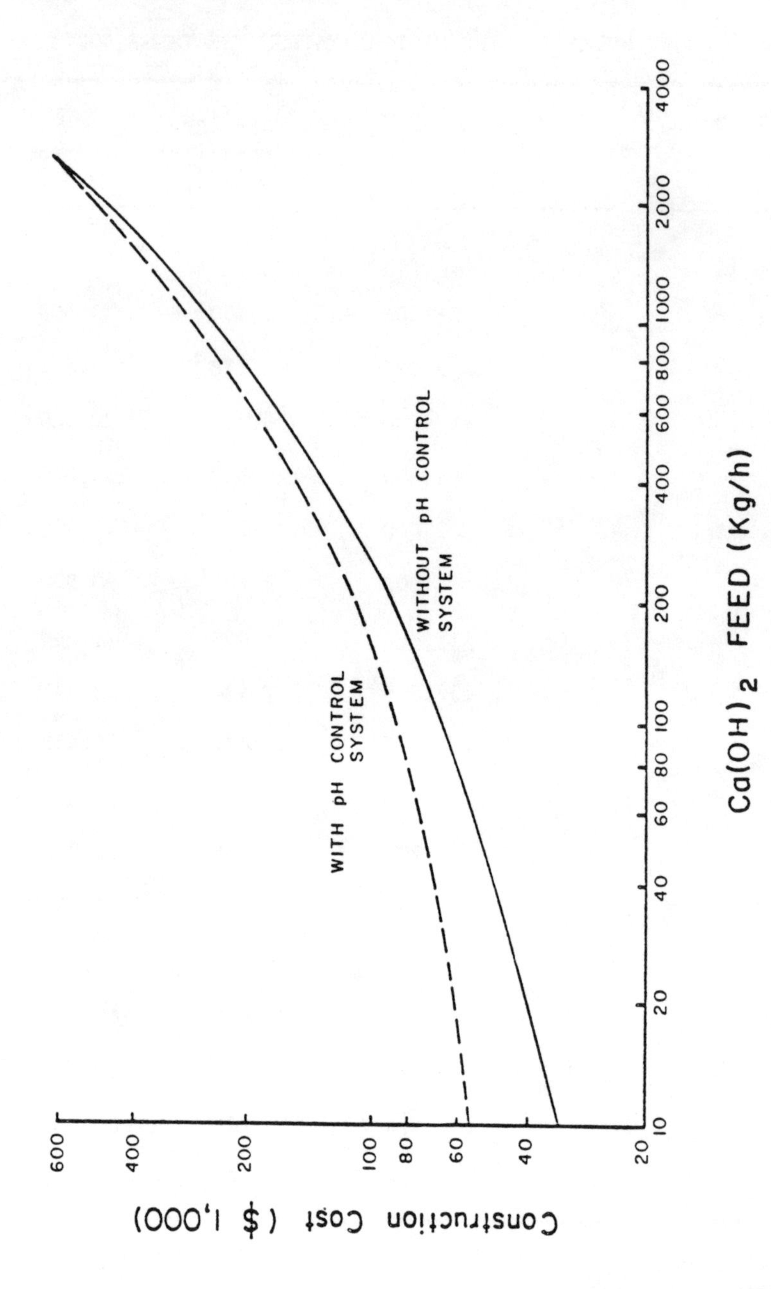

Figure 4.4.3. Investment costs for hydrated lime feed systems.

Source: Reference 16.

valves, instrument panel, miscellaneous hardware and installation. The reagent demand for the neutralizer system was based on a basicity factor of 0.710 and a 10 percent insoluble feed. Sludge handling equipment sizing was estimated on a sludge generation factor of 1.322 lbs of dry solids/pound of sulfuric acid neutralized.

As with limestone, hydrated lime sludge generation and subsequent disposal costs are a significant portion of the overall treatment application. In the advent of a land disposal ban, the increased costs and liability in disposing of wastewater treatment sludges may outweigh any possible benefit gained from the reduced chemical costs associated with calcium based reagent systems. Therefore, when evaluating neutralization systems careful consideration of options such as acid recovery or reagents with soluble end products should be performed to reduce total sludge volumes.

4.4.4 Process Status

Lime slurry treatment is a widely used technology for neutralizing dilute and concentrated acidic waste streams. Its ability to treat a wide variety of manufacturing waste streams has been well demonstrated in bench, pilot, and full-scale systems. Environmental impacts can result from emissions during the neutralization process and the production of large volume of potentially hazardous sludge.[18] Exit gases can be scrubbed by using a control system, however, sludge reduction methods (seeding, dilution, vacuum filtration, etc.),[19,20] have only partially offset the problems associated with sludge generation. Therefore, new methods of sludge disposal and reduction and recycle/reuse options (such as agricultural liming) should be considered.

The advantages and disadvantages of lime-slurry neutralization are summarized in Table 4.4.10.

TABLE 4.4.10 ADVANTAGES AND DISADVANTAGES OF LIME-SLURRY NEUTRALIZATION

Advantages
Proven technology with documented neutralization efficiencies
No temperature adjustments normally necessary
Modular design for plant expansion
Can be used in different configurations
Able to overneutralize and precipitate metals lowering space requirements
Low-cost and widely available reagent
Disadvantages
Relatively high capital and operating costs when operating slaking equipment
Forms voluminous sludge product
Requires large and moisture-proof storage facilities when using quicklime
Will foul and clog process equipment unless continually recirculated

Source: References 1, 3, 4, 12, 21

REFERENCES

1. Camp, Dresser, and McKee. Technical Assessment of Treatment Alternatives for Wastes Containing Corrosives. Contract No. 68-01-6403, September 1984.

2. Besselievre, E.B. The Treatment of Industrial Wastes. McGraw-Hill Book Company, New York, NY. 1967.

3. Boynton, R.S. Chemistry and Technology of Lime and Limestone. Interscience Publishers, New York, NY. 1966.

4. Cushnie, G.C. Removal of Metals from Wastewater: Neutralization and Precipitation. Pollution Technology Review, No. 107. Noyes Publications, Park Ridge, NJ. 1984.

5. Kirk-Othmer Encyclopedia of Chemical Technology. Vol. 14, 3rd Edition, John Wiley & Sons, New York, NY. 1981.

6. U.S. EPA. Process Design Manual: Sludge Treatment and Disposal. EPA-625/1-79-011. September 1979.

7. Mace G.R. Lime vs. Caustic for Neutralizing Power. Chemical Engineering Progress. August 1977.

8. HSU D.Y., et al. Soda Ash Improves Lead Removal in Lime Precipitation Process. 34th Industrial Waste Conference, Purdue University. 1977.

9. USEPA. Characterization of Carbide Lime to Identify Sulfite Oxidation Inhibitors, Interagency Energy/Environment RPD Program Report. U.S. EPA-600/7-78-176. September 1978.

10. USEPA. Disposal and Utlilization of Waste Kiln Dust From Cement Industry. U.S. EPA-670/2-75-043. May 1975.

11. U.S. Department of Transportation. Kiln Dust-Fly Ash Systems for Highway Bases and Subbases. Report No. FHWA/RD-82/167. 1983.

12. Haynes B.W., and G.W. Kramer. Characterization of U.S. Cement Kiln Dust. Bureau of Mines Information Circular No. 8885. 1982.

13. Faust, S.D. Sludge Characteristics Resulting from Lime Neutralization of Dilute Sulfuric Acid Wastes. 13th Industrial Waste Conference, Purdue University. 1958.

14. Hugget et al. Automatic Continuous Acid Neutralization. 23rd Industrial Waste Conference, Purdue University. 1968.

15. Mooney, G.A., et al. Two-Stage Lime Treatment in Practice. Environmental Progress, Vol. 1, No. 4. November 1982.

16. MITRE Corp. Manual of Practice for Wastewater Neutralization and Precipitation. EPA 600/2-81/148. August 1981.

17. Peters M.S., and K.D. Timmerhaus. Plant Design and Economics for Chemical Engineers. McGraw-Hill Book Company, New York, NY. 1980.

18. U.S. EPA. Reducing Water Pollution Control Costs in the Electroplating Industry. U.S. EPA-625/5-85/016. September 1985.

19. Levine, R.Y. et al. Sludge Characteristics of Lime Neutralized Pickling Liquor. 7th Industrial Waste Conference, Purdue University. 1952.

20. Dickerson, B.W. et al. Neutralization of Acid Wastes: Industrial Engineering Chemistry 42:599-605. 1950.

21. Arthur D. Little Inc. Physical, Chemical, and Biological Treatment Techniques for Industrial Wastes. U.S. EPA SW-148. November 1976.

4.5 CAUSTIC SODA TREATMENT

4.5.1 Process Description

Pure anhydrous sodium hydroxide (NaOH) or caustic soda is a white crystalline solid manufactured primarily through the electrolysis of brine. Caustic soda is a highly alkaline, water soluble compound especially useful in reactions with weakly acidic materials where weaker bases such as sodium carbonate are less effective.[1] It is also useful in the precipitation of heavy metals and in neutralizing strong acids through the formation of sodium salts. Table 4.5.1 lists the properties of pure anhydrous sodium hydroxide.

Although available in either solid or liquid form, NaOH is almost exclusively used in water solutions of 50 percent or less.[2] The solution is marketed in either lined 55-gallon drums or in bulk; i.e., tank car or truck. As a solution, caustic soda is easier to store, handle, and pump relative to either lime or limestone. In comparison to lime slurries, caustic soda will not clog valves, form insoluble reaction products, or cause density control problems. However, when sodium hydroxide is stored in locations where the ambient temperature is likely to fall below 12°C, heated tanks should be provided to prevent reagent freezing.

After lime, sodium hydroxide is the most widely used alkaline reagent for acid neutralization systems. Its chief advantage over lime is that, as a liquid, it rapidly disassociates into available hydroxyl (OH-) ions. Holdup time is minimal, resulting in reduced feed system and tankage requirements. Caustic soda's main disavantage is reagent cost.[3] As a monohydride, in neutralizing diprotonated acids such as sulfuric, two parts base are required per part of acid neutralized. In contrast, dihydroxide bases such as hydrated lime, only require one part base per part of acid neutralized:

This increase in reagent requirements combined with a higher cost/mole (approximately five times that of hydrated lime), makes caustic soda more expensive on a neutralization equivalent basis. Generally in high volume applications where reagent expenditures constitute the bulk of operating expenses, lime is the reagent of choice. However, in low volume applications where low space requirements, ease of handling, and rapid reaction rates are

TABLE 4.5.1 PHYSICAL CONSTANTS OF PURE SODIUM HYDROXIDE

CAS Registry No.	1310-73-2
Molecular Weight	39.99
Specific gravity 20°/4°C	2.13
Melting Point °C	318
Boiling Point °C	1390
Index of Refraction	1.358
Latent Heat of Fusion (Cal/g)	40.01
Heat of Formation (kcal/mol)	
Alpha Form	100.97
Beta Form	101.96
Transition Temperature °C	299.6
Solubility at 20°C (g/100 g water)	109

Source: Reference 2

the deciding factor in reagent selection, caustic soda is clearly superior. Also, in any system where sludge disposal costs will be high, caustic soda will compete more favorably with lime.

The higher solubility of NaOH in water (approximately 100 times that of lime at 25°C) reduces or eliminates the need for complex slaking, slurrying, or pumping equipment. In a typical system, caustic is added through an air-activated valve controlled by a pH sensor.[4] Reagent is demanded as long as the pH of the acidic waste stream remains below the controller setting. Agitation is provided by a mechanical mixer to prevent excessive lag time between the addition of the reagent and the first observable change in the effluent pH. The neutralized solution is then pumped to a large settling tank for liquid/solid separation.

Caustic soda is a particularly versatile alkali reagent that can be used in either over or under-neutralization applications. The neutralization reaction is typically carried out under standard operating temperatures and pressures. The reaction is almost instantaneous since caustic soda reacts vigorously with water. At concentrations of 40 percent or greater, the heat generated by dilution can raise the temperature above the boiling point. Handling precautions are required when performing dilution or other reagent handling since even moderate concentrations of NaOH solution are highly corrosive to skin.[2]

Process configurations for caustic soda treatment are a function of waste type, volume, and raw waste pH level and variability. For example, the neutralization of concentrated acidic waste streams with low dead times depends on pH as follows: one reactor system for feeds with pH ranging between 4 and 10, a reactor plus a smoothing tank for feeds with pH fluctuations of 2 and 12, and two reactors plus a smoothing tank for feeds with pH less than 2 or greater than 12.[5] Retention times vary with the rate of reaction and mixing, however, 15 to 20 minutes appears to be optimal for complete neutralization in most systems.[6] The interval between the addition of sodium hydroxide and the first observable change in effluent pH (dead time) should be less than 5 percent of the reactor residence time in order to maintain good process control.[5] A summary of typical operating parameters is provided in Table 4.5.2.

TABLE 4.5.2. SODIUM HYDROXIDE NEUTRALIZATION: SUMMARY OF TYPICAL OPERATING PARAMETERS

Parameter	Unit(s)	Operating range	Ideal range
Sodium Hydroxide Concentration	% NaOH	12 - 50	40 - 50
Dead Time	% Retention Time	3 - 10	3 - 5
Influent pH	H_2O = 7.0	0.1 - 6	2 - 4
Retention Time	Min	5 - 30	15 - 20
Batch Treatment Throughput	gal/min	1 - 20	20
Continuous Treatment Throughput	gal/min	>15	20
Suspended Solids	Weight %	3 - 10	10
Storage Temperature (40 - 50% NaOH)	°C	12 - 20	16 - 20

Source: References 3, 5, 7, 8

A typical caustic system is designed to add most of the reagent in a preliminary neutralization stage, while a second stage acts as a smoothing and finishing tank. In this manner, the second reactor is able to compensate for pH control overshoots or concentrated batch dumps which may temporarily overwhelm the primary neutralization system.[7]

Overshoot is due primarily to the lack of sodium hydroxide solution buffering capacity. For example, Figure 4.5.1 illustrates the titration curve for the neutralization of a ferric chloride etching solution (pH 0.5) with a 5 Molar caustic soda solution. The steep slope of the titration curve beginning at pH 2.0 combined with a strong demand for alkali prior to that point, often make over- or under- correction unavoidable. For continuous neutralization applications of greater than 20 gpm, pH control in the portion of the titration curve which is nearly vertical (between pH 2.0 and 9.0) is achieved in a second reactor to prevent excess reagent usage or effluent discharge violations.

Sodium carbonate (soda ash) is an alternative sodium alkali for acidic wastestreams lacking buffering capacity such as deionized acid-bath rinsewaters. The use of sodium carbonate (a weak base) with strong acids, such as sulfuric, will impart a buffer to the wastewater stream, thereby facilitating pH control within the neutral range. These buffering reagents will produce a smaller change in pH per unit addition than comparable unbuffered, strong bases such as high calcium lime or caustic soda. This phenomena can be seen in Figure 4.5.2, which illustrates the neutralization of a 1 percent sulfuric acid solution with caustic soda and soda ash. A small incremental addition of caustic soda caused the pH to change from 2 to 11 standard units. Alternatively, approximately 3 times the quantity of soda ash, resulted in a modest pH change from 6 to 9 units.

Due to its carbonate-based reaction mechanism, the neutralization of acidic rinsewaters with soda ash (as with limestone) proceeds at a much slower pace than comparable hydroxide-based reagent systems such as lime or caustic soda. Accordingly, continuous flow reactors must be sized to provide a minimum of 45 minutes hydraulic retention in each stage.[6] In addition, soda ash is commercially available only in a dry form. Consequently, onsite batch mixing and solution preparation facilities, similar to those of hydrated lime, are manadatory when using this chemical as a neutralizing agent. The

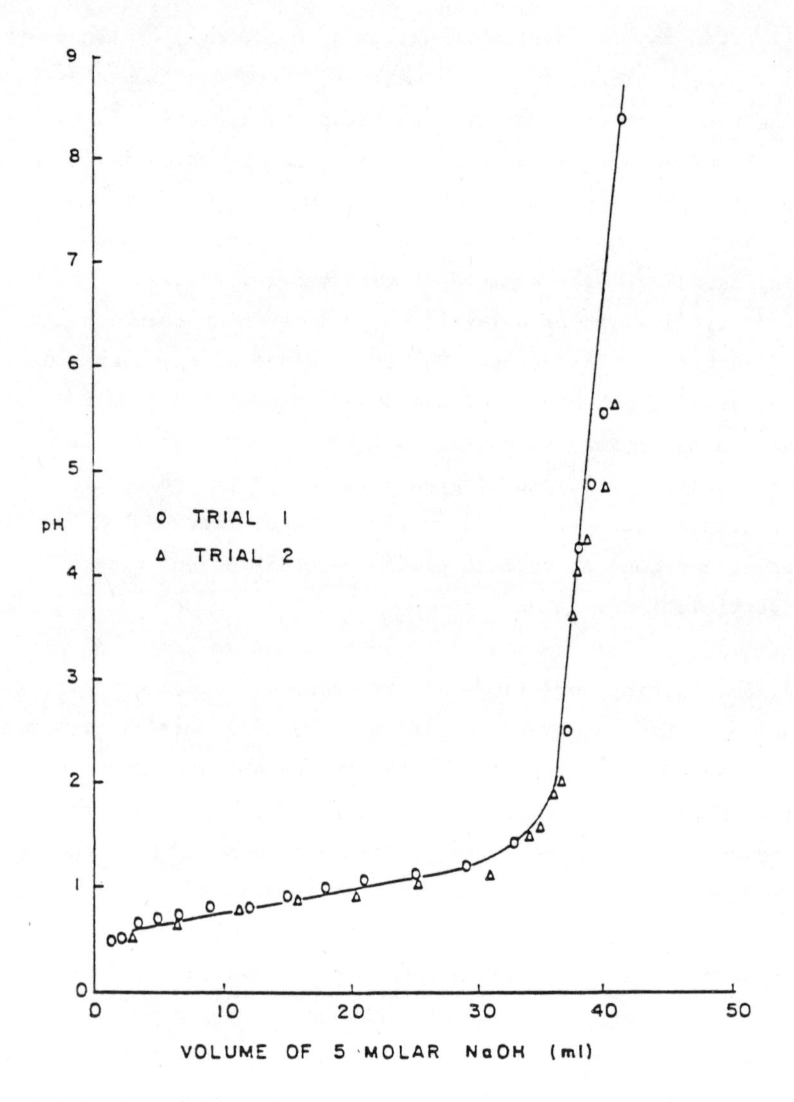

Figure 4.5.1 Neutralization of ferric chloride etching waste by sodium hydroxide

Source: Reference 9

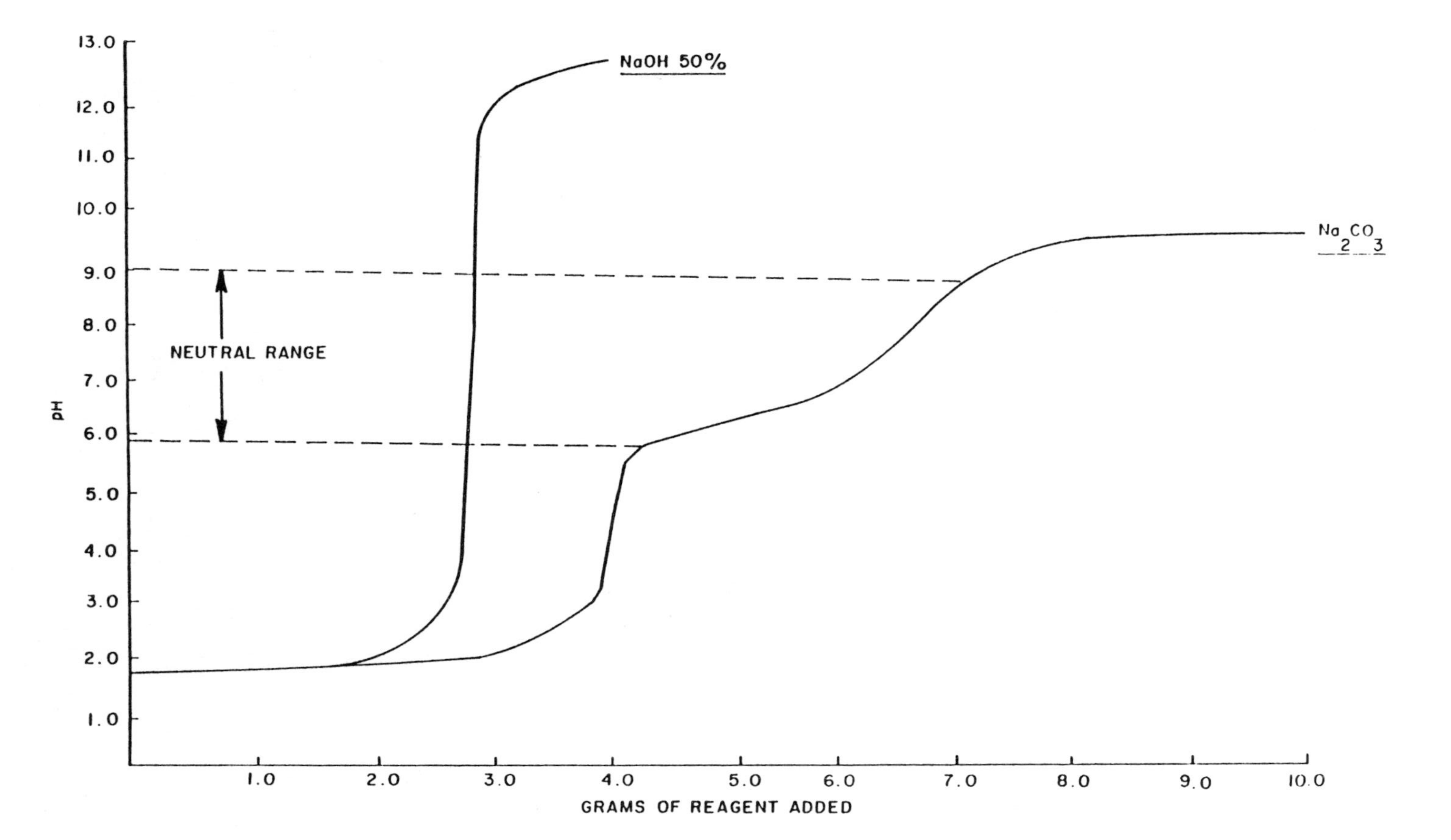

Figure 4.5.2 Titration curve for the neutralization of a 1% H_2SO_4 solution with sodium hydroxide and sodium carbonate.
Source: Reference 6

solubility of soda ash also limits its use since a chemical solution feed strength of only 20 percent by weight can be maintained at ambient temperatures without salt recrystallization. Continuous mixing of the prepared solution is recommended to maintain homogeneity.

An advantage of soda ash is the generation of less sludge since sodium-based end products are more soluble than calcium-based products. However, sodium-based sludges do not filter as readily or to as high solids content as calcium based sludges.[4] In addition, the clarified liquid effluent may not be as low in metals content or total dissolved solids as insoluble end product systems such as lime. All these factors must be carefully weighed before selecting sodium carbonate or any other alkaline reagent as a neutralizing agent.

4.5.2 Process Performance

While design constraints such as deadtime, reagent splashing, and pH control system overshoot will determine minimum system retention time, parameters such as influent flowrate and concentration will usually establish overall system volume and configuration. This is partially as a result of the availability of caustic soda pH control systems which rapidly respond to instantaneous changes in acid concentration. Thus, the following three case studies summarized in Table 4.5.3, are intended to illustrate caustic soda neutralization system design as a function of influent flowrate.

The first application involved two-stage continuous neutralization of acidic wastewater from semiconductor wafer processing operations.[6] Rinsewaters and process bath dumps are collected in an equalization sump prior to the first stage neutralization as shown in Figure 4.5.3. The influent averages a pH of less than 2.0 and a flowrate of 30 gal/minute. The neutralization system components include a neutralization tank with pH control instrumentation, a mixer (1/2 hp) and mounting bracket, one chemical feed tank with mixer, one chemical metering pump, and an electrical control panel.

In the first neutralization tank (150 gallons), the pH is raised to 5 or 6 with a 3 to 5-minute retention time. In the second neutralization tank (750 gal), the pH is raised to 9.0 with a 15 to 25-minute retention time. Upon complete neutralization, the effluent is checked by a final pH probe and then discharged to the sewer.

TABLE 4.5.3 SUMMARY OF SODIUM HYDROXIDE NEUTRALIZATION DATA

Parameter	Treatment system		
Type of waste (manufacturer)	Printed circuit[a] board	Integrated ciruits[b] (semi-conductors)	Cutting oils[c] (automobile)
Influent pH	2.0	2.0	1.5
Influent Flow rate (gpm)	60	30	700-4000
NaOH Concentration (%)	45	50	50
Mode of Operation	Continuous	Continuous	Batch
Process			
Sump			
Capacity (gal)	250	-	-
Retention Time (Min)	4	-	-
Effluent pH	5.0	-	-
Primary Neutralization			
Capacity (gal)	1500	150	500,000
Retention Time (Min)	25	3-5	125
Effluent pH	7.0	5.0-6.0	6.0-9.0
Secondary Neutralization			
Capacity (gal)	1500	750	-
Retention Time (Min)	25	15-25	-
Effluent pH	8.5	9.0	-

Sources: a. Reference 8
b. Reference 6
c. Reference 10

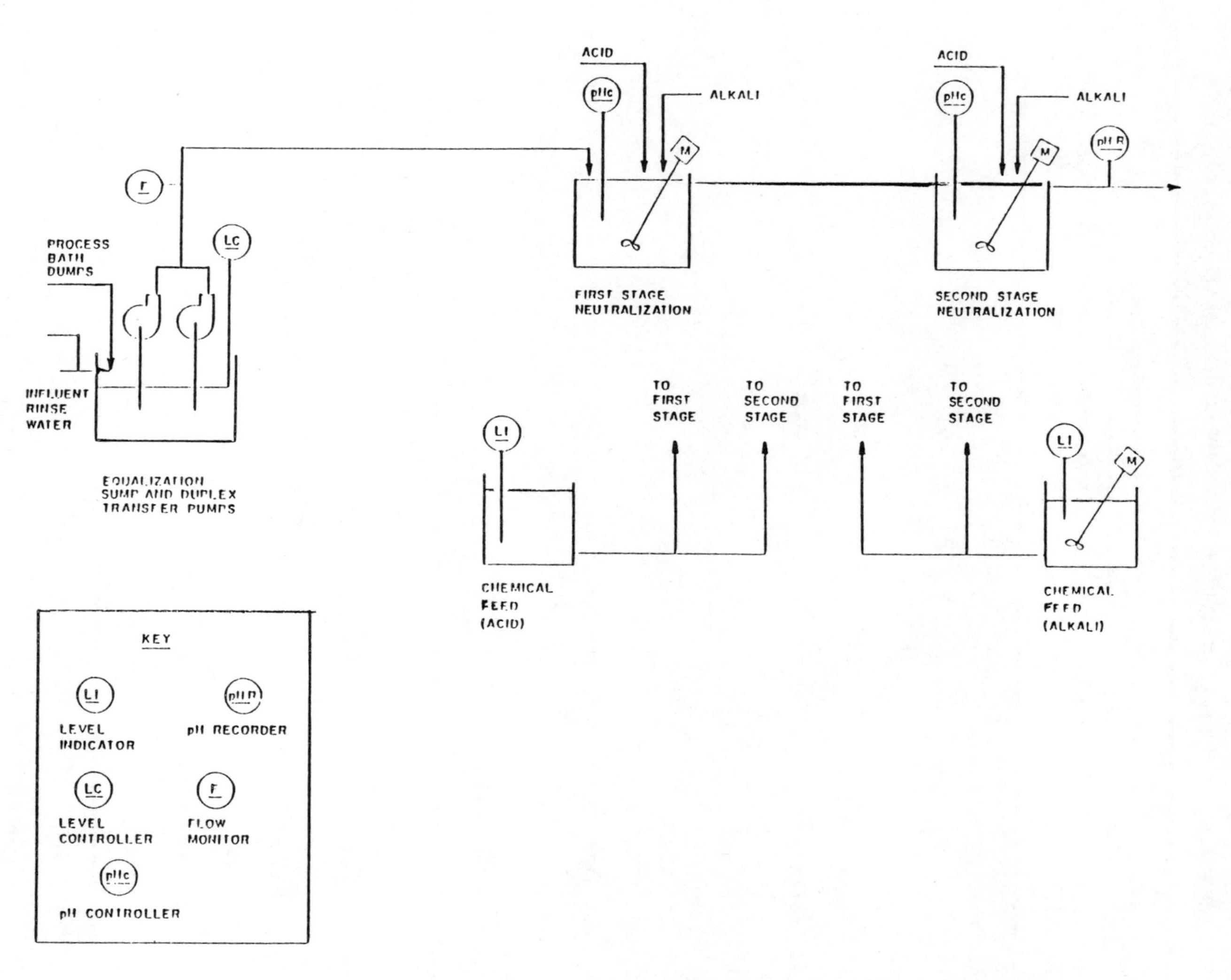

Figure 4.5.3 Process schematic: two-stage neutralization system.

Source: Reference 6

The second application of caustic soda involves the continuous neutralization of acidic wastewater from a large printed circuit board facility.[8] The facility operates on a 24-hour, 7-day/week basis. The average influent flow rate is 60 gal/minute with a pH of less than 2.0. The reagent solution is stored inside in a 3,500 gallon tank at 45 percent strength to prevent freezing. Reagent usage is approximately 3,000 gallons each week. The system configuration is very similar to the previous case study, however, a third stage and increased retention times have been added to accomodate a flow rate which has essentially doubled. This expanded system allows for a greater safety margin in insuring effluent compliance and minimal reagent wastage. It also provides more flexibility in the event that future expansion is required.

The influent collects in a 250 gallon sump where the pH is raised to 5 during a four minute retention time. Primary and secondary neutralization takes place in two 1,500 gallon tanks (25-minute retention time) and the pH is raised to 7 and 8.5, respectively. Agitation in provided in all tanks with a 3/4 hp mechanical agitator and reagent is regulated through feedback control. Following caustic neutralization, liquid/solid separation is achieved through two activated carbon filters arranged in series which remove suspended solids and trace organics. Following separation, the supernatant is discharged to the sewer while the absorption columns are backwashed and the resulting solution is dewatered in a filter press.

The third application consists of the large scale neutralization of acid-emulsion breaking wastes.[10] The physical plant consists of a 44 ft x 60 ft treatment building, three 500,000 gallon treatment tanks, and 10,000 gallon reagent storage tanks. The waste stream (pH 1 to 2) consists of an oily waste that has been acidified with sulfuric acid, heated to 120°F, and then skimmed to remove the oil in a batch emulsion-breaking process. The remaining contents are clarified with a flocculating agent (alum), neutralized by the addition of caustic soda or lime, and then discharged to the city sewer system (pH 6-9). Air for agitation is provided by a six-stage, 60 hp blower at 7 psig. Plant air is available for standby through a pressure reducing station. When the control panel starts from the bottom for the blower is activated, a control valve closes to prevent surging.

The reagent system (Figure 4.5.4) consists of 10,000 gallon liquid caustic storage tank and a 100 gal/minute recycle pump system. Reagent is added as needed and overdose is prevented by time interlocking of the reagent injection operation. Most of the piping is fiberglass reinforced polyester and all controls are pneumatic.

While the influent flow rate is greater than any of the other case studies (700-4,000 gpm), the flow from the emulsion breaking system requires that the neutralization process be carried out in a batch mode.

4.5.3 Process Costs

Table 4.5.4 presents the process costs developed for the construction and operation of 50 percent sodium hydroxide continuous neutralization systems. Capital investment and operational costs are based on the methodologies that were presented in previous sections.[5,11,12] Reactor sizing was based on influent flow rate. The individual treatment configurations for the 400, 2500, and 3,500 gpm systems consisted of respectively; a single-stage reactor, a primary reactor in series with one finishing tank, and a primary reactor in series with two finishing tanks.

Reagent demand was estimated to be 1.18 lbs of 50 percent caustic solution per pound of sulfuric acid neutralized. The reagent usage ratio is based on a basicity factor of 0.69 and a 50 percent molar excess of water. Sludge generation was assumed to be negligible since the neutralization of pure sulfuric acid will result in the formation of a soluble (44 grams/100 milliliters) sodium sulfate dihydrate reaction product. However, the presence of metallic species or other reducible compounds in the wastestream would necessitate the additional investment of sludge handling equipment. Table 4.5.5 presents sludge generation factors for seven common metallic contaminants.

4.5.4 Process Status

Sodium hydroxide neutralization is widely applied technology for treating corrosive waste streams. The rapid reaction rate of NaOH treatment facilitates the use of smaller tanks and shorter retention times. Also,

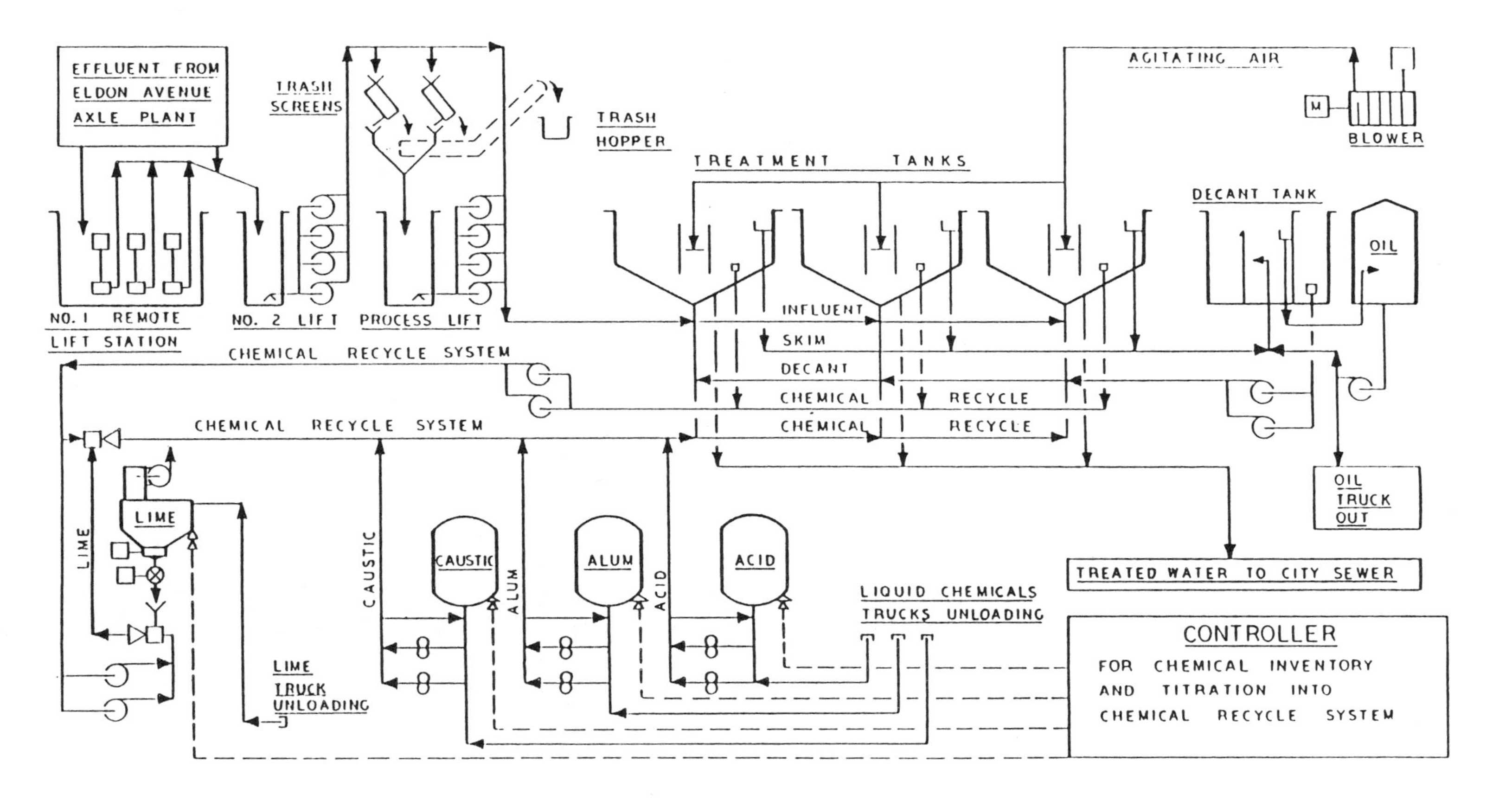

Figure 4.5.4 Process schematic sodium hydroxide batch treatment system.

Source: Reference 10

TABLE 4.5.4. CONTINUOUS CAUSTIC SODA NEUTRALIZATION TREATMENT COSTS

	Flow rate (gph)		
	400	2,500	3,500
Capital investment ($)			
Collection sump	3,100	3,600	3,600
Piping system	2,838	2,838	2,838
Reactor system	19,000[a]	48,000[b]	74,000[c]
Total capital cost	24,938	54,438	80,438
Annulized capital	4,414	9,635	14,238
Operating costs[d] ($)			
Taxes and insurance (7%)	309	675	997
Maintenance (5%)	221	482	712
Labor ($20/hr)	24,000	24,000	24,000
Reagant cost ($205/ton)	35,493	221,908	311,146
Total cost/yr ($)	64,439	256,700	351,093
Cost/1000 gal	67	43	42

[a]Includes single stage continuous reactor with feed system.
[b]Includes single stage continuous reactor in series with 25 minute finishing reactor with feed systems.
[c]Include single stage continuous reactor in series with two 25 minute finishing reactors with feed systems.
[d]Does not include utilities.

Source: Adapted from References 5, 11, 12 using August 1986 CPI index cost data.

TABLE 4.5.5 SODIUM HYDROXIDE SLUDGE GENERATION FACTORS

Metal ion	lb dry solids generated / lb of metal precipitated
Cr	1.98
Ni	1.58
Cu	1.53
Cd	1.30
Fe	1.61
Zn	1.52
Al	2.89

Source: Reference 12

sodium hydroxide minimizes the usual requirement to install large flow/pH equalization basins due to the flexibility of proportional control. It also results in lower quantities of sludge generated relative to calcium based reagents. However, these sludges may be more difficult to dewater. The advantages and disadvantages of caustic soda neutralization are summarized in Table 4.5.6.

TABLE 4.5.6 ADVANTAGES AND DISADVANTAGES OF CAUSTIC SODA NEUTRALIZATION

Advantages

- Proven technology with documented neutralization efficiencies
- Strong base with rapid reaction rate
- Smaller tanks and retention times than comparitive reagents
- Inventory and storage handling procedures are less complicated due to liquid form
- Storage does not require continuous agitation to maintain homogeneity
- Does not require complex slaking or slurrying equipment
- Produces more soluble by products in low pH applications

Disadvantages

- Chemical costs are significantly higher ($205/ton vs. $46/ton for hydrated lime)
- Does not impart any buffering capacity to industrial waste streams
- Close attention must be given to the design of the pH control
- Caustic soda precipitation will result in a fluffy gelatinous floc increasing the size of the clarification chambers and sludge dewatering equipment.
- Cannot effectively precipitate sulfate waste streams due to solubility of sodium sulfate.

Source: References 1, 2, 3, 6, 13

REFERENCES

1. Cushnie, G.C. Removal of Metals from Wastewater: Neutralization and Precipitation. Pollution Technology Review, No. 107, Noyes Publication, Park Ridge, NY. 1984.

2. Kirk-Othmer Encyclopedia of Chemical Technology. Vol. 1, 3rd. Edition, John Wiley & Sons, New York, NY. 1981.

3. Camp, Dresser, and McKee. Technical Assessment of Treatment Alternatives for Wastes Containing Corrosives. Contract No. 68-01-6403. September 1984.

4. Mace, G.R. Live vs. Caustic for Neutralizing Power. Chemical Engineering Process. August 1977.

5. MITRE Corp. Manual of Practice for Wastewater Neutralization and Precipitation. EPA-600/2-81-148. August 1981.

6. Mabbett, Cappacio & Associates. Industrial Wastewater Pretreatment Study: Preliminary Engineering Design Report. January 1982.

7. Hoffman, F. How to Select a pH Control System for Neutralizing Waste Acids. Chemical Engineering. October 30, 1972.

8. Leedberg, T., Honeywell Corporation. Telephone conversation with Steve Palmer, GCA Technology Division, Inc. August 13, 1986.

9. Okey, R.W. Neutralization of Acid Wastes by Enhanced Buffer. Journal of the Water Pollution Control Federation. July 1978.

10. Hoad, J.G. How One Pollution Problem was Solved - Simply Industrial Waste. November 1971.

11. Peters, M.S., and K.D. Timmerhaus. Plant Design and Economics for Chemical Engineers. McGraw-Hill Book Company, New York, NY. 1980.

12. U.S. EPA. Economics of Wastewater Treatment Alternatives for the Electroplating Industry: Environmental Pollution Control Alternatives. EPA-625/5-79-016. 1979.

13. Arthur D. Little, Inc. Physical, Chemical, and Biological Treatment Techniques for Industrial Wastes. U.S. EPA SW-148. November 1976.

4.6 MINERAL ACID TREATMENT

4.6.1 Process Description

Mineral acid treatment of corrosive waste streams is the most widespread of neutralization processes for alkaline wastes. The acid-base reaction is essentially the analog of those discussed in previous sections; i.e., a proton (H^+) is donated by the neutralization reagent. The two primary mineral acid reagents are sulfuric and hydrochloric which are characterized by their highly reactive nature, complete miscibility with water, and rapid disassociation rates. In concentrated form, application may result in the generation of an acid mist or toxic fumes. Therefore, the choice of an acidic reagent is typically based on ease of handling, as well as cost per unit basicity, and in some cases such as food processing; end product characteristics.

Sulfuric acid (H_2SO_4) is the most widely used of all mineral acid reagents[1]. Ease of manufacture, diprotonated reaction chemistry and concentrated nature, combine to make it the least expensive mineral acid on a neutralization equivalent basis. It is supplied in concentrated liquid form, (93 to 98 percent), is highly reactive, strongly hygroscopic and presents a burn hazard to personnel.[2] Dilute solutions are highly corrosive to iron and steel, whereas concentrated solutes (greater than 93 percent) are not corrosive.[3] Protection from freezing during storage and transport is required since sulfuric acid exhibits a maximal freezing point of 8°C (47°F) at a concentration of 85 percent. In addition, pH control overshoot and fuming characteristics frequently require diluting the acid to 30 percent concentration prior to application[4]. In diluting operations, the acid should be slowly added to the water, with provisions for agitation, adequate ventilation and protective clothing. A summary of sulfuric acid physical data is provided in Table 4.6.1.

When sulfuric acid is uneconomical or otherwise inapplicable as a neutralization reagent, hydrochloric acid (HCl) is often used as a substitute.[5] Hydrochloric (Muriatic) acid is supplied in aqueous solutions of 35 to 37 percent acid. Although it is somewhat more reactive than sulfuric, the most concentrated commercial grade contains at least 63 percent water, increasing transportation costs and limiting most major uses to a

TABLE 4.6.1 SULFURIC ACID PHYSICAL PROPERTIES

Chemical name	Sulfuric acid
Reagent grade concentration	78 - 96.5%
CAS No.	7664-93-9
Formula	H_2SO_4
Molecular weight	98.08
Boiling point	310°C
Melting point	-14 to -9°C
Specific gravity	1.842
Nonvolatiles (max) %	0.025 - 0.05
As (max) ppm	1.0 - 2.5
SO_2 (max) ppm	40 - 80
Fe (max) ppm	50 - 100
Nitrate (max) ppm	5 - 20
APHA turbidity (max)	100 - 150
APHA color (max)	100 - 200

Source: Reference 2

radius of 300 to 500 km from the producing source[5]. In addition, its higher unit cost and monoprotonated reaction chemistry result in an overall reagent cost approximately double that of sulfuric acid on a neutralization equivalen basis. Hydrochloric acid is highly disassociated (a 10,000 mg/L HCl solution will result in pH of 0.9), and extremely corrosive, attacking most metals through surface dissolution. Commonly used plastics and elastomers are recommended as materials of construction when designing neutralization system using hydrochloric acid as reagent.

The main advantage of using HCl is that it generates soluble end products. However this attribute may not be beneficial in some cases, since it may cause the waste to exceed dissolved solids and metal effluent standards. As with sulfuric, hydrochloric acid will react vigorously with water, sometimes evolving an acid mist which can destroy the mucous membrane and cause choking, coughing, headache and dizziness. In addition, hydrochloric acid will decompose in the presence of heat into toxic hydrogen chloride gas. Therefore, caution must be exercised when storing or handling concentrated hydrochloric acid to minimize splashes, spills or mist generation. A summary of hydrochloric acid physical property data is provide in Table 4.6.2.

Conventional mineral acid treatment systems dispense a highly reactive aqueous reagent and are, therefore, similar to those of caustic soda. Equipment consists of chemical storage and reaction vessels, mixers, chemical-feed metering pumps or valves, and a pH control system. Typically the reagent storage vessel is constructed of fiberglass reinforced plastic (FRP) and sized for one week to one months supply, depending on reagent usag and/or reagent delivery schedules.[1] For low volume systems, the reagent can be metered directly from the barrel to the reaction system. Reaction vessel can be constructed of FRP or reinforced concrete, depending on corrosion resistance and flexibility for operation and expansion. Since mineral acids disassociate rapidly into solution, reaction vessels are generally sized for 15 to 20 minutes of retention time.

Mixers should be of the type that generate a top-to-bottom flow pattern with fluid redistribution outward from the center to the walls of the tank[1]. This helps avoid "short circuiting", in which the wastewater exits the tank without sufficient dispersion and treatment. Metering pumps should

TABLE 4.6.2 HYDROCHLORIC ACID PHYSICAL PROPERTIES

Trade name Chemical name	Muriatic acid Hydrochloric acid
Reagent grade concentration	35 - 37%
CAS No.	7647 - 01-0
Formula	HCl
Molecular weight	36.46
Boiling point	110°C
Melting point	--
Specific gravity	1.1885
Evaporation rate (Butyl acetate = 1)	10
Vapor pressure (20°C)	160 mm Hg
Vapor density (air = 1)	1.25
Percent volatiles by volume	37
pH 1.0 N (aqueous) 0.1 N (aqueous) 0.02 N (aqueous) 0.002 N (aqueous)	0.1 1.1 2.02 3.02

Source: Reference 5

be of the positive displacement type, (large turndown ratio) which allow for accuracy and flexibility in regulating the chemical dosage to the system. In systems where precipitation reactions occur, solids adhere to the pH sensing electrode and eventually build up a film that interferes with the measurement. Proper cleaning and maintenance are thus necessary to maintain accurate pH control.

Due to the similar nature of mineral acid and sodium hydroxide treatment systems, process equipment and operating parameters for each system are essentially the same and, in most cases, interchangeable. The primary difference between these treatment systems is the inclusion of a ventilation and an acid mist scrubber. The scrubbing system is necessary when neutralizing concentrated alkaline wastes with strong mineral acids. A significant amount of heat can be generated raising the temperature of the neutralization tanks and causing emissions of acid fumes and hazardous gases. The scrubbing system captures these fumes and neutralizes them in a packed tower. A caustic scrubbing solution is typically used in the packed tower, since a lime slurry would rapidly coat the packing medium and plug the tower.[6]

4.6.2 Process Performance

When utilized as a final treatment, mineral acid systems are used singularly or in tandem with an alkali reagent depending on the variability of influent pH. They are also used in pretreatment systems to reduce the alkalinity of highly caustic waste streams prior to secondary treatment. In addition, mineral acids are used as an emergency pH control for mutual neutralization systems in which the concentration and flow rate of the alkali waste stream can exceed the neutralization capacity of the acid waste stream. Finally, although sulfuric acid is usually chosen as the acid reagent, there are situations for which hydrochloric acid achieved better overall performance. The following case studies demonstrate the aforementioned used mineral acid neutralization systems and provide an example where hydrochloric acid was found to be a more suitable reagent than sulfuric acid.

The first case study[7] illustrates a neutralization system in which sulfuric acid was used singularly to partially neutralize a highly alkaline waste stream prior to activated sludge treatment (Figure 4.6.1). The pH reduction was necessary to prevent toxic shock and maintain optimal growth conditions for the bio-culture.

The coal-liquefication process wastewater was segreated into a high strength caustic waste stream of pH 9.8 to 13.5 and a low strength liquid waste steam of pH 3 to 11.

Waste characteristics are summarized in Table 4.6.3. The high strength waste stream, which has been non-chemically treated for oil removal, is treated on a fill and draw basis in one of two 10,000 gallon chambers in the pretreatment tank. The solution is batch catalyzed with manganese sulfate (400 mg/L) and diffused air (2,500 SCFM) to oxidize the sulfide groups. The chambers are used alternately to maintain aeration in one while the other is being filled. When the sulfide content falls below 20 mg/L, the pH is adjusted to between 9.5 and 10.5 with sulfuric acid. The pH is controlled in the moderately basic range since the bio-reaction system tends to drive the pH to the acid side. The high strength waste stream is then combined with the low strength waste in a 55,000 gallon, concrete equalization storage tank. The drawoff point is protected by a baffle to allow a 2.5 ft buildup of sludge in the bottom. In addition, a 55,000 gallon emergency storage basin is provided to hold and dilute high-strength spills and retain bio-food during extended shutdowns.

As previously stated, singular mineral acid neutralization systems are usually only applicable when alkaline influent pH characteristics are fairly uniform in nature. However, the wide fluctuations in pH normally encountered in industrial manufacturing wastewaters often necessitate the use of dual reagent system. The following case study illustrates the most commonly used form of modular, dual reagent neutralization systems.

In 1986, Alliance Technologies Corporation evaluated a 2-stage, 22,000 gpd wastewater treatment system for the neutralization and precipitation of a variable pH effluent from a printed circuit board manufacturing operation.[8] The wastewater streams consisted of alkaline etching solution (pH 11.8 to 13.7), acidic plating rinsewaters (pH 2-3), or a combination of the two, depending on the process operations. The wastewaters were neutralized to a final endpoint of 8.5 with 93 percent sulfuric acid or 50 percent sodium hydroxide, as required to facilitate sodium borohydride

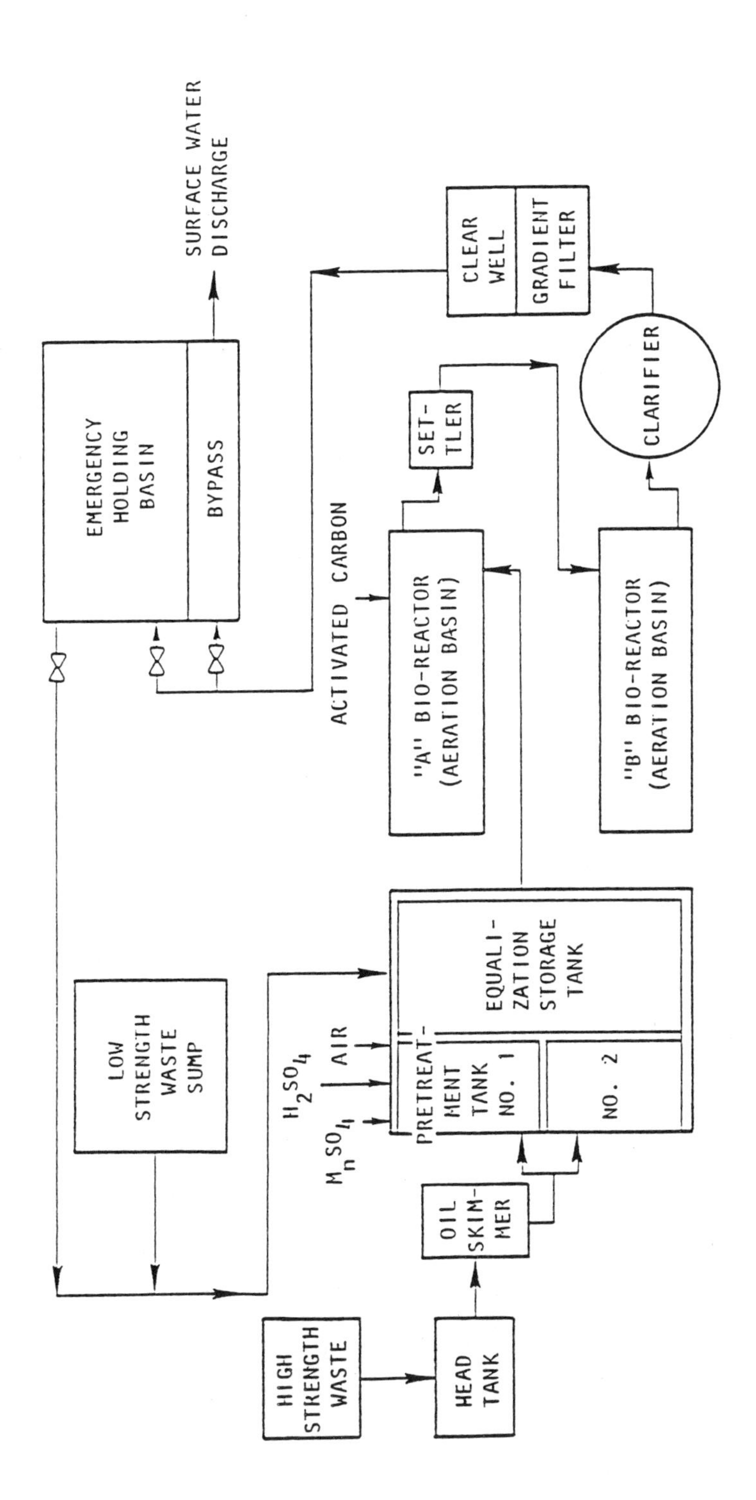

Figure 4.6.1 Coal-liquefaction facility wastewater treatment system.

Source: Reference 7.

TABLE 4.6.3 SUMMARY OF COAL-LIQUEFACATION FACILITY WASTEWATER CHARACTERISTICS

	High strength waste stream	Low strength waste stream	Combined[a]
Flow, gpd	4,000-5,000	5000-21,000	9,000-26,000
pH	9.8-13.5	3-11	9.5-10.5
BOD, mg/L	250-2,500	7-23	100-1,100
COD, mg/L	900-5,500	NA	400-8,000
Phenolics, mg/L	30-800	0.1-1.0	20-600
Oil Grease, mg/L	20-110	1-15	5-70
Sulfides, mg/L	70-3,400	1.0	1.0
NH_3, N, mg/L	75-200	--	15-75

[a]Streams combined after catalyzed air oxidation of CWS and flow equalization.

Source: Reference 7

metals reduction. On the day of testing, the influent to the waste treatment system primarily consisted of dilute alkaline etching solution (average pH 12.28). In addition, the metals content of the waste stream was 790 mg/L of which approximately 99 percent was divalent copper. After neutralization, metals reduction occurred through the addition of a 12 percent sodium borohydride solution. Liquid/solid separation of the precipitate was achieved through the use of a Memtek ultrafiltration unit with subsequent dewatering in a Delta unifilter low-pressure filter press. A schematic of the wastewater treatment system is provided in Figure 4.6.2. Sludge production was approximately 185 lbs/day of dry solids, of which approximately 78 percent was reduced copper. Table 4.6.4 contains a summary of alkaline wastestream and final effluent characteristics.

The wastewater treatment system consisted of a 250 gallon polypropylene collection sump and two 825 gallon carbon steel, FRP lined reactors in series. The wastewater collection sump was automatically emptied into the neutralization tank through an adjustable 35 gpm Gould feed pump. The reaction tanks were vertical, rectangular, and flat bottomed with open tops. Agitation was provided through an electric propeller mechanical mixer with stainless steel shaft and props, driven by a 2 hp motor. The neutralization pH controller/recorder was a single position, three set point unit which controlled pH and sulfuric acid/caustic soda addition. The chemical feed system consisted of a duplex pump system which drew reagent directly from 55 gallon barrels to minimize hazardous transport. The chemical feed pumps were positive displacement, diaphragm type with adjustable output of 0 to 20 gph.

Since industrial processes rarely produce only one type of corrosive waste stream, the dual reagent system tends to predominate over the singular reagent system. Regardless of the alkaline reagent selected, sulfuric acid is the overwhelming mineral acid of choice, except in specialized cases. An exception to these general rules of thumb is the neutralization of food processing wastewaters from caustic peeling operations, as described below.

A peeling operation to remove skin materials is a necessary step in the processing of numerous fruits and vegetables.[9,10,11] One such process involved submerging tomatoes for about 30 seconds in a 200°F, 16 to 20 percent sodium hydroxide bath.[11] In the subsequent washing step which removed the softened outer peel and residual caustic, the pH of the washwater ranged from

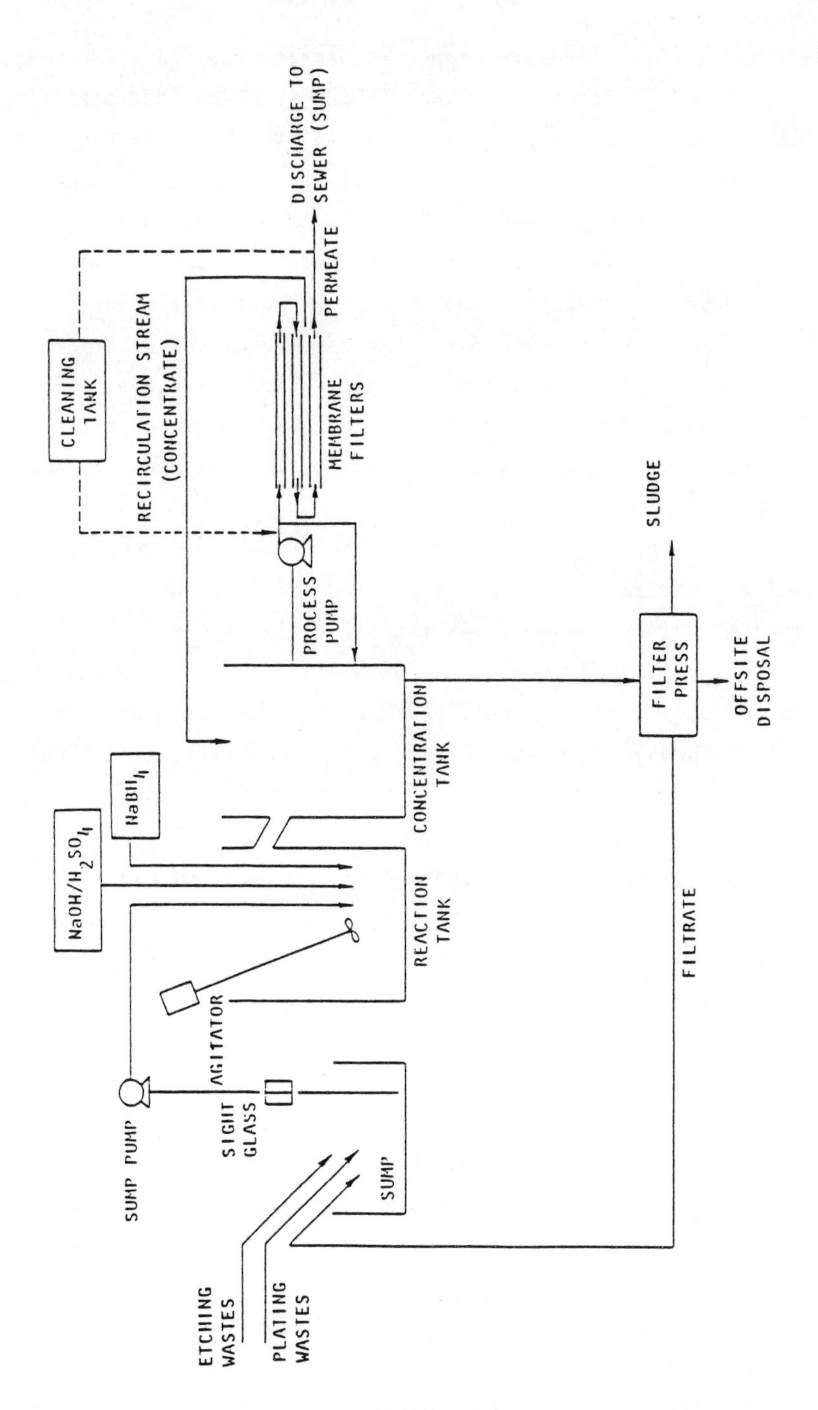

Figure 4.6.2. Process schematic showing plating/etching waste treatment system.

Source: Reference 8.

TABLE 4.6.4. SUMMARY OF NEUTRALIZATION/PRECIPITATION TEST DATA

Parameter	Stream 3 Influent waste (mg/L)	Stream 5 Effluent wastewater (mg/L)	Stream 7a Sludge[a] (μg/g)
pH	12.28	8.5	NA
Total organic carbon	40.0	36.1	184.8
Total organic halide	1.76	1.75	--
Total trace metals:			
Cu	786.0	1.49	780,000
Ni	0.055	0.03	58.7
Pb	0.57	0.10	300
Zn	3.86	0.028	1,430
EP toxic metals:			
Ar	--	--	0.03
Ba	--	--	0.522
Cd	--	--	0.002
Cr	--	--	0.003
Pb	--	--	1.8
Ug	--	--	0.0002
Se	--	--	0.04
Ag	--	--	0.56

[a]Results given on a dry weight basis for sludge.

Source: Reference 8.

13 to 14. The caustic peeling operation processed 35,000 tons of tomatoes/year, generating approximately 24 gallons/minute of caustic sludge.[11] In addition, over 100 gallons/minute of washwater containing 9.1 lbs of 50 percent sodium hydroxide had to be subsequently neutralized.

Sulfuric acid was originally used as the neutralizing agent (4 lbs/min), but resulted in the loss of approximately 120 tons/day of tomato constituents. By substituting food grade hydrochloric acid for sulfuric acid, the company expect to realize an estimated 50 percent reduction in sludge production as shown in Table 4.6.5. The use of hydrochloric results in the formation of sodium chloride (table salt) since this is permitted as a food ingredient under the FDA standards (21 CFR 53-10, 53.20, and 53.30), a substantial fraction of the neutralized waste can be recycled. A flowsheet for the proposed recovery of the waste peeling sludge is presented in Figure 4.6.3.

4.6.3 Process Costs

Table 4.6.6 presents the process costs developed for the construction and installation of two 93 percent sulfuric acid neutralization systems and two 37 percent hydrochloric acid systems. Each mineral acid system was evaluated on the basis of its ability to cost effectively neutralize: (1) a 2 percent sodium hydroxide wastestream; and (2) a 2 percent calcium carbonate wastestream. The reaction chemistries for each mineral acid neutralization reaction are presented below:

$$NaOH + H_2SO_4 \rightleftharpoons NaSO_4 + H_2O \quad (1)$$

$$NaOH + 2\ HCl \rightleftharpoons NaCl + H_2O \quad (2)$$

$$CaCO_3 + H_2SO_4 \rightleftharpoons CaSO_4 + H_2O + CO_2 \quad (3)$$

$$NaCO_3 + 2\ HCl \rightleftharpoons CaCl + H_2O + CO_2 \quad (4)$$

Capital costs for each of the four waste treatment scenarios have been adapted from cost information presented previously for sodium hydroxide neutralization (Section 4.5.3) due to a similarity in process equipment and operational parameters. Reagent demand was estimated to be 1.32 lbs of

TABLE 4.6.5. SUMMARY DATA ON CAUSTIC TOMATO PEELING OPERATION AND ACID USAGE

Parameter	Caustic waste stream	Reagent system	
		H_2SO_4	HCl
Number of caustic peelers	4		
Loading, rate, tons/min, each	0.33		
Caustic solution in peeler, %	7-10		
Total season throughput, tons	35,000		
Caustic usage, tons	121		
Unit caustic usage, lbs/ton	6.9		
Acid usage, tons		53	99
Sludge production, tons/day		120	64

Source: Reference 11

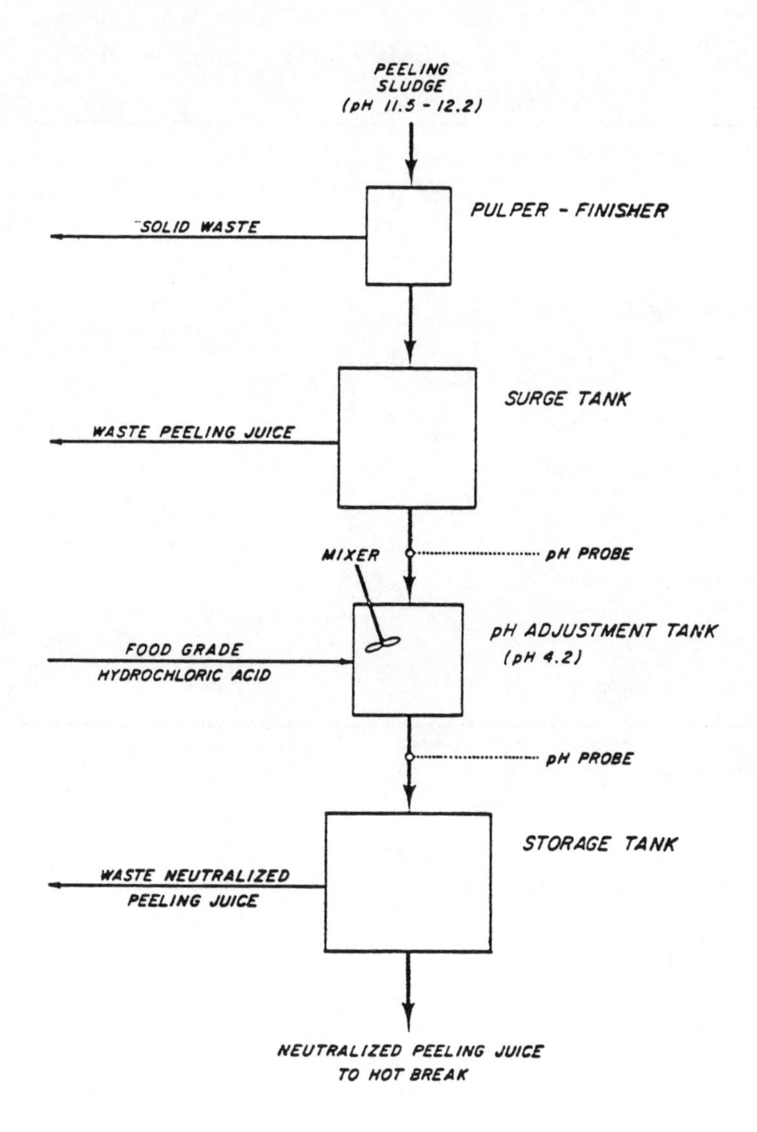

Figure 4.6.3 Flowsheet for proposed recovery of waste peeling sludge.

Source: Reference 11.

TABLE 4.6.6. CONTINUOUS MINERAL ACID TREATMENT COSTS USING SULFURIC AND HYDROCHLORIC ACIDS AS THE NEUTRALIZING AGENT

		Treatment cost ($)								
Waste stream	Reagent (gph)	Total capital	Annualized capital	Taxes & insurance (7%)	Maintenance (5%)	Labor ($20/hr)	Reagent costs/yr	Disposal ($200/ton)	Total cost/yr	Cost/ 1,000 gal
2% NaOH	H_2SO_4									
	400	24,938	4,414	309	221	24,000	15,943	--	44,887	47
	2,500	54,438	9,635	675	482	24,000	99,639	--	134,431	22
	3,500	80,438	14,238	997	712	24,000	139,496	--	179,443	21
2% NaOH	HCl									
	400	24,938	4,414	309	221	24,000	31,085	--	60,029	63
	2,500	54,438	9,635	675	482	24,000	194,286	--	227,078	38
	3,500	80,438	14,238	997	712	24,000	271,999	--	311,946	37
2% $CaCO_3$	H_2SO_4									
	400	59,938	10,609	743	530	24,000	16,833	104,475	157,190	163
	2,500	109,938	19,459	1,362	973	24,000	100,942	626,504	773,240	129
	3,500	148,438	26,273	1,839	1,314	24,000	141,318	877,100	1,071,844	128
2% $CaCO_3$	HCl									
	400	24,938	4,414	309	221	24,000	31,492	--	60,436	63
	2,500	54,438	9,635	675	482	24,000	197,094	--	231,886	39
	3,500	80,438	14,238	997	712	24,000	275,556	--	315,503	38

Source: Adapted from References 13 and 14 using September 1986 CPI index cost data.

93 percent sulfuric acid and 2.47 lbs of 37 percent hydrochloric acid/pound of sodium hydroxide neutralized. Similarly, for the 2 percent calcium carbonate wastestream, reagent demand was estimated to be 1.05 lbs of 93 percent sulfuric acid and 2.02 lbs of 37 percent hydrochloric acid/pound of calcium carbonate neutralized. The reagent usage ratios were based on stoichiometric equivalents with a 7 percent molar excess of water for sulfuric acid and a 63 percent excess for hydrochloric.

Based on end-product solubilities and reagent requirements, sludge generation sludge generation is negligible in all cases except for the reaction involving sulfuric acid and calcium carbonate. The reaction product (calcium sulfate dihydrate) was estimated to generate sludge at approximately 0.75 pounds of sludge/pound of calcium carbonate neutralized.[12,13] The sludge product was assumed to be dewatered to a filter cake containing 60 percent solids and disposed of in a secure landfill at a cost of $200/ton. In addition, the cost of both a sludge storage unit and filter press were included in total capitol cost for this system.

As shown in Table 4.6.6, sulfuric acid is more cost-effective than hydrochloric on a neutralization equivalent basis in situations where sludge generation does not occur or is not required; e.g., precipitation of heavy metals. However, in situations where sludge generation cannot be avoided, hydrochloric may be the mineral acid of choice.

4.6.4 Process Status

Mineral acid treatment is the most widely used and demonstrated technology for the neutralization of corrosive alkaline waste streams. Both sulfuric and hydrochloric have very high acidities, so that quantities required for neutralization are relatively low in comparison to other acids.[6] Consequently, reactor volumes and handling/storage facilities are smaller. Sulfuric acid, being the most widely available and lowest in cost on a neutralization equivalent basis, is the most prevalent acidic reagent. It is typically used in combination with an alkali reagent to control pH fluctuations in both the acidic and alkaline ranges. Hydrochloric acid is generally used in situations requiring rapid reaction rates and soluble reaction products.

The primary environmental impact from the use of these regeants is the generation of potentially hazardous sludge (sulfuric acid) or generation of an acid mist or toxic/hazardous fumes (hydrochloric acid). The highly corrosive nature of mineral acids presents a burn hazard to personnel and increases the likelihood of a possible catastrophic release during bulk transport or storage. A summary of both the advantages and disadvantages of sulfuric and hydrochloric acid are presented in Tables 4.6.7 and 4.6.8, respectively.

TABLE 4.6.7 ADVANTAGES AND DISADVANTAGES OF SULFURIC ACID NEUTRALIZATION

Advantages

- Highly reactive, concentrated acid with rapid disassociation rate
- Widely available and low in cost
- Unlikely to evolve noxious or hazardous fumes
- Proven technology with documented neutralization efficiencies
- Easy to store and can be used in a variety of configurations

Disadvantages

- Strongly hydroscopic and presents burn hazard to personnel
- Highly corrosive to metals when in dilute form
- Will freeze when stored at temperatures below 47°F and concentrations greater than 85 percent
- Can possibly form voliminous, insoluble endproducts
- Can introduce sulfates into the effluent wastestream

Source: References 1, 2, 3, and 6.

TABLE 4.6.8 ADVANTAGES AND DISADVANTAGES OF HYDROCHLORIC ACID NEUTRALIZATION

Advantages

- Has a more rapid reaction rate than sulfuric acid
- Will form insoluble endproducts and thereby minimize sludge production
- Is a proven technology which can be applied in solution form
- Not as hydroscopic as sulfuric acid

Disadvantages

- Highly corrosive and represents burn hazard
- Higher unit cost than sulfuric acid
- Can evolve both noxious and hazardous fumes
- Can possibly exceed effluent standards due to solubility of reaction products
- Will decompose in the presence of heat to form hydrogen chloride gas

Source: References 3, 5, and 6.

REFERENCES

1. Capaccio, R.S., and R. Sarnelli. Neutralization and Precipitation. Plating and Surface Finishing. September 1986.

2. Kirk-Othmer Encyclopedia of Chemical Technology. Volume 12, 3rd Edition. pp. 983-1011. John Wiley & Sons, New York, NY. 1981.

3. Cushnie, G.C. Removal of Metals from Wastewater: Neutralization and Precipitation. Pollution Technology Review No. 107, Noyes Publication, Park Ridge, NJ. 1984.

4. Francini, F. Honeywell Corporation. Telephone conversation with Stephen Palmer, GCA Technology Division, Inc. September 12, 1986.

5. Kirk-Othmer Encyclopedia of Chemical Technology. Volume 22, 3rd Edition, pp. 226-229. John Wiley & Sons, New York, NY. 1981.

6. Camp, Dresser, & McKee. Technical Assessment of Treatment Alternatives for Wastes Containing Corrosives. Contract No. 68-01-6403. September 1984.

7. Sapp, J.B. Wastewater Treatment at a Coal-Liquefaction Facility. Environmental Progress (Volume 2, No. 3). August 1983.

8. Palmer, S. Case Studies of Existing Treatment Applied to Hazardous Waste Banned from Landfill, Phase II. U.S. EPA Contract No. 68-03-3243. July 1986.

9. Graham, R.P. et al. Double-Dip Caustic Peeling of Potatoes. Proceedings of the Sixth National Symposium on Food Processing Wastes. EPA-600/2-76-224. December 1976.

10. Schultz, W.G., Graham, R.P., and M.R. Nant. Pulp Recovery from Tomato Peel Residue. Proceedings of the Sixth National Symposium on Food Processing Waste. EPA-600/2-76-224. December 1976.

11. Fernbach, E. et al. Wastewater Management of Hickmott Foods Inc. Proceedings of the Sixth National Symposium on Food Processing Wastes. EPA-600/2-76-224. December 1984.

12. CRC Handbook of Chemistry and Physics. 54th Edition 1973-1974, CRC Press, Cleveland, Ohio.

13. MITRE COrp. Manual of Practice for Wastewater Neutralization and Precipitation. EPA-600-81-148. August 1981.

14. US EPA Economics of Wastewater Treatment Alternatives for the Electroplating Industry. EPA 625/5-79-016.

4.7 CARBONIC ACID TREATMENT

4.7.1 Process Description

Carbonic acid neutralization of alkaline waste streams is a relatively old, but as of yet, undeveloped treatment technology. As early as 1931, Curtis and Copson patented a process using a reaction product carbonic acid to neutralize a cotton waste (Kier liquor) treated with caustic soda.[1] The inherent problem with carbonic acid treatment is that carbonic acid, a weak acid disassociates slowly in solution, retards reaction rates, and limits pH reduction applications to the pH 7 to 8 range.[2] In addition, carbonic acid reaction products are slightly alkaline in nature and tend to act as buffers in the neutralization of concentrated alkaline wastes. For example:

H_2CO_3	+	$NaOH$	$\rightleftharpoons$	$NaHCO_3$	+	H_2O	(1)
Carbonic acid		Sodium hydroxide		Sodium bicarbonate pH 8.4 (0.1 normal)			

H_2CO_3	+	$Ca(OH)_2$	$\rightleftharpoons$	$CaCO_3$	+	$2H_2O$	(2)
Carbonic acid		Hydrated lime		Calcium carbonate pH 9.4 (saturated)			

Typically, carbonic acid is generated directly in the neutralization chamber by injecting carbon dioxide into the wastewater solution. Upon hydration, the carbon dioxide will form carbonic acid and neutralize excess alkalinity.[2] Carbon dioxide is available as either a compressed gas or the by-product of a combustion process. Table 4.7.1 contains a summary of carbon dioxide physical property data.

Compressed (liquid) carbon dioxide is stored and transported at ambient temperatures in cylinders containing up to 22.7 kilograms. Larger quantities are stored in refrigerated, insulated tanks maintained at -18°C and 20 atmospheres.[3] Transportation is by insulated tank truck and rail car.

TABLE 4.7.1. SUMMARY OF CARBON DIOXIDE PHYSICAL PROPERTY DATA

Trade name	Carbon dioxide
Molecular formula	CO_2
CAS number	124-38-9
Sublimation point (°C)	-78.5
Latent heat of vaporization (Btu/lb at 0°C)	101.03
Gas density (g/L)	1.976
Liquid density (g/L)	914
Viscosity (cp)	0.015
Heat of formation (Btu/mol at 25°C)	373.4
pH (saturated solution)	
1 atm	3.7
23.4 atm	3.2

Source: Reference 3

The standard method of applying compressed carbon dioxide for pH control is to vaporize carbon dioxide in a heat exchanger or across a flash valve. The pressurized gas is forced through porous diffuser tubes placed along the bottom of a batch treatment tank. Carbon dioxide gas is released from the diffusers as fine bubbles (15 microns) which are preferentially absorbed by the surrounding wastewater. This type of treatment requires a slow-moving effluent stream with a treatment tank of sufficient depth to ensure that the carbon dioxide is fully absorbed before reaching the surface.[4] Figure 4.7.1 shows the solubility of carbon dioxide in water as a function of temperature and pressure. Since hydration of carbon dioxide forms carbonic acid, it is recommended that the diffuser assembly be constructed of a corrosion-proof material.

The primary advantages of compressed carbon dioxide are minimal capital requirements, uncomplicated piping, and the inability to over-acidify the wastewater. Its primary disadvantages are a low dissolved oxygen content (4.5 percent) at the point of injection, and a high reagent cost on a neutralization equivalent basis (approximately \$200 to \$300/ton). However, for large volume users of 200 tons or more per year, the unit cost per ton of compressed carbon dioxide drops to \$90 to \$100/ton.[5]

A secondary source of carbon dioxide is flue or stack gas from furnaces using fossil fuel. The resultant flue gas, which may contain up to 14 percent carbon dioxide, can be used to neutralize waste caustic solutions in a manner analogous to that of compressed carbon dioxide.[2] In plants with a ready source of available boiler flue gas, dramatic savings in reagent purchases can be realized. At a minimum, seven times the SCFM of boiler flue gas will be required on a neutralization equivalent basis in comparison to compressed carbon dioxide. However, since flue gas is usually available in sufficient quantities, this does not generally create a problem.

One problem with using boiler flue gas as the neutralization reagent is that it is typically available only at low pressure and high temperature (450°F). It contains moisture, carbon dioxide, carbon monoxide, oxygen, nitrogen, and sulfur dioxide in varying amounts, depending on the fuel and the efficiency of the combustion process. Hot flue gas is not especially corrosive, but when the gas is cooled, the water vapor forms liquid droplets in which the various gases can dissolve and form compounds which are corrosive to metallic surfaces.[6]

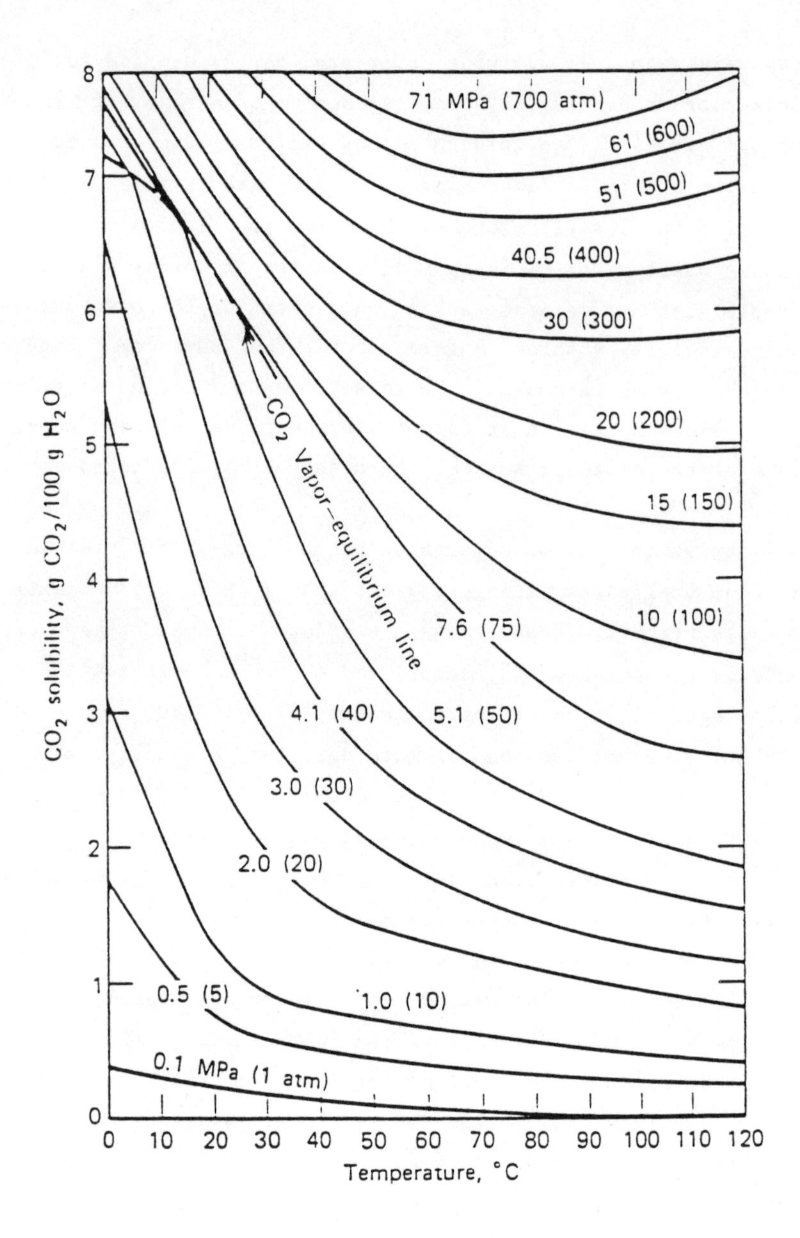

Figure 4.7.1. Solubility of carbon dioxide in water.
Source: Reference 3.

Since boiler flue gas must be pressurized prior to use in order to overcome the liquid head of the treatment tank, a compressor must be supplied. This adds substantially to system capital costs since the compressor, sparger, and associative piping must be constructed of corrosion resistant materials. Alternatively, gas can be dissolved in a pressurized sidestream prior to introduction into the waste treatment tank (see Section 4.7.2).

The primary drawback which has limited the use of flue gas neutralization to a few industrial applications is the presence of both reduced and oxidized sulfur groups in the gas. These sulfur groups will result in an atmospheric pollution hazard from hydrogen sulfide evolution. They also create a potential water pollution hazard due to the incorporation of sulfate groups into the neutralized effluent stream. Therefore, at this point in time, boiler flue gas neutralization of alkaline wastestreams is not considered to be a technically viable treatment process.

A third method of producing carbon dioxide is the underwater combustion of natural gas (see Figure 4.7.2). In 1951, this method was used to reduce the alkalinity of a sulfur dye waste through a submerged combustion process.[7] However, due to high fuel costs, this type of treatment is more economically attractive as an evaporative acid-recovery process than as a neutralization technology.

4.7.2 Process Performance

Although research involving carbonic acid neutralization of alkaline wastestreams has been ongoing for over 50 years, practical applications and literature citations are limited. Only recently have compressed carbon dioxide neutralization systems become competitive with sulfuric acid in select applications. These include plants which use over 200 tons/year of reagent or have flow rates greater than 100,000 gpd.[8]

The following case study (see Table 4.7.2 for summary of process data), illustrates an automatic, compressed carbon dioxide neutralization process which reduces the pH of a 17 MGD chlor-alkali plant effluent from 11.8 to 8.3.[4] The caustic wastestream is fast flowing (1 ft/sec) and shallow (3 ft by 12 ft) and is composed primarily of waste sodium hydroxide. Reagent usage is approximately 45 tons every 2 to 3 weeks. Compressed CO_2 is stored onsite in a leased liquid storage tank.

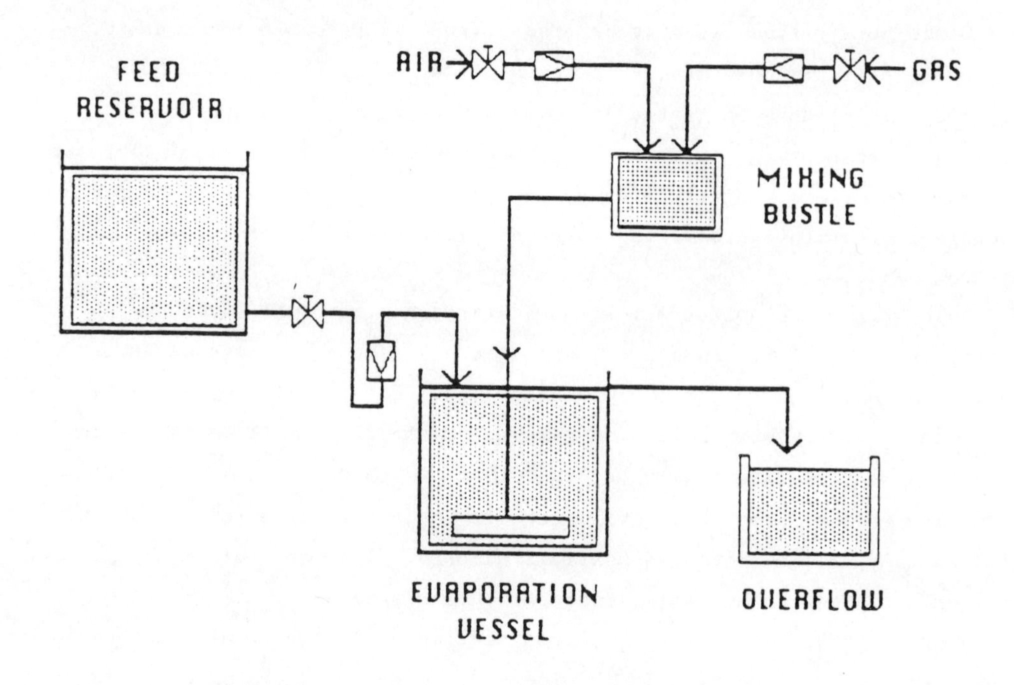

Figure 4.7.2. Submerged combustion pilot unit.
Source: Reference 2

TABLE 4.7.2. SUMMARY OF AUTOMATIC CARBON DIOXIDE NEUTRALIZATION PROCESS DATA

Parameter	Range
Wastewater flow rate (MGD)	17
Wastewater pH (S.U.)	11.8
Sidestream flow rate (MGD)	0.85
Sidestream pH (S.U.)	3.2 - 8.3
Carbon dioxide (psig) vapor pressure	60 - 120
Carbon dioxide (tons/yr) usage	700
Number of discharge orifices	20
Diameter of discharge orifices (in.)	1/8 - 5/15
Depth of discharge manifold (ft)	3
Effluent pH	8.3

Source: Reference 4

A sidestream pH control system is used to increase reagent efficiency (Figure 4.7.3). Five percent of the total waste stream flow is pressurized to 50 to 150 psig and mixed with CO_2 gas. The high pressure in the sidestream greatly increases the carbon dioxide solubility. This increased solubility allows the total CO_2 requirement for the system to be completely dissolved in a small sidestream flow. The sidestream water ranges in pH from 8.3 when no CO_2 is required, to as low as 3.2 at the maximum gas feed rate. Since waste effluent pH is controlled with a supersaturated CO_2 water solution, little gas is released from the discharge manifold into the effluent stream. Thus, losses are negligible.

The carbonated sidestream water is injected into the mainstream through a submerged injection manifold. The orifices in the manifold can serve two purposes. They restrict the flow of the sidestream pump so that the correct water pressure and flow rate are maintained, and they thoroughly mix the acid sidestream water with the caustic mainstream.

The high absorption efficiency of the system results from the fact that the CO_2 gas is in a supersaturated solution and can only be lost to the atmosphere by coming out of solution in the form of bubbles. In the event of an unusually high alkalinity or reagent system electronic failure, an operator can manually open the CO_2 control valve and overtreat the effluent. The excess gas will then slowly bubble out downstream and neutralize the effluent to an 8.3 pH equilibrium.

The sidestream pH control system includes two water pumps (20 and 40 hp) which are automatically operated on the basis of required CO_2 flow. The smaller pump is sized to handle the average carbon dioxide flow requirement of 0 to 7 lb/min., while the larger pump is sized to handle peak flows of up to 35 lb/min. The controls monitor the CO_2 valve position and turn on the pumps as required for a particular effluent flow.

4.7.3 Process Costs

Table 4.7.3 summarizes the process costs for the construction and installation of a continuous, automatic, compressed carbon dioxide neutralization system. Equipment sizing and operating costs were based on the

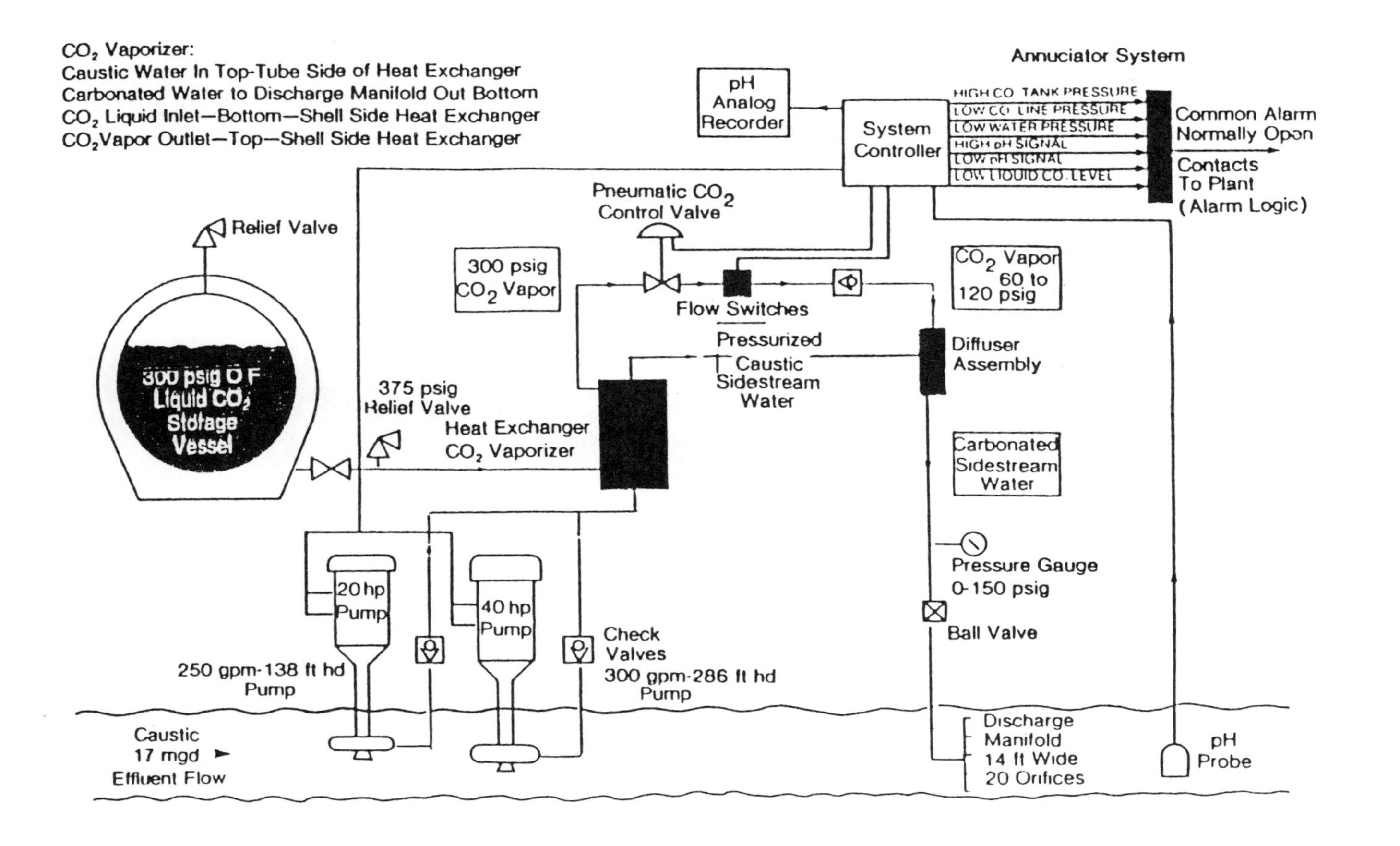

Figure 4.7.3. Compressed carbon dioxide treatment system.

Source: Reference 4.

TABLE 4.7.3. COMPRESSED CARBON DIOXIDE TREATMENT COSTS

	NaOH waste Flow rate (gph)[a]			$Ca(OH)^2$ waste Flow rate (gph)[b]		
	12,500	15,000	20,000	12,500	15,000	20,000
Capital investment ($)						
Liquid CO_2 storage tank	32,728	43,125	48,650	32,728	43,125	48,650
pH control system	6,800	6,800	6,800	6,800	6,800	6,800
Pump system	1,976	2,233	2,576	1,976	2,233	2,576
Heat exchanger	5,000	6,000	8,500	5,000	6,000	8,500
Sparger system	574	656	820	574	656	820
Flocculation/clarification	--	--	--	100,000	112,000	120,000
Filter press	--	--	--	600,000	630,000	700,000
Total capital cost	47,078	58,814	67,346	747,078	800,814	887,346
Annualized capital	8,333	10,410	11,920	132,233	141,744	157,060
Operating cost ($)[c]						
Taxes and investment (7%)	583	729	834	9,256	9,922	10,994
Maintenance & overhead (5%)	417	521	596	6,612	7,087	7,853
Labor & overhead ($20/hr)	48,000	48,000	48,000	48,000	48,000	48,000
Sludge disposal ($200/ton)	--	--	--	416,667	500,000	666,667
Reagent cost ($100/ton)	55,140	66,165	88,220	55,140	66,165	88,220
Total cost/year	112,473	125,825	149,570	667,908	772,918	978,794
Cost/1,000 gallon	4	3.5	3	22	21	20

[a]0.1 N NaOH wastestream.

[b]1.7 percent $Ca(OH)_2$ wastestream.

[c]Does not include utilities.

Source: Cardox quote September 1986, Reference 9, 10 using September 1986 CPI index cost data.

neutralization of a 0.1 N (4 g/L) sodium hydroxide wastestream and a 1.7 percent hydrated lime wastestream. The flow rates were assumed to be 12,500, 15,000, and 20,000 gal/hr, operating 2,400 hrs/yr. Capital costs include: a refrigerated, liquid carbon dioxide storage tank, a pH control system, sidestream pump, liquid carbon dioxide vaporizer, diffuser assembly, and sludge separation and handling equipment. Operational expenses were assumed to include: taxes and insurance, maintenance and overhead, labor and overhead, sludge disposal, and reagent costs.

The liquid CO_2 storage tank is sized to hold a 3-week supply under normal operating conditions. The pH control system includes a 30-day recorder, pH probe, alarm system, control valves, and switches. The quoted price for this particular pH control system was $6,800,[8] although costs for comparable equipment may be as high as $12,000.[5] The sidestream pump is assumed to remove 5 percent of the total effluent flow at a pressure of 150 psi. The pumps are API-610, cast steel casing, horizontal, in-line centrifugal pumps with in-line vertical motors. Pump costs are based on updated pump capacity curve data contained in the literature.[9] The heat exchanger/vaporizer system costs are based on those supplied for fixed-tube-sheet heat exchangers with 3/4 in. O.D. x 1 in. square pitch with carbon-steel shells operating at 150 psi.[9] The diffuser system cost is represented by vendor-supplied manifolds, each of which can displace 600 ft^3/hr of carbon dioxide gas at $82/sparger.[5]

Flocculation/clarification unit costs are based on information contained in Figure 4.1.6, with multiple units provided when maximum capacities are exceeded. Sludge generation for the sodium hydroxide system is assumed to be negligible based on a sodium bicarbonate formation of 8.4 g/L and a minimum solubility of 69 g/L.[11] Sludge generation for the hydrated lime system is assumed to be 1.35 lbs of calcium carbonate precipitate per pound of lime neutralized, based on a maximum solubility of 0.018 g/L.[11] Clarifier underflow is based on a sludge containing 2 percent solids which is subsequently dewatered to 60 percent solids in a recessed plate filter press. Filter press capital investment costs are based on Figure 4.7.4 which includes: the filter press (maximum capacity 224 ft^3), feed pumps (including one standby), a sludge conditioning and mixing tank, an acid wash system, and housing.[10] Housing costs are for a two-story, concrete block

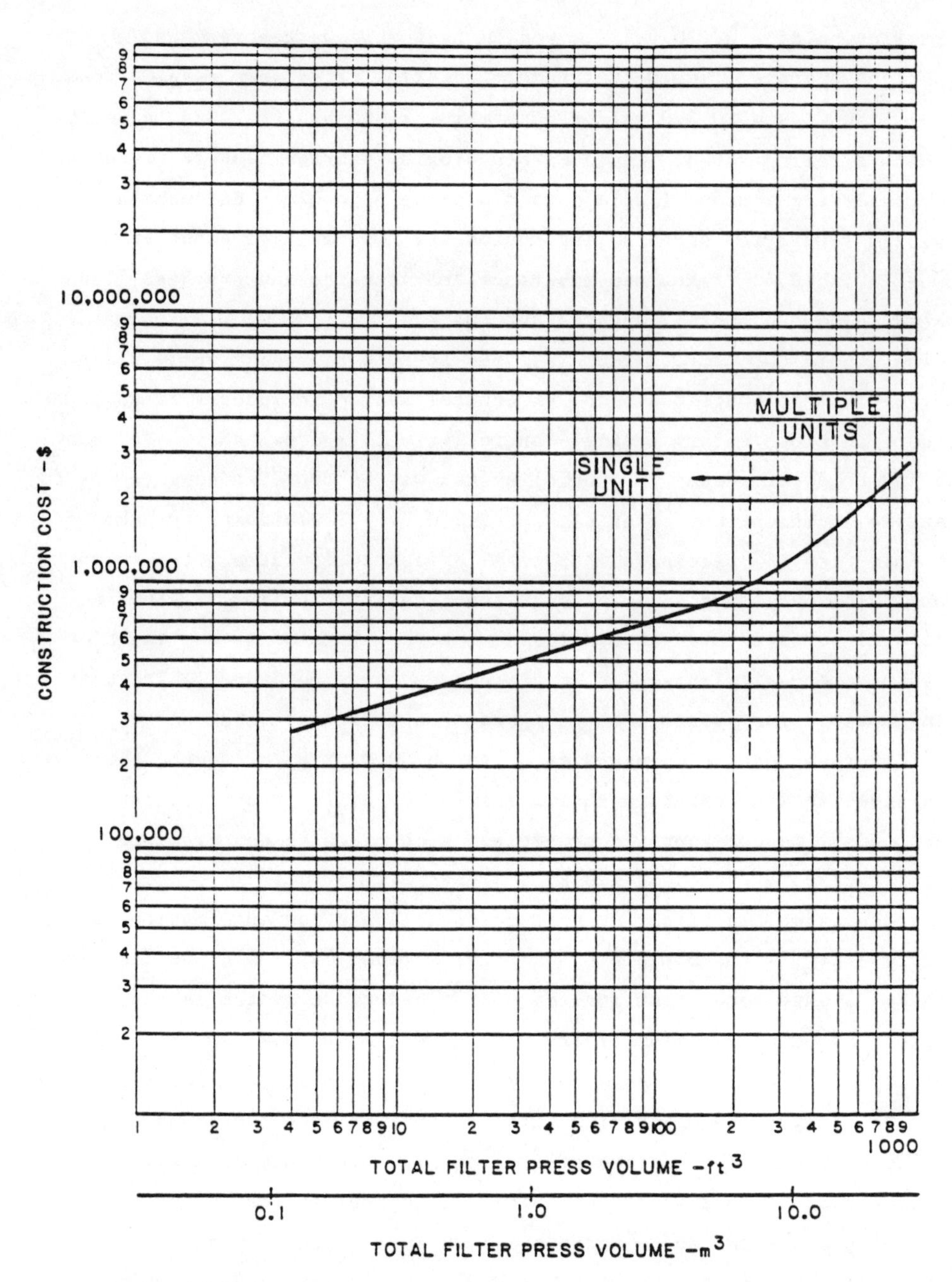

Figure 4.7.4. Construction cost for recessed plate filter press.

Source: Reference 10.

building with the filter press located on the upper floor, and discharging through a floor opening to a disposal truck. Also included in the Figure 4.7.4 cost estimates are a lime storage bin and feeders and ferric chloride solution storage and feeders. However, these costs have been deducted from the final cost analysis since they are not applicable to the example waste streams. Cycle times were estimated to be 2.25 hours with a 20-minute turnaround time between cycles.[10]

Operating and maintenance costs are based precepts presented previously, except that labor has been increased to 8 hrs/day at a rate of $20/hr, and reagent costs are $100/ton based on reagent usage greater than 200 tons/yr.[8] Reagent consumption is assumed to be 0.25 percent above stoichiometric requirements due to atmospheric losses.

Table 4.7.3 demonstrates that, in high volume applications, reagent costs can constitute up to 50 percent of total costs in non-sludge generating systems. In non-carbon dioxide producing regions such as the east coast, the cost of compressed carbon dioxide can effectively double. Sulfuric acid has a 30 percent minimum price advantage over carbon dioxide on a neutralization equivalent basis. Combined with its proven performance and widespread availability, sulfuric acid is typically the reagent of choice in the majority of applications. However, when an available source of carbon dioxide is nearby, space limitations are critical, and rapid reaction rates are necessary (e.g., in-line neutralization), liquid carbon dioxide may be the reagent of choice.

4.7.4 Process Status

Liquid carbon dioxide treatment of alkaline wastestreams is a promising but not widely applied technology. Improved facilities for the transportation, storage, and handling of liquid carbon dioxide have contributed to the recent emergence of this technology as a viable treatment in the last 5 years.[3] While application of this process is limited to less than 300 facilities nationwide,[5] increasing concern over the hazardous aspects of mineral acid treatment (burn dangers, acid mists, etc.) may increase utilization of this technology in high volume applications. A summary of the advantages and disadvantages of liquid carbon dioxide treatment is provided in Table 4.7.4.

TABLE 4.7.4. ADVANTAGES AND DISADVANTAGES OF CARBON DIOXIDE NEUTRALIZATION

Advantages:

- Capital requirements are minimal.
- Wastewater cannot be over-acidified
- No complex piping required.
- 1 to 1-1/2 minute retention time allows for in-line neutralization of alkaline wastes which are fairly uniform in nature.
- No danger of acid burns as with sulfuric or hydrochloric

Disadvantages:

- Low dissolved oxygen content (4.5 percent) at point of injection.
- High reagent cost limits applications to facilities with minimum flow rates of 100,000 gpd or minimum reagent usage of 200 tons/year.
- May be an asphyxiant at high concentration.
- Care must be exercised when contacting liquid CO_2 with wastewater in the vaporizer to prevent freezing.

Source: References 2, 3, 4, 5, and 8.

REFERENCES

1. Curtis, H.A., and R.L. Copson. Treating Alkaline Factory Waste Liquors Such as Kier Liquor from Treating Cotton with Caustic Soda. U.S. Patent No. 1,802,806. April 28, 1931.

2. Camp, Dresser, and McKee. Technical Assessment of Treatment Alternatives for Wastes Containing Corrosives. Contract No. 68-01-6403. September 1984.

3. Kirk-Othmer Encyclopedia of Chemical Technology. Vol. 4, 3rd Edition. pp. 725-741. John Wiley & Sons, New York, NY. 1981.

4. Griffith, M.J. et al. Carbon Dioxide Neutralization of an Alkaline Effluent Industrial Waste. March 1980.

5. Ponzevik, D. Liquid Air Products. Telephone conversation with Stephen Palmer, GCA/Technology Division, Inc. September 6, 1986.

6. Beach, C.J., and M.G. Beach. Treatment of Alkaline Dye Waste with Flue Gas. Proceeding - Fifth SMIWC. 1956.

7. Murdock, H.R. Stream Pollution Alleviated-Processing Sulfur Dye Wastes. Industrial and Engineering Chemistry. 43:77A (1951).

8. Berbick, D., Cardox Corporation. Telephone conversation with Stephen Palmer, GCA Technology Division, Inc. September 25, 1986.

9. Peters, M.S. et al. Plant Design and Economics for Chemical Engineers. 3rd. Edition. McGraw-Hill Book Company, New York, NY. 1980.

10. U.S. EPA. Design Manual: Dewatering Municipal Wastewater Sludges. EPA-625/1-82-014. October 1982.

11. CRC Handbook of Chemistry and Physics. 58th Edition. CRC Press, West Palm Beach, FL. 1978.

5. Recovery/Reuse Technologies

5.0 INTRODUCTION

The processes discussed in this section are used to recover or reuse corrosive wastes. Information is presented on process description, performance, costs, and current status for the following technologies:

- Evaporation/Distillation
- Crystallization
- Ion Exchange
- Electrodialysis
- Reverse Osmosis
- Donnan Dialysis & Coupled Transport
- Solvent Extraction, and
- Thermal Decomposition.

Crystallization and evaporation/distillation involve the use of temperature changes to effect a separation of contaminants and recovery of corrosive solutions. Ion exchange methods are based on the use of an anionic or cationic selective resin to remove ionic contaminants (i.e., metal ions) from corrosive wastes. Electrodialysis, reverse osmosis, Donnan dialysis, and coupled transport processes involve the use of a membrane to separate contaminants from corrosive solutions. Solvent extraction uses the differential distribution of constituents between the aqueous phase waste and an organic phase solvent to separate constituents from a mixed solution of metal salts and acid wastes. Thermal decomposition involves decomposing metal salts (present in spent acid wastes) in a roaster and collecting vaporized acid in a condensor.

In addition, a brief discussion of the role of waste exchanges in reuse of corrosives is presented at the end of this section. Waste exchangese involve the transfer of unwanted corrosive waste material generated by one company to another company capable of using it.

5.1 EVAPORATION AND DISTILLATION

5.1.1 Process Description

Evaporation is a concentration process used in the metal finishing and electroplating industries to recover plating solutions, chromic acid, nitric acid/hydrofluoric acid pickling liquors, and metal cyanides from spent baths and rinsewaters.[1,2,3] Distillation techniques are commonly used in conjunction with evaporation to recover water vapors by condensing them and returning them to the rinse tank, and to recover acids for return to acid baths.

Evaporation can be performed using atmospheric or vacuum techniques. Atmospheric evaporation occurs by boiling the liquid at atmospheric pressure (14.7 psi). The evaporation temperature can be lowered by spraying the liquid on a heated surface, and blowing air over this surface.[1,4] Thus, atmospheric evaporation occurs by humidification of the air stream.[1,4] With vacuum evaporation, the system pressure is lowered which causes the liquid to boil at a lower temperature. The vapor is subsequently condensed. A vacuum pump is used to maintain the vacuum condition during this process. The basic types of evaporation systems include the following:

1. Spray Evaporator;
2. Rising (or Climbing) Film Evaporator;
3. Submerged Tube Evaporator; and
4. Atmospheric Exhaust Evaporator.

Process flow diagrams for these systems are presented in Figure 5.1.1. The spray and rising film evaporators involve covering the heating surface with a thin film of waste liquid, whereas the submerged tube and the atmospheric exhaust evaporation systems transfer heat to a reservoir (tank) of waste liquid.[3] The distillate is returned to the rinse tanks for each of these systems, with the exception of the atmospheric exhaust evaporator which

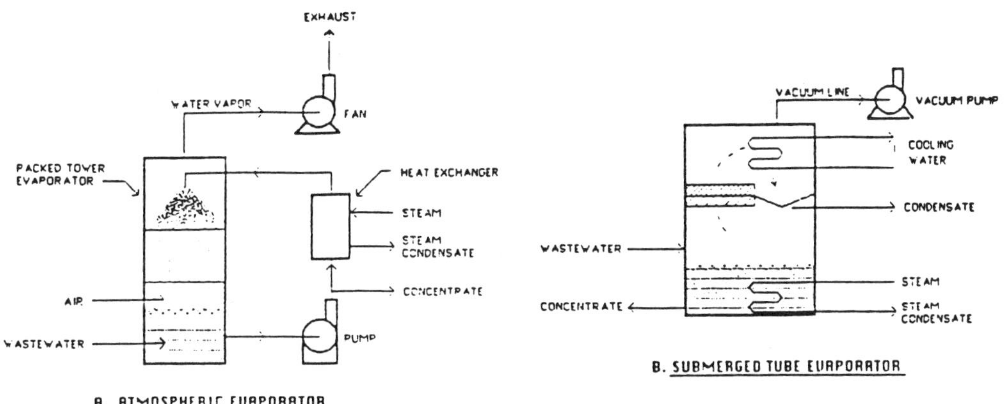

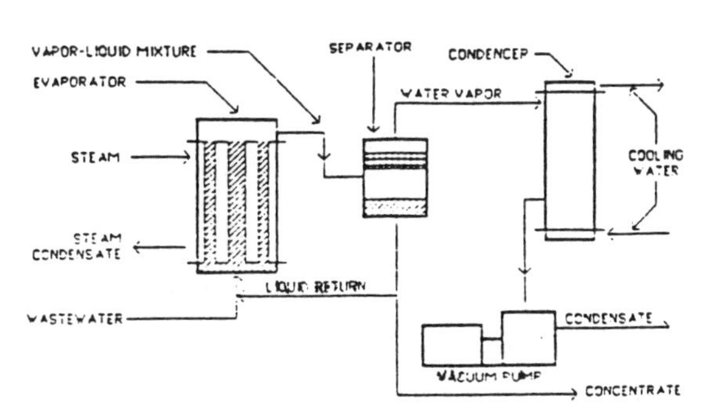

Figure 5.1.1. Process flow diagrams for commonly used evaporation systems.

Source: References 1 and 4.

vents the distillate to the air.[3] The surface film evaporators (nos. 1 and 2 above) are more efficient due to their higher heat transfer coefficients.[3] Therefore, smaller corrosive-resistant surface areas are required for surface film evaporators, which in turn reduces capital costs.[3]

Heat energy requirements for the evaporation system will depend upon the method used to supply heat. Heat is generally supplied in the form of steam. According to the basic laws of thermodynamics, it takes 970 BTU to evaporate 1 lb of water at atmospheric pressure. However, the amount of energy required to produce sufficient steam can be reduced if heat is reused at successively lower temperatures by multi-effect evaporation.[5] For example, a double-effect evaporator uses half the heat of a single-effect evaporator. However, capital costs will increase with increasing number of effects. The use of low-grade waste heat achieves the optimum amount of energy savings.[5] If sufficient waste heat is not available, a vapor compression (VC) evaporator can be employed.[5]

The VC technique involves the use of a mechanical compressor to heat the plating solution and to increase the pressure and temperature of the separated water vapor.[3] VC units operated under atmospheric conditions are only applicable to alkaline wastes due to the corrosive carryover of acid fumes under these conditions.[3] However, VC systems operated under vacuum conditions have lower temperature requirements, and can also make wider use of waste heat.[5] Thus, the vapor compression (VC) system reduces the operating costs of the evaporation process by lowering energy consumption.[3]

Although, distillation can be performed using either atmospheric or vacuum techniques, the high temperatures associated with atmospheric techniques can cause degradation in plating chemicals.[3] Also, carryover of corrosive fumes can occur with atmospheric techniques.[3] Vacuum evaporation systems do not require cooling towers or external steam sources, and therefore can be more cost-effective.[3] Additionally, vacuum systems allow the use of lower distillation temperatures and therefore have the advantages of greatly reduced corrosive action and lower-cost construction materials.[2,3]

Evaporation/distillation techniques are typically used in the recovery of acids and bases from spent rinsewaters. The spent solution is heated to evaporate water from the solution, thereby increasing the concentration of solute in the remaining solution. During distillation, the water vapor resulting from the evaporation process is condensed. The distillate is returned to the rinse tanks and the concentrate is used in the fresh bath make-up.[3]

Currently, the only application of vacuum evaporation-distillation techniques for the recovery of corrosive wastes directly from the spent bath is in the recovery of spent nitric acid/hydrofluoric acid pickling liquor. The process can be operated using either a one-stage evaporation system or a two-stage system. The one-stage system is suited for treating undiluted wastes, and the two-stage system is applicable to diluted waste acids obtained during pickling.[6]

A two-stage system is diagrammed in Figure 5.1.2. With the two-stage system, the spent pickle liquor is initially vacuum distilled to reduce the total liquid volume as much as possible before appreciable loss of acid in the distillate occurs.[2] Following this step, sulfuric acid is added to the residual liquid causing precipitation of metal sulfates.[2] Hydrofluoric and nitric acids have a higher vapor pressure than sulfuric acid and will therefore boil off with the remaining water.[8] The hydrofluoric and nitric acids are later condensed and recycled to the pickling tank (Stephenson et al., 1984). The sulfuric acid/metal sulfate slurry is passed through a solid-liquid separator, which generates a liquid consisting of 50 percent sulfuric acid and a metal sulfate sludge.[2] The 50 percent sulfuric acid solution can be reused in the second stage of the evaporation process. The metal sulfate sludge must either be disposed or subjected to further treatment.

A one-stage system is diagrammed in Figure 5.1.3. With this process, the waste liquid flows continuously into the evaporation system without concentration and evaporates under vacuum.[6] Concentrated sulfuric acid is fed with the waste acid into the evaporator. Upon boiling, more sulfuric acid is added to maintain the liquid level in the evaporator. Nitric and

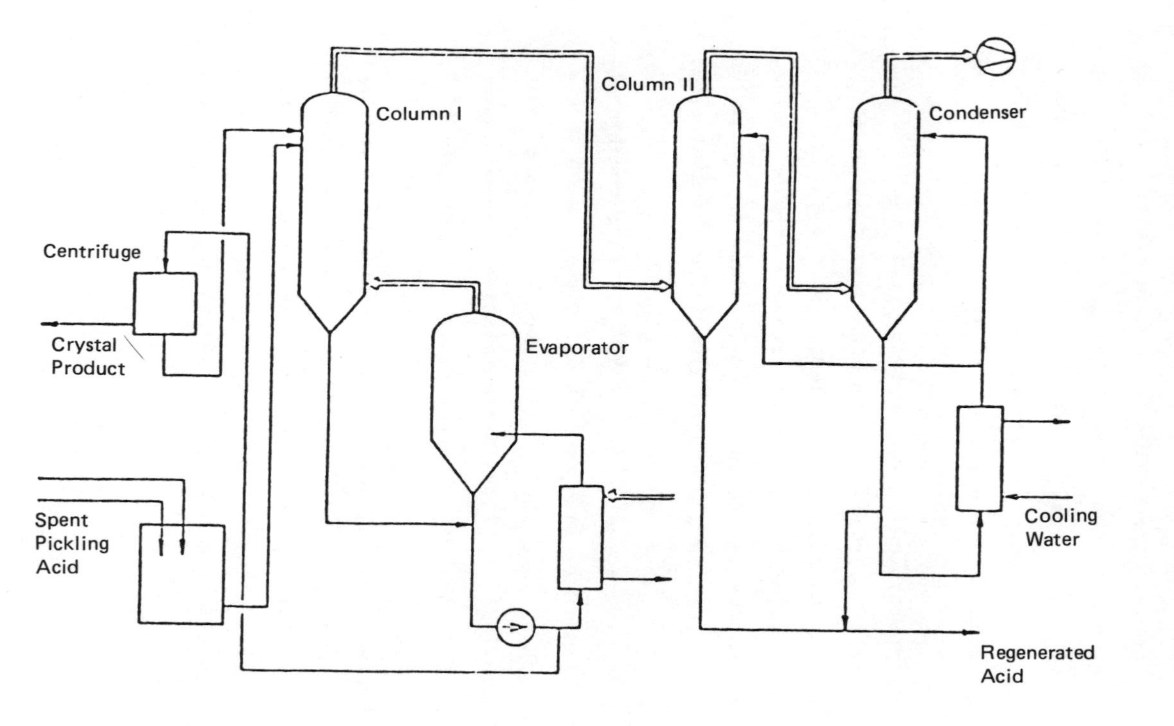

Figure 5.1.2. Process flow diagram for a two-stage vacuum evaporation/distillation system to recover spent HNO_3/HF pickling liquors.

Source: Reference 7.

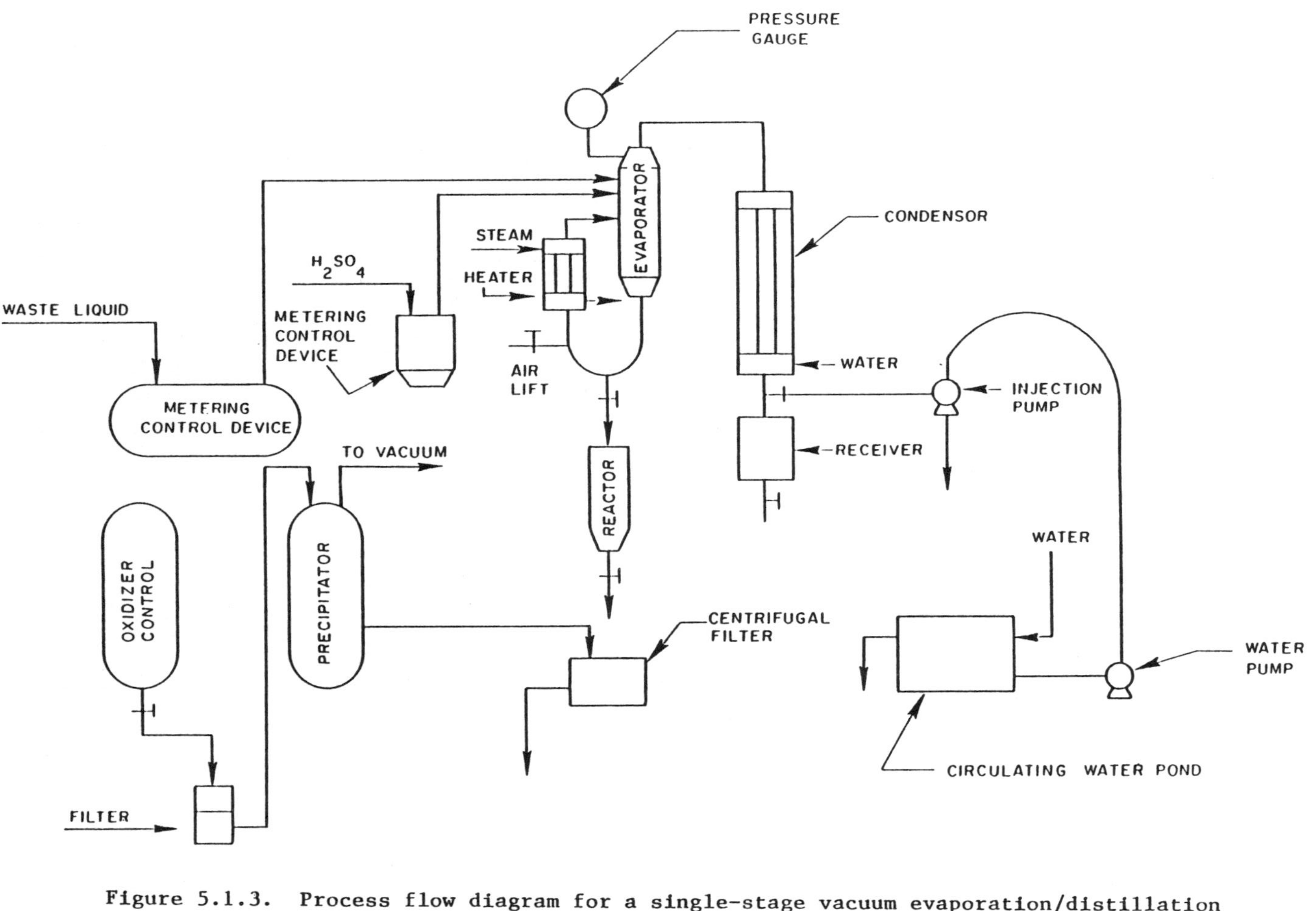

Figure 5.1.3. Process flow diagram for a single-stage vacuum evaporation/distillation system to recover spent HNO_3/HF pickling liquors.

Source: Reference No. 6.

hydrofluoric acid evaporate continuously and condense in the condenser. The distilled liquid consists of the regenerated acid. Metal salts in the waste liquid are continuously converted into metal sulfates which remain in the evaporator. As with the single-stage operation, the metal sulfates must either be disposed or treated.

Operating Parameters--

Important parameters which affect the operation of an evaporation/distillation system include: temperature, pressure, acid concentrations, and construction materials.

For direct treatment of spent acid, the acid concentration is a critical control factor in the proportions of constituents distilled. For a two-stage system, it is important to drive off most of the water while retaining the acid in the residual liquid. As shown in Figure 5.1.4, when the mixed-acid is being distilled, the acid concentration in the distillate remains relatively low until approximately 50 percent of the original acid volume is distilled.[8] As distillation continues beyond this point, the concentrations of HF and HNO3 in the distillate increase sharply.

Sulfuric acid is added to the system to transform the nitrates and fluorates to sulfates. The amount of sulfuric acid added to the system will affect the acid recovery ratio. If the concentration of circulating sulfuric acid is too low, the acid recovery ratio will decrease.[6] Also, if the sulfuric acid concentration is too high, iron salts will precipitate too soon, which will lower the acid recovery ratio. The concentration of sulfuric acid should be maintained at 12.5 N for proper control.[6]

The pressure of the system also effects distillate losses. As can be seen from Table 5.1.1, the distillate losses decrease with decreasing pressures. The permeability and compression strength of the graphite heater determine the pressure of the heating steam. It usually does not exceed 2 kg/sq cm, and is generally controlled at 1 kg/sq cm.[6]

Lower operating temperatures are desirable in order to minimize corrosion of evaporation/distillation equipment. Operating temperatures should not exceed 65°C.[6] By lowering the pressure the boiling point

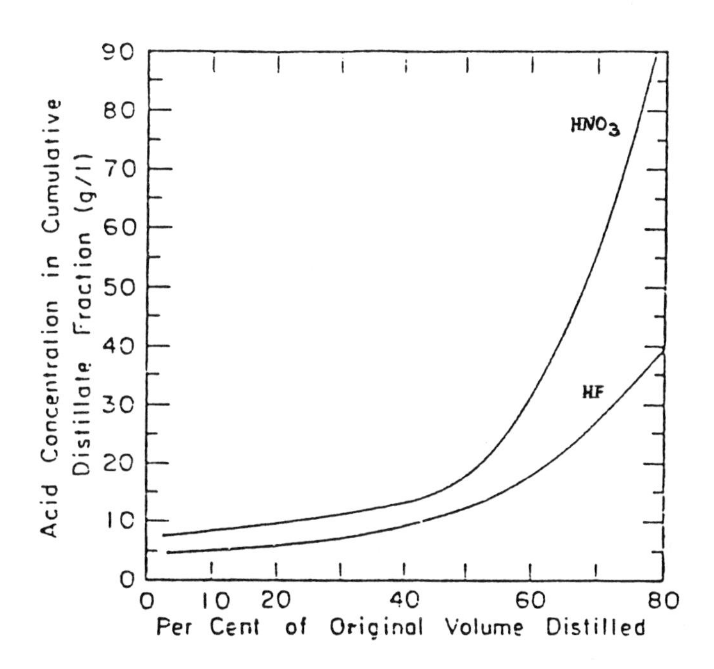

Figure 5.1.4. Nitric and hydrofluoric acid concentrations in cumulative distillate fraction versus original volume distilled.

Source: Reference 8.

TABLE 5.1.1 EFFECTS OF PRESSURE ON DISTILLATE LOSSES.

Pressure (psia)	Final Temp. (°F)	Percent-Original Volume Distilled	Distillate Losses (%)	
			Nitrate	Fluoride
14.7	224	51.2	4.0	6.4
14.7	221	50.0	3.5	4.6
1.8	150	49.0	1.8	1.3

Source: Reference 8.

of a liquid increases. Thus, the use of a vacuum allows lower operating temperatures. The vacuum pressure typically ranges from 660 to 680 mm Hg.[6]

Materials used in the evaporation/distillation system need to be able to withstand the corrosive properties of the waste. Commonly used materials include titanium, tantalum, borosilicate glass, fiberglass-reinforced plastic, and polyvinyl chloride.[4] The materials used will depend upon the specific application. Metallic materials will be dissolved by the waste acid solution. Ceramics will be degraded by the hydrofluoric acid. Certain plastics will withstand the corrosive properties of the waste if the temperature is not too high. Fiber-reinforced PVC plastics are typically used in the construction of the evaporator and peripheral piping (Delu, et al., 1980). However, plastics are not good heat conductors. Therefore the use of an impermeable graphite heat exchanger is recommended due to its good heat conductivity, impermeability, strong oxidation resistance, and non-fouling property.[6] It has been successfully used in several plants since 1974.[6]

Additional factors to consider when designing an evaporation system for the recovery of corrosives from rinsewaters include: rinse ratios, mixing techniques, rinse flow rates, and rinse concentrations. Each of these parameters must be optimized in order to design an efficient, properly-sized evaporation system.

The rinse ratio quantifies the amount of rinse water available to a particular load.[9] For example, if the dragout amounted to a workload of 2 gal/hr, and the flow rate of the rinse water was 100 gal/hr, the rinse ratio would be 50 (100/2). With a lower rinse ratio, a smaller evaporator is required, which in turn lowers the costs.[9] However, the rinse ratio must be high enough to assure product quality.

Effective mixing in the rinse water tank will aid in maintaining uniformly good quality when fresh water additions are made. Some methods of achieving proper mixing include: maintaining a high water flow rate in the rinse tank, air agitation using a low-pressure blower, mechanical mixing using electric or air driven stirrers, and the use of countercurrent rinsing (using two or more rinse tanks in series).

If effective rinsing techniques are used, the evaporator can generally be sized to approximately 15 times the dragout rate from the plating tank.[9] The evaporator size is equivalent to the water flow rate in the rinse ratio formula. In order to determine if a rinse ratio is appropriate for a particular application, the following formula can be used:

$$C_i = \frac{C_o}{R_i}$$

where Ci=concentration in rinse tank, Co=concentration of the plating tank, and Ri=rinse ratio raised to a power equal to the number of rinse tanks.[9] The total quantity of concentrate that can be recovered can be calculated using the following formula:

$$\% - \text{Capture} = (1 - 1/R_i) \times 100$$

These two formulas can be used to optimize an evaporative recovery system for various rinse ratios, rinse tank numbers, and plating bath concentrations.

It is also important to consider the energy efficiency of a system when designing an evaporation system. Energy requirements contribute the most to operating costs for these systems. Efficiencies are typically measured using the coefficient of performance (COP). The COP is equal to the energy output divided by the energy input. Higher heat transfer efficiencies will contribute to a higher thermal efficiencies. In addition, effective reuse of waste heat will lower the overall energy requirements.

Pre-Treatment--

A filtration system may be required to remove suspended particulates present in the spent solution to avoid clogging of the distillation system. Also, if any impurities are present in solution or suspension in the spent solution, pretreatment prior to evaporation may be necessary because these impurities will be concentrated in the evaporator.[5] Chemical treatment or filtration techniques may be employed to remove these impurities. For continuous operation, it is recommended that deionized water be used in makeup additions in order to avoid buildup of impurities.[5]

Post-Treatment--

The evaporation/distillation process generates a metal sulfate sludge. The sludge can either be treated and disposed, or subjected to further treatment via roasting techniques to recover the metal oxides (see section 5.8).

Also certain impurities may be recovered along with the concentrate. A filtration system may be required prior to returning the concentrate to the bath. A Freon-cooled crystallization/filtration system is commonly required to remove carbonates from the concentrated solution.

5.1.2 Process Performance

The performance of an evaporation/distillation system is typically evaluated on the basis of the percent removal of contaminants, the percent product concentration achieved, the product quality, the amount of waste generated, the energy requirements, and the economics of the process.

Evaporation/distillation systems can be used effectively to recover acidic waste streams. A system using a vapor compression evaporator operated under vacuum conditions was tested at the Naval Facilities Engineering Command (NFEC) Charleston, South Carolina to recover chromic acid from a hard chrome plating line rinse tank.[5] Testing was performed over a 9-month period with more extensive data collected between March 23 and April 23, 1984.[5] Typical operating parameters and results during this period are summarized in Table 5.1.2. As shown this table, the unit was able to successfully recover chromic acid at relatively low operating temperatures (i.e., lower energy requirements). An economic evalution determined that the system was only cost-effective for processes generating more than 100 gal/hr dragout.[5] A smaller capacity system was then developed which could operate economically at boiling temperatures in the range of 100° F.[5]

TABLE 5.1.2. SUMMARY OF OPERATING PARAMETERS AND RESULTS DURING TESTING OF HIGH VACUUM VAPOR COMPRESSION EVAPORATION SYSTEM AT THE CHARLESTON NAVY YARD.

Parameter	Result
Compressor Efficiency	
Coefficient of Performance (COP)	10.3
Adiabatic Efficiency	25 %
Capacity	25 gph @ 700 rpm speed 40 gph @ 1170 rpm speed
Total Chrome Recovered (70 gals x 54,900 mg/l)/7484	32.1 lb (513.5 oz.)
Dragout Rate (32.1 lb/320 hrs/month) x (1 gal/2 lb Cr+6)	0.05 gal/hr
Rinse Ratio (Ratio of plating bath concentration to final rinse concentration using 3 countercurrent rinse tanks)	20,000
Rinse Flow Rate	27 gph per 1 gph dragout
Evaporator Capacity (Required Rinse Rate) 27 gph x 0.05 gph = 1.35	1.35 gph
Recovered Process Water	
Quantity	8.75 gph (33,600 gpy)
Conductance	10 mmho
Operating Temperatures	95 - 122°F
Electrical Requirements	9 kw

Source: Reference 5.

The performance of the smaller capacity system was tested at the Superior Plating Division of Florida Plating, Inc. in St. Petersburg, Florida for recovery of a cadmium cyanide plating solution. The process line consisted of a manual hoist and barrel (10 x 18 in.) with two plating tanks and three countercurrent rinse tanks. The average dragout was 0.44 gal/hr.

The evaporation system employed is diagrammed in Figure 5.1.5. The system consists of single-effect, high-vacuum, climbing-film evaporator, a 20-ton Freon heat-pump, and a Freon crystallizer. Operating and design parameters are summarized in Table 5.1.3. The evaporator uses bayonet augmented tube (BAT) heat exchangers to recover heat from the heat pump, which allows the evaporation temperature to be maintained below 110°F so that cyanide breakdown is minimized.[10] The Freon crystallizer also employs BAT heat exchangers to extract heat from the cadmium cyanide concentrate before returning it to the plating bath in order to raise the temperature of the incoming rinsewater from the cadmium cyanide operation.[10] The purpose of the crystallizer is to remove carbonates (impurities that interfere with product quality) from the bath by chilling the concentrated cadmium cyanide complex to 28°F, which precipitates sodium carbonate crystals that can be removed by settling.[10] The distilled rinsewater is returned to the last countercurrent rinse tank).

Superior Plating Division is satisfied with the performance of the system for their application. Detailed monitoring of the performance of the system was conducted for an 11-day period in April 1985.[10,11,12,13,14] It was found that during 90 hours of operation, 40 gal of cadmium cyanide complex solution was recovered. The average concentrations in the recovered solution were 2.14 oz/gal cadmium and 15.3 oz/gal sodium cyanide.[10] The system is currently operated during the day shift. Approximately 1 to 2 hours per day are required for start-up, data collection and recording, and shut-down.[10] An economic evaluation of the system is presented in Table 5.1.4.

Performance data on the use of evaporation/distillation system to recover corrosives directly from the spent solutions (as opposed to recovery from rinsewaters) is limited to research conducted on the recovery of nitric and hydrofluoric acids from spent pickling liquors. The performance of this

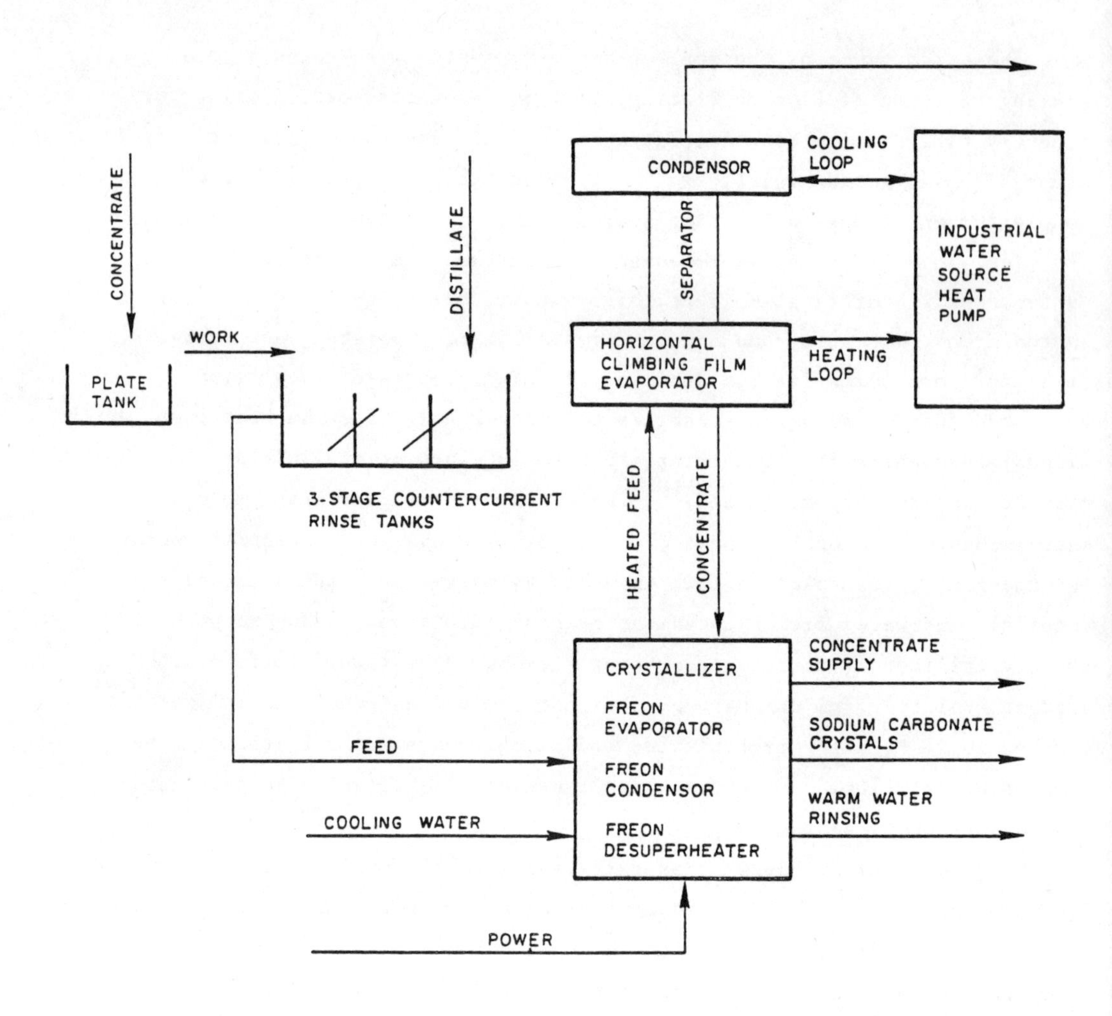

Figure 5.1.5. Flow diagram of Evaporative Recovery System installed at Superior Plating, Inc.
Source: Reference 5.

TABLE 5.1.3. SUMMARY OF OPERATING PARAMETERS AND RESULTS USING EVAPORATION TO RECOVER A CADMIUM CYANIDE PLATING BATH AT SUPERIOR PLATING, INC.

Parameter	Result
Heat Pump Capacity	300,000 BTU/hr
Heat Pump Exit Temperature	125°F
Evaporator Capacity	200 - 250 Btu per lb water distilled
Evaporation Temperature	110°F
Chiller Exit Temperature	95°F
Freon Condensor Exit Temperature	140°F
Coefficient of Performance (COP)	4.35
Recovered Cadmium Conc.	2.14 oz./gal.
Recovered Sodium Cyanide Conc.	15.3 oz./gal.

Source: Reference 10.

TABLE 5.1.4. ECONOMIC EVALUATION OF VACUUM COMPRESSOR EVAPORATOR UNIT EMPLOYED AT SUPERIOR PLATING, INC.

Item	Cost
Capital Equipment Cost	$ 24,000/yr
Operating Costs	$ 2,000/yr
Cost Savings	$ 11,560/yr
Net Savings (Cost Savings - Operating Costs)	$ 9,560/yr
Payback Period (Capital Costs/Net Savings)	2.5 yrs

Source: Reference 10.

process is typically evaluated on the basis of the percent of nitric and hydrofluoric acid recovered, the percent of metal contaminants removed, and acid additions required in the makeup solution. Pilot-scale evaporation/distillation systems for the recovery of nitric/hydrofluoric acid pickling liquors have not been installed in the United States.

A single-stage system was employed at a steel plant in China.[6] Monitoring of this pilot-scale system was conducted over a 3-year period as part of an acid recovery research project. Typical results are shown in Table 5.1.5.

A two-stage system developed by Rosenlew has been installed at the NYBY Steel Works in Sweden.[7] Typical operating parameters and results are summarized in Table 5.1.6. In all tests, the evaporation of hydrofluoric acid went to near completion.[7] However, the percent recovery of nitric acid was dependent upon the sulfuric acid addition.[7]

Commercial-scale evaporation/distillation systems for recovery of corrosives directly from the spent bath have not been tested in the United States. Thermal decomposition (see Section 5.8) appears to be a more cost-effective process for this application.

5.1.3 Process Costs

Capital costs for an evaporative recovery system will vary with the waste type, waste quantity, process flow rates, type of heat exchanger employed, and system size. Table 5.1.7 presents costs for various system sizes.

Operating costs generally include 1-2 hours labor for system maintenance and operation (labor requirements will be reduced if the system is operated continuously), electrical and fuel energy requirements for heat supply, taxes and insurance, and depreciation costs. Approximately 10 lbs of low pressure steam (15 psig) is required for every gallon of liquid evaporated.[9]

Evaporation/distillation processes require large amounts of heat energy, which can make the process quite costly. However, efficient use of energy systems can lower these costs significantly. Waste heat from other industrial processes (diesel generators, incinerators, boilers, and furnaces) within the plant can be recovered for use in the evaporation/distillation system. The use of multi-effect evaporators and vapor compression systems can also improve thermal efficiencies.

TABLE 5.1.5. SUMMARY OF RESULTS OF PILOT-SCALE UNIT INSTALLED AT A CHINESE STEEL PLANT.

Item	Volume (liters)	Concentration (g/l)					
		H^+	F^-	NO_3^-	Fe^{+2}	Ni^{+2}	Cr^{+3}
Spent Acid	1140	2.58	2.67	1.70	21.50	3.58	4.27
Residual Liquid	350	13.47	0.88	0.23	80.05	12.30	13.7
Regenerated Acid	858		3.23	2.12			
Sulfuric Acid	150	32.4					
Recovery Ratio (%)		92.9	93.9				

Source: Reference 6.

TABLE 5.1.6. TYPICAL OPERATING PARAMETERS DURING TESTING OF THE PILOT-SCALE EVAPORATION/DISTILLATION SYSTEM AT THE NYBY STEEL WORKS IN SWEDEN.

Parameter	Result
Design Capacity	1.5 cu. meters/hr
Equipment Size	9m x 9m x 15m
Vacuum Evaporation Temp.	80°C
Nitric Acid Concentration in Recovery Evaporator	10 to 20 wt-%
Hydrofluoric Acid Conc. in Recovery Evaporator	2 to 8 wt-%
Sulfuric Acid Conc.	60 wt-%

Source: Reference 7.

TABLE 5.1.7 TYPICAL CAPITAL EQUIPMENT COSTS FOR VARIOUS EVAPORATION SYSTEM CAPACITIES

Evaporator Capacity (gph)	Capital Costs ($)
20	25,000
40	33,800
55	39,199
120	44,129
300	115,000

Source: References 15 and 16 (August 1986).

Cost savings will be realized in reduced neutralization costs, reduced sludge disposal costs, and reduced purchase requirements for fresh bath makeup solutions.

5.1.4 Process Status

Evaporation/distillation is one of the oldest recovery techniques, and is widely used in industry.[4] Over 600 units are currently in operation in the United States.[9,15] They are most commonly used in metal finishing and electroplating industries to recover plating solutions, chromic acid and other concentrated acids, and metal cyanides. In addition, water recovered from the evaporation process is of high purity and can be reused in process waters. The percentage of these units used in various plating applications is presented in Table 5.1.8. These systems are most effective in recovering acids, bases, and metals from rinsewaters. Systems can be designed cost-effectively with capacities ranging from 20 gph to 300 gph.[9] These systems are cost-competitive with conventional neutralization and disposal technologies. Greater cost savings are realized with larger operations.

The use of evaporation/distillation systems to recover concentrated streams directly from the spent solution is limited. Pilot-scale evaporation/distillation systems for recovery of nitric/hydrofluoric acid pickling liquors have been tested at facilities in Europe. However, cost-effective systems for direct recovery of spent solutions via evaporation/distillation have not been developed at the commercial-scale for application in the United States. Other technologies, such as thermal decomposition (see Section 5.8) appear to be more cost-effective for this purpose.

In summary, evaporation/distillation systems are effective in recovering corrosives from rinsewaters. Cost-effective systems are commercially available for a wide range of spent rinsewater constituents and process sizes. Application to direct treatment of spent corrosive solutions is limited by costs.

TABLE 5.1.8. PERCENTAGE BREAKDOWN BY PLATING TYPE OF EVAPORATION UNITS CURRENTLY IN OPERATION.

Plating Chemical	Percent of Units
Chrome	50
Chrome Etch	10
Nickel	20
Cyanide	10
Other	10

Source: Reference 9.

REFERENCES

1. Cushnie, G. C. Centec Corporation, Reston, Virginia. Navy Electroplating Pollution Control Technology Assessment Manual. Final report prepared for the Naval Civil Engineering Laboratory, Port Hueneme, California. NCEL-CR-84.019. February 1984.

2. Stephenson, J.B., J.C. Hogan, R.S. Kaplan. Recycling and Metal Recovery Technology for Stainless Steel Pickling Liquors. Environmental Progress, (3)1: 50-53. February 1984.

3. Chacey, K., L. Mellichamp, and W. Williamson. Chrome Electroplating Waste BAT. Pollution Engineering. April 1983.

4. Camp, Dresser, and McKee, Inc. Technical Assessment of Treatment Alternatives for Wastes Containing Corrosives. Prepared for the U.S. EPA Office of Solid Waste under EPA Contract No. 68-01-6403 (Work Assignment No. 39). September 1984.

5. Williamson, R. C., and W. R. Williamson. Licon, Inc., Pensacola, Florida. Energy Effective Systems for Closed Loop Treatment of Electroplating Waste Water. Final Report prepared for the U.S. Department of Energy under DOE Contract No. DE-AC07-79CS40290. March 17, 1986.

6. Delu, H., L. Xiuchung, and W. Chingwen. The Regeneration of Nitric and Hydrofluoric Acids From Waste Pickling Liquid. In: Symposium on Iron and Steel Pollution Abatement Technology for 1980 held in Philadelphia, Pennsylvania. November 18-20, 1980.

7. Solderman, J. New Method for Recovery of Spent Pickling Acids. In: Third International Congress on Industrial Wastewaters and Wastes, Stockholm, Sweden. February 6-8, 1980.

8. Dasher, J., and D. Goldstein. Recovery of Nitric and Hydrofluoric Acids. Metal Finishing, 61(5): 60-63. May 1963.

9. Constantine, D. Corning Process Systems, Corning, New York. Technical Data Sheet No. RT-1: Rinse Theory. May 12, 1980.

10. Williamson, R. C., and S. Natof. Evaporative Recovery for Cadmium Cyanide Plating. Plating and Surface Finishing. November 1985.

11. Brandvwine, P. Mixed Rinses Treated by Evaporation. Product Finishing. August 1984.

12. Industrial Finishing Staff. Recovering Brass Cyanide Plating Solution. Industrial Finishing. June 1978.

13. Industrial Finishing Staff. Evaporator/Chiller System Recovers Brass Plating Solution. Industrial Finishing. January 1980.

14. Rose, B. A. Design for Recovery. Industrial Finishing. May 1979.

15. Licon, Inc. Pensacola, Florida. Product Literature. Received August 1986.

16. Constantine, D. Corning Process Systems. Letter to J. Spielman, GCA Technology Division, Inc. August 6, 1986.

5.2 CRYSTALLIZATION

5.2.1 Process Description

Crystallization is a recovery technique whereby metal contaminants in a spent corrosive solution are crystallized and removed by settling or centrifugation. Crystallization techniques for the recovery of corrosive wastes are most applicable to spent acid pickling liquors, and spent caustic soda aluminum etching solutions.

Crystallization techniques are similar differing only in the methods used to crystallize the metal salts and to separate the crystals from the recovered solution. Crystallization can be induced by cooling or evaporating water from the pickling solution, or a combination of the two.[1] The process may be operated in either a batch or a continuous mode.[1,2]

Cooling crystallization techniques are used for the recovery of sulfuric acid pickling liquors by removing iron contaminants. Typically, the process involves crystallizing the iron salts in a cooling chamber, removal of the crystals in a settling/drainage chamber, and addition of fresh sulfuric acid to restore the pickling solution to its original strength.[1,3] Thus, the free sulfuric acid remaining in the spent pickling solution is able to be reused.

A batch-mode process using cooling to induce crystallization is diagrammed in Figure 5.2.1. Spent pickling solution is pumped to the crystallizer. Recovered sulfuric acid from a previous batch operation replaces the spent solution. Fresh sulfuric acid is added to the crystallizer to increase the yield of ferrous sulfate crystals. Cooling water flows through teflon-cooling coils around the crystallization chamber and is cycled through a refrigeration unit. The temperature of the spent liquor is slowly reduced (over a 6 to 10 hour period) to a temperature of 35 to 50°F, causing ferrous sulfate heptahydrate crystals to form.[1,2]

The resulting crystal/acid slurry is then transferred to a drainage chamber which retains the ferrous sulfate crystals, but allows the acid solution to pass through to an acid recovery tank[1]. The recovered

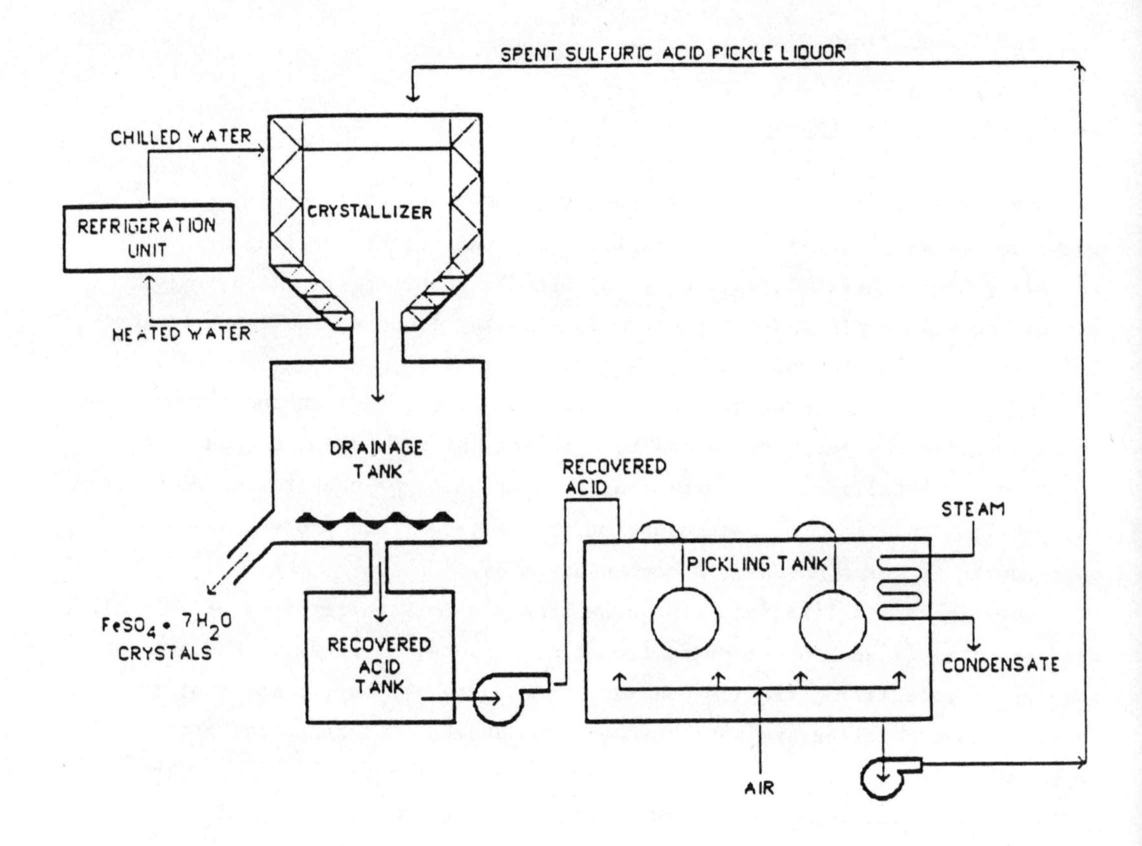

Figure 5.2.1. Flow diagram of crystallization system for recovery of sulfuric acid pickling liquor.

Source: Reference 1.

acid is preheated with steam and then returned to the pickling liquor tank during the next batch cycle. The ferrous sulfate crystals are washed with water to remove any remaining free acid, air-dryed, and then removed for disposal (or marketed, if possible).

Greater contaminant removal can be achieved using a combination of evaporation and cooling techniques in a two-stage system.[4,5] A flow diagram of a two-stage process to recover nitric hydrofluoric acid pickling liquors is presented in Figure 5.2.2.

During the first stage, approximately half of the waste pickle liquor is evaporated in order to concentrate the dissolved metals (supersaturated solution). The vapor resulting from the evaporation process is condensed, and the acid condensate (the initial portion of regenerated acid) is directed to a storage tank.[4]

During the second stage of the process, the remaining solution is sent to a crystallizing chamber. Cooling is used to induce crystallization of metal fluoride crystals. The crystals are then separated from the solution in a drainage chamber. The filtered concentrate, which contains 90 percent of the nitric acid and the free hydrofluoric acid, is directed to the recovered acid storage tank.[5] The metal fluoride crystals can be disposed, or thermal decomposition techniques (discussed in Section 5.8) can be used to regenerate hydrofluoric acid from the crystals and to form reusable metal oxides. The recovered acid in the storage tanks is then metered to the pickling bath in appropriate proportions.

A crystallization process has also been developed to recover sodium hydroxide from spent aluminum etch solutions. Aluminum contaminants need to be removed in order to recover the sodium hydroxide (caustic soda). Figure 5.2.3 presents a flow diagram for a continuous crystallization process for recovery of caustic soda. In this process, a vacuum chamber is used to induce crystallization of aluminum hydroxide.[1] The pressure in the chamber is reduced, which causes cooling of the spent etchant and some vaporization of the water. The recovered acid, which has a high caustic concentration and a low aluminum concentration, is returned to the etching tank.[1] A centrifuge is used to separate and wash the crystals formed in the crystallization chamber.[1,3,6] The filtrate from the centrifuge can also be returned to the etchant tank. The dewatered crystals are of commercial-grade quality and

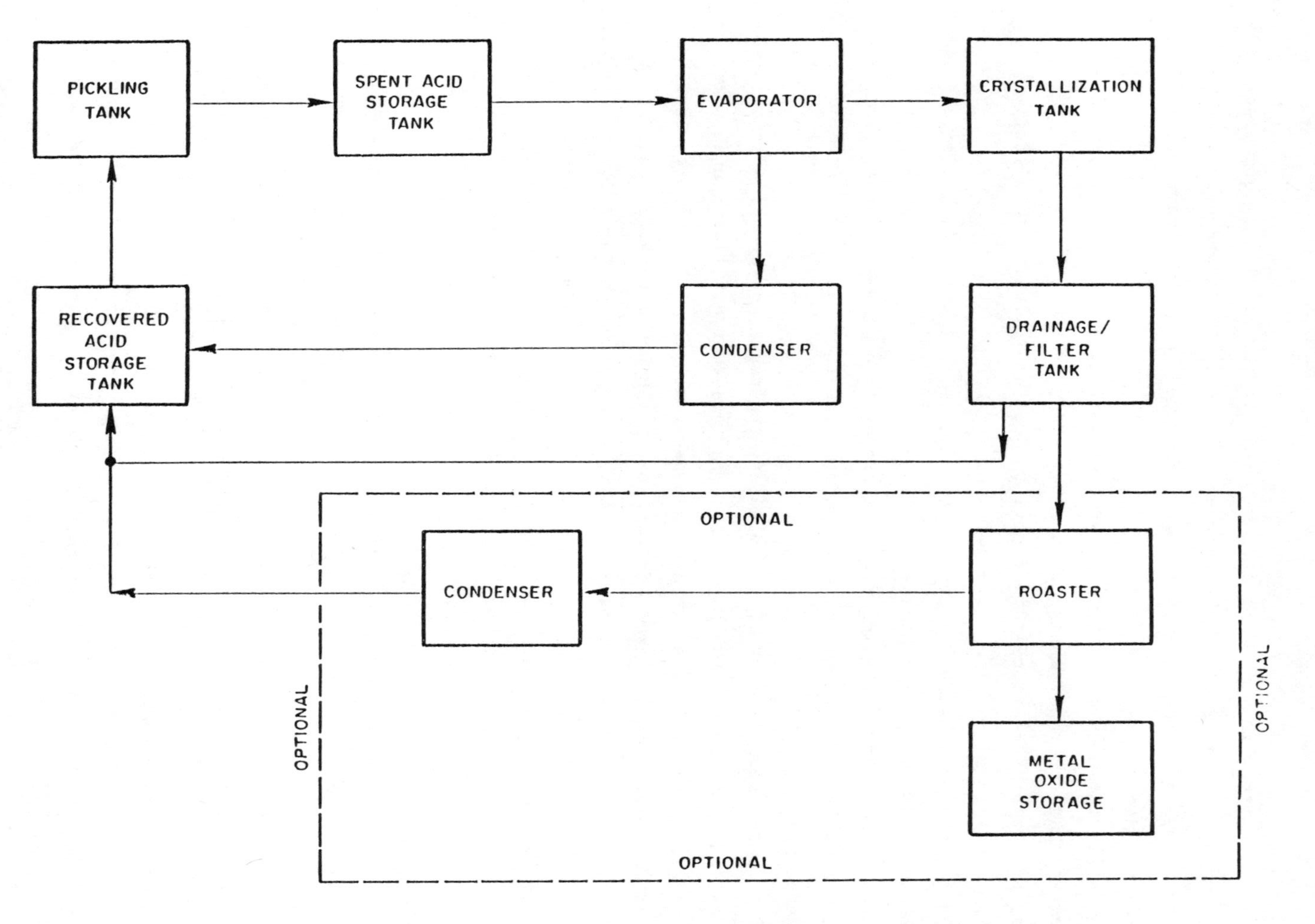

Figure 5.2.2. Flow diagram of two-stage recovery system for nitric-hydrofluoric acid pickling liquor.

Source: Reference 4.

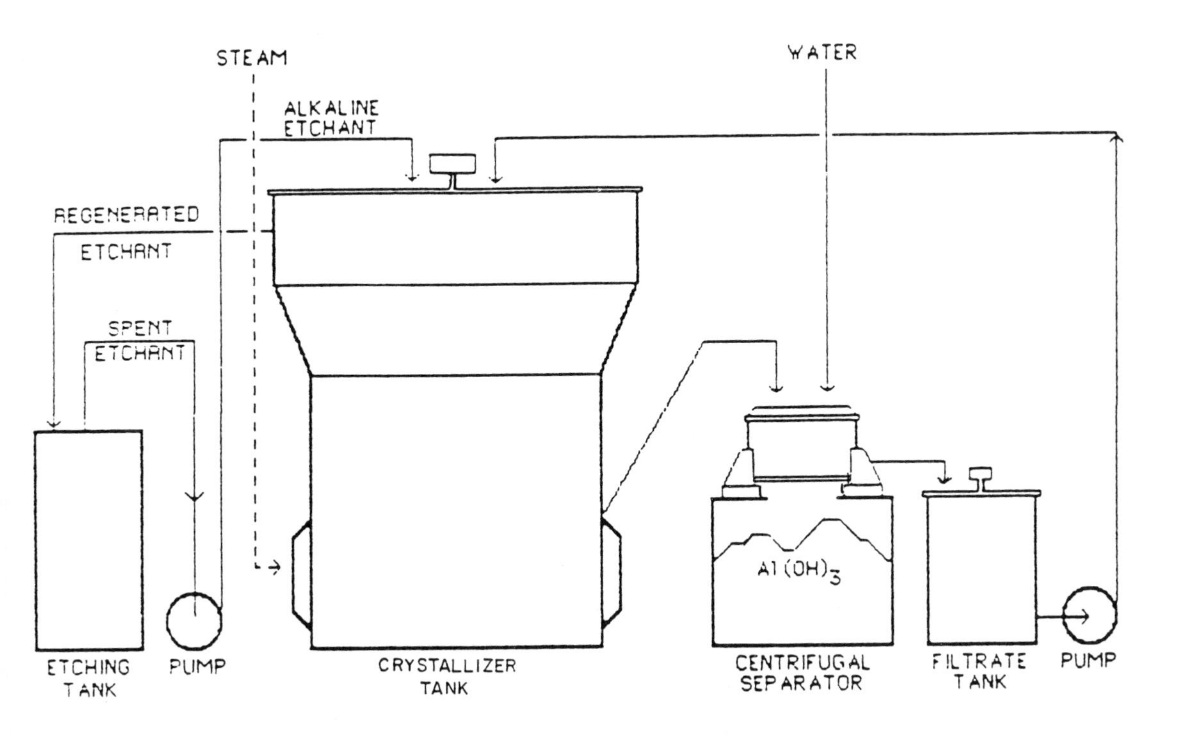

Figure 5.2.3. Flow diagram of crystallization system for the recovery of caustic soda aluminum etching solution.

Source: Reference 1.

can be traded or sold depending on the available market. Additions of fresh sodium hydroxide are only required to replace dragout losses.

Operating Parameters--

Crystallization techniques for sulfuric acid recovery are based on the solubility of ferrous sulfate decreasing with decreasing temperatures.[3] Figure 5.2.4 shows that the solubility of ferrous sulfate also decreases with increasing acid concentrations.[3] Therefore, more efficient operation of a crystallization process is achieved when pickling lines have a high acid concentration and a relatively low temperature.

Sulfuric acid recovery systems using crystallization techniques typically cool the pickle liquor to approximately 40°F under controlled conditions to form the ferrous sulfate heptahydrate crystal.[3,8] At this low temperature, most of the ferrous sulfate will be in the heptahydrate form rather than the monohydrate form, which allows for easier removal. This crystallization system works most efficiently in treating solutions with high iron concentrations and high free acid concentrations.[3] High acid concentrations minimize the solubility of ferrous sulfate in the pickling solution, which produces a lower iron content in the recovered acid. Also, the high iron concentration allows less operating time for the acid recovery system and/or the use of a smaller recovery system.[3]

Typical operating parameters for a two-stage evaporation-crystallization system to recover nitric-hydrofluoric acid pickling liquors are presented in Table 5.2.1. The metal salt crystals are formed during the cooling stage.[4] The metal removal efficiencies achieved with this process are directly dependent on the availability of fluoride ions in an amount sufficient to combine with the metal ions.[9] Therefore, instead of adding the makeup hydrofluoric acid at the end of the process (to achieve the proper pickling concentrations), it is desirable to add the makeup hydrofluoric acid following the evaporation stage and prior to the crystallization stage.[9] However, nitric acid should not be added at this stage because a portion of the acid will be lost with the crystals.[9] The effect of nitric and hydrofluoric acid concentrations on the solubility of iron fluoride crystals over various temperature ranges is shown in Figure 5.2.5. Also, as illustrated

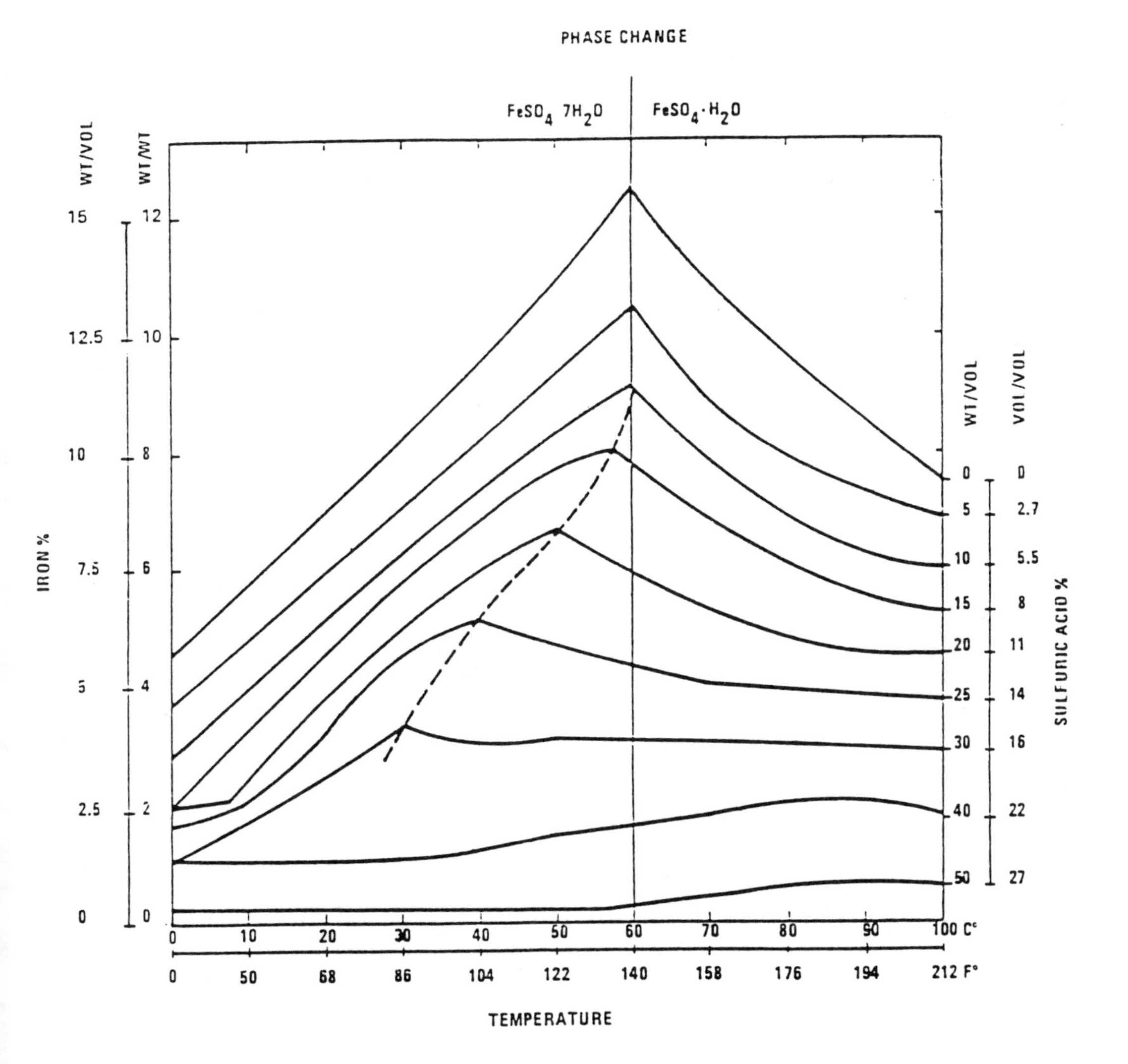

Figure 5.2.4. Solubility of ferrous sulfate in various sulfuric acid concentrations.

Source: References 3 and 7.

TABLE 5.2.1. EVAPORATION-CRYSTALLIZATION SYSTEM FOR RECOVERY OF NITRIC-HYDROFLUORIC ACID

Parameter	Result
Operating temperatures	
Spent pickle liquor	40 to 60°C
Evaporator	105°C
Vaporization	107.5°C
First crystallizer	70 to 78°C
Second crystallizer	40 to 60°C
Residence times	
Evaporator	15 minutes
First crystallizer	4 to 12 hrs
Second crystallizer	4 to 12 hrs

Source: References 9 and 10.

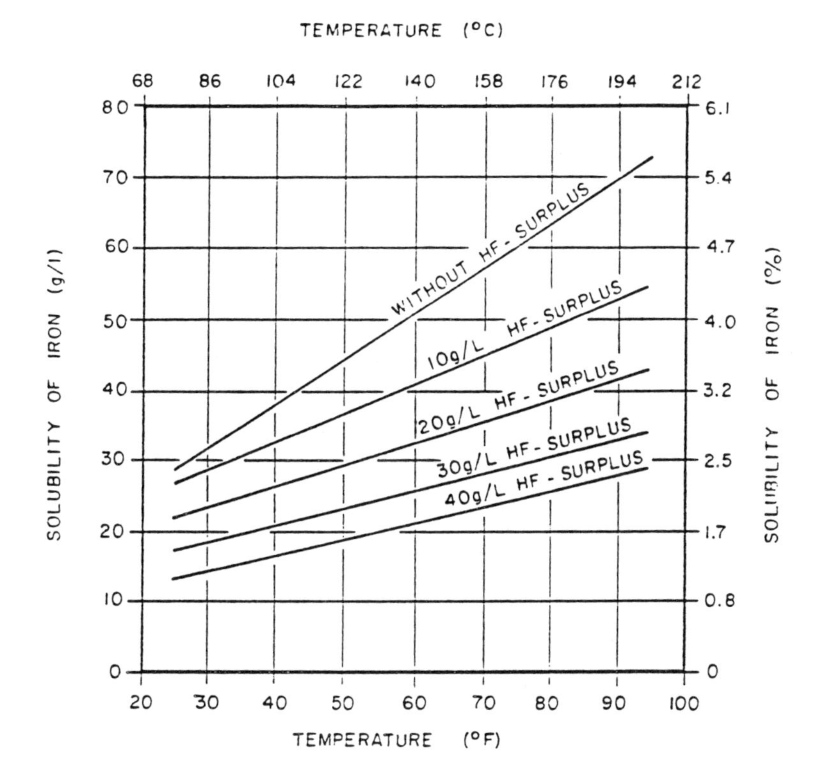

Figure 5.2.5 Solubility of iron in mixed acid containing 150 g/L (12.5 percent) nitric acid and different amounts of free hydrofluoric acid at varying temperatures.
Source: Reference 4.

in Figure 5.2.6, the rate of crystallization of the iron fluoride crystals is dependent on the initial concentration of iron present in the solution, which is controlled by the evaporator.

Equipment used in each of these crystallization processes should be resistant to corrosion. Commonly used materials include Teflon, polyfluorohydrocarbons, sintered corundum, and structural graphite.[8,9]

Pretreatment Requirements--

For efficient operation of the crystallization recovery process, the amount of water added to the pickling or etching system should be minimized. Under optimum conditions, the water added to the system should not significantly exceed the water exiting the system through evaporation, free moisture, and water of hydration of the crystals.[11] Otherwise, the recovered solution becomes too dilute for pickling/etching purposes. Some methods which can be used to reduce water in the pickling/etching system include: counter-current rinsing operations, indirect steam heating of the pickling/etching tank, and air agitation of the bath to promote evaporation.[11]

Post-Treatment Requirements--

Upon completion of the acid recovery process, the purified acid or etchant is returned to the pickling bath. The removed metals are in the form of crystal salts, which will require further handling.

The iron removed in the single-stage cooling crystallization system for recovery of sulfuric acid pickling liquor is in the form ferrous sulfate heptahydrate crystals. These crystals can either be disposed, traded locally, or marketed. The value of ferrous sulfate crystals on the market varies widely. Crown Technology will purchase the iron sulfate crystals from their clients.[3,8] Crown dries the crystals, bags them, and then ships them for use in water and sewage treatment facilities (flocculation processes), fertilizer industry, and the animal feed industry.[8]

The metal fluoride crystals formed in the two-stage evaporation-crystallization system for recovery of nitric-hydrofluoric acid can be treated by thermal decomposition techniques (discussed in Section 5.8) to recover additional hydrofluoric acid. The thermal decomposition process generates a metal oxide product which can be reused in the steel process.

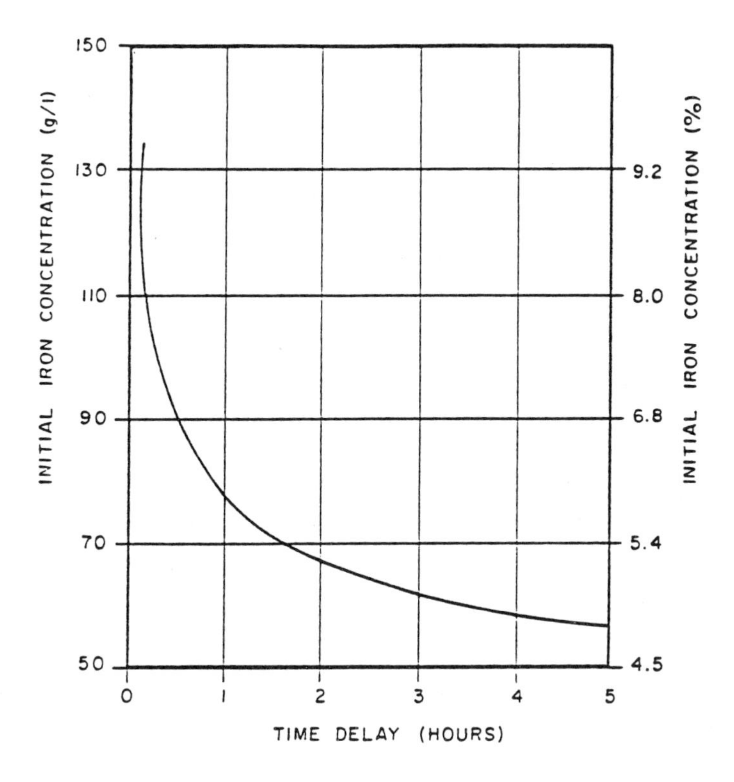

Figure 5.2.6. Delay of formation of visible crystals in oversaturated mixed acid versus initial concentration of iron (150 g/L HNO_3 and 20 g/L free hydrofluoric acid).

Source: Reference 4.

5.2.2 Process Performance

The performance of a crystallization system is typically evaluated on the basis of percent metal removal, percent acidity loss, product quality, processing time, and economics.

Case studies demonstrating the performance of single-stage cooling crystallization systems for the recovery of sulfuric acid are being prepared by Acid Recovery Systems, Inc. in Lenexa, Kansas, but they are currently not available.[2] Typical performance data for single-stage acid recovery systems are summarized in Table 5.2.2. Improvements are generally noted in product quality when a continuous recovery process is employed due to the consistency of the bath concentrations. Although the acid recovery process is typically operated in a batch mode, interruptions to the pickling line are minimal since bath dumping is no longer required.

Typical performance data for a two-stage evaporation-crystallization system for the recovery of nitric-hydrofluoric acid are presented in Table 5.2.3. Better performance is achieved with this system than the evaporation-distillation systems described in Section 5.1 for nitric-hydrofluoric acid recovery because of reduced sludge generation (i.e., sulfuric acid additions with subsequent neutralization treatment are not required).

Limited performance data is available for crystallization systems used to recover caustic soda. However, recoveries of up to 80% have been reported for this process.[1]

5.2.3 Costs

Sulfuric acid recovery systems are available with throughput rates ranging from 600 gpd to 30,000 gpd.[3,8] Capital costs will include crystallization equipment, two tanks (customer-supplied); one to hold the high iron content solution, and one to hold the low iron content solution. Additional requirements include: connections for hooking up the system to the product line, electrical requirements, and heating coils for the tanks.

TABLE 5.2.2. TYPICAL OPERATING PARAMETERS AND RESULTS FOR SULFURIC ACID RECOVERY SYSTEM USING CRYSTALLIZATION

Parameter	Result
Optimum iron content in the waste feed	10 to 14%
Iron removal efficiency	80 to 85%
Acidity losses in recovered acid	2 to 3%
Average cycle time	6 hrs

Source: Reference 2.

TABLE 5.2.3. TYPICAL PERFORMANCE OF A TWO-STAGE CRYSTALLIZATION SYSTEM FOR THE RECOVERY OF NITRIC-HYDROFLUORIC ACID

	Concentration, weight-percent (lbs/hr)					
Parameter	Fe	Cr	Ni	NO_3	F	Water
Feed to evaporator	3.4 (26.5)	1.1 (8.6)	1.6 (12.5)	12.0 (93.6)	6.0 (46.8)	75.9 (592)
Feed to crystallizer	6.5 (26.5)	2.1 (8.6)	3.1 (12.5)	22.1 (89.9)	10.1 (41.2)	56.1 (228.3)
Condensed vapor	-	-	-	1 (3.7)	1.5 (5.6)	97.5 (363.7)
Residue from crystallizer	25 (20.0)	4.6 (3.7)	0.8 (0.6)	6.0 (4.8)	30.8 (24.6)	32.9 (26.3)
Filtrate from crystallizer	2.0 (6.5)	1.5 (4.9)	3.5 (11.9)	26.0 (15.1)	5.1 (16.6)	61.7 (202.0)
Total concentration recovered	0.9 (6.5)	0.7 (4.9)	1.7 (11.9)	12.7 (88.8)	3.2 (22.2)	80.8 (565.7)
Total required additions	-	-	-	(43)	(32)	(261.1)

Source: Reference 9.

Table 5.2.4 presents cost estimates for three acid recovery systems with varying throughput rates. The economics of sulfuric acid recovery varies considerably with the costs of acids, the market for ferrous sulfate, and the costs for disposal.[1,6]

Two-stage systems are generally not cost-effective for recovering sulfuric acid pickling liquors due to the high capital equipment costs. However, two-stage systems can be cost-effective for recovering hydrofluoric-tric acid pickling liquors, because of the higher acid purchase costs for these acids.[12] An economic evaluation of the two-stage system for recovery of nitric-hydrofluoric pickling liquors is presented in Table 5.2.5. Costs are given for a system with a regeneration capacity of 2,000 L/hr (530 gal/hr), which was designed for a steel plant with a pickling capacity of 40 mL/hr (400 gal/hr).

5.2.4 Process Status

Crystallization is a demonstrated and commercially available technology for the recovery of acid pickling liquors and caustic etching solutions.[2,4,5,8,11] Due to the large capital investment costs, the process is more economically feasible for use in operations that generate larger quantities of spent solutions.

The use of crystallization techniques for the recovery of sulfuric acid pickling liquors and caustic aluminium etching solutions is limited by economics due to the small quantities of these solutions used by individual manufacturers, the costs for plan modifications, and the varying demand for the crystal product.[1,6] Despite these limitations, these processes are currently being used in the metal finishing industry.

Nitric-hydrofluoric pickling liquors are used in larger quantities by individual manufacturers in the steel industry than sulfuric acid pickling liquors. Therefore, crystallization techniques would have wider application for this waste type.

Due to economics, acid and alkaline wastes are normally treated by neutralization techniques and land disposal. With increasing restrictions on land disposal, recovery using crystallization techniques may become more economically viable.

TABLE 5.2.4. ECONOMIC EVALUATION OF ACID RECOVERY SYSTEM USING CRYSTALLIZATION TECHNIQUE

Item	Small unit ($)	Medium unit ($)	Large unit ($)
Flow rate (gal/day)	2,400	16,000	30,000
CAPITAL COSTS			
Equipment	175,000	460,000	850,000
Tank (2 tanks @ $1.25/gal)	5,000	40,000	75,000
Installation (10% of investment)	17,500	46,000	85,000
Total capital costs:	197,500	546,000	1,010,000
OPERATING COSTS			
Maintenance (6% of investment)	10,500	27,600	51,000
Taxes & insurance (0.5% of investment)	875	2,300	4,250
Utilities (@ $0.02/KW-h)	8,000	10,000	12,000
Depreciation (10% of investment)	17,500	46,000	85,000
Total operating costs:	36,875	85,900	152,250
COST SAVINGS			
Neutralization savings	22,653	139,400	261,375
Disposal savings	51,025	314,000	588,750
Process water savings	2,633	16,200	30,375
Acid makeup savings	16,250	100,000	187,500
Total cost savings:	92,560	569,600	1,068,000
NET SAVINGS: (Gross savings-Operating costs)	55,685	483,700	915,750
PAYBACK PERIOD (Capital costs/Net savings)	3.59 yrs (43 months)	1.16 yrs (14 months)	1.14 yrs (14 months)

Source: References 1, 2, 3 and 12 (July, 1986 cost data).

TABLE 5.2.5. ECONOMIC EVALUATION OF TWO-STAGE SYSTEM FOR RECOVERY OF NITRIC-HYDROFLUORIC ACID

Item	Cost ($)
Capital costs	10,000,000
Operating costs	2,064,000/year
Cost savings	5,222,000/year
Net savings (Gross savings-Operating costs)	3,158,000/year
Payback period (Capital costs/Net savings)	3 years

Source: References 4 and 12.

REFERENCES

1. Camp, Dresser, and McKee, Inc. Technical Assessment of Treatment Alternatives for Wastes Containing Corrosives. Prepared for the U.S. EPA Office of Solid Waste under EPA Contract No. 68-01-6403 (Work Assignment No. 39). September 1984.

2. Luhrs, R. Acid Recovery Systems, Inc., Lenexa, Kansas. Telephone conversation with L. Wilk, GCA Technology Division, Inc. Re: Sulfuric Acid Recovery System. September 4, 1986.

3. Crown Technology, Inc. Product Literature: Crown Acid Recovery Systems. Received July 1986.

4. Krepler, A. Total Regeneration of the Waste Pickle Liquor for Stainless Steel. Ruthner Industrieanlagen Aktiengesellschaft Technical Report No. 3, Vienna, Austria. 1980.

5. Smith, I., Cameron, G. M., and H. C. Peterson. Chemetics International Co., Toronto, Canada. Acid Recovery Cuts Waste Output. Chemical Engineering. February 3, 1986.

6. Versar, Inc. National Profiles Report for Recycling/A Preliminary Assessment. Draft Report prepared for the U.S. EPA Waste Treatment Branch under EPA Contract No. 68-01-7053, Work Assignment No. 17. July 8, 1985.

7. Peterson, J. C. Crown Technology, Inc. Closed Loop System for the Treatment of Waste Pickle Liquor. Prepared for the U.S. EPA-Industrial Environmental Research Laboratory, Research Triangle Park, North Carolina. EPA-600/2-77-127. July 1977.

8. Peterson, J. C. Crown Technology Inc., Indianapolis, Indiana. Telephone conversation with L. Wilk, GCA Technology Division, Inc. Re: Sulfuric Acid Recovery System. July 10, 1986.

9. Krepler, A. Apparatus for Recovering Nitric Acid and Hydrofluoric Acid from Solutions. U.S. Patent No. 4,252,602. Assigned to Ruthner Industrianlagen-Aktiengesellschaft, Vienna, Austria. February 24, 1981.

10. Krepler, A. Process for Regenerating a Nitric Acid-Hydrofluoric Acid Pickling Liquor. U.S. Patent No. 4,144,092. Assigned to Ruthner Industrieanlagen-Aktiengesellschaft, Vienna, Austria. March 13, 1979.

11. Micheletti, W. C., Nassos, P. A., and K. T. Sherrill. Radian Corporation. Spent Sulfuric Acid Pickle Liquor Recovery Alternatives and By-Product Uses. Draft Report prepared for U. S. EPA-IERL, Research Triangle Park, North Carolina. DCN-80-203-001-17-07. November 3, 1980.

12. Chemical Marketing Reporter. Spot Market Prices. Volume 230, No. 3, Pages 32-40. July 21, 1986.

5.3 ION EXCHANGE

5.3.1 Process Description

Ion exchange has been used to recover corrosive wastes from the metal finishing, electroplating, and fertilizer manufacturing industries by removing metal contaminants and recycling the treated solution.[1] Industrial process waters, plating baths, and acid pickling baths contain dissolved metal salts which dissociate to form metal ions. Ion exchange is a reversible process which involves an interchange of these ions between the solution and an essentially insoluble, solid resin in contact with the solution. Ions from the solution are exchanged for similarly charged ions attached to the solid resin.

The maximum quantity of exchanges per unit of resin is set by the number of mobile ion sites (exchangeable ions) attached to the resin. This is dependent on the type of resin, which can be composed of either naturally occurring materials (e.g., clays or zeolites) or synthetic organic polymers. Synthetic resins are more frequently employed because they can be designed for specific applications.

Typical equipment used in an ion exchange treatment system includes: a waste storage tank, prefilter system, cation and/or anion exchanger vessels, and caustic or acid regeneration equipment.[2] The ion exchange system may be operated in a batch or flow-through (column) mode; the latter mode is generally preferred due to greater exchange efficiencies.

With the batch mode of operation, the ion exchange resin and the waste solution are mixed in a batch tank. Upon completion of the exchange reaction (i.e., equilibrium is reached), the resin is separated from the treated solution by filtering or settling. The spent resin is then regenerated and reused. Unless the resin has a very high affinity for the contaminant ion, the batch mode of operation is chemically inefficient and thus has limited applications.

Flow-through operation involves the use of a bed or packed column of the exchange material (resin). These systems are typically operated in cycles consisting of the following four steps:

1. Service (Exhaustion) - Waste solution is passed through the ion exchange column or bed until the exchange sites are exhausted.

2. Backwash - The bed is washed (generally with water) in the reverse direction of the service cycle in order to expand and resettle the resin bed.

3. Regeneration - The exchanger is regenerated by passing a concentrated solution of the ion originally associated with it through the resin bed or column; usually a strong mineral acid or base.

4. Rinse - Excess regenerant is removed from the exchanger; usually by passing water through it.

A flow-through (column) system can be designed with cocurrent or countercurrrent flow of the waste and regenerant (steps 1 and 3 above). In cocurrent systems, the feed and the regenerant both pass through the resin in a downflow mode. Figure 5.3.1 illustrates the cocurrent flow process. Each ion exchange unit consists of a cylindrical vessel having distributors or collectors at the top and bottom. Resin is loaded into approximately half of the vessel to accommodate resin expansion during the backwash cycle. Cocurrent systems are most cost-effective for weak acid or base exchangers which do not require highly concentrated regenerant solutions. However, regeneration of strong exchangers (high exchange capacity) requires strong acid and base solutions which can be more costly.

Often it is too costly to fully regenerate a bed. In order to avoid carry over of contaminants into the next service run, two or more sets of fixed columns arranged in parallel series can be used. Also, to avoid excessive down-time during the regeneration cycle, dual sets of fixed columns are generally used. While one set of columns is being regenerated, the second set of columns will be switched on line. This technique allows continuous operation of the system. As illustrated in Figure 5.3.2, improved regenerant efficiency can also be accomplished by reusing the last portion of the regenerant solution that flows through the resin. For example, if 5 lb/cu ft (80 g/L) of regenerant were used for the system shown in the figure, the first 50 percent of spent regenerant would only contain 29 percent of the original acid concentration, whereas the remaining regenerant would contain 78 percent of the original acid.[4] If the last portion of the regenerant is

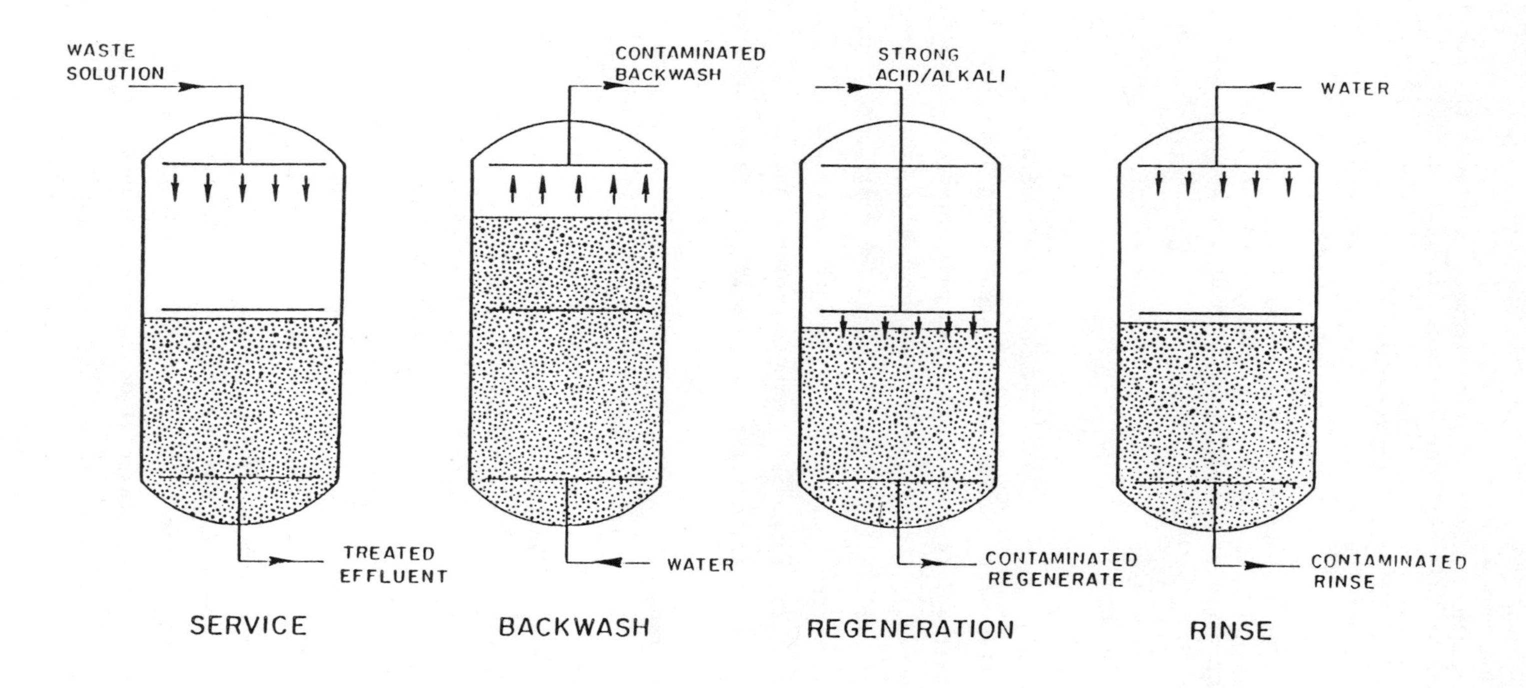

Figure 5.3.1. Cocurrent ion exchange cycle.

Source: Reference 3.

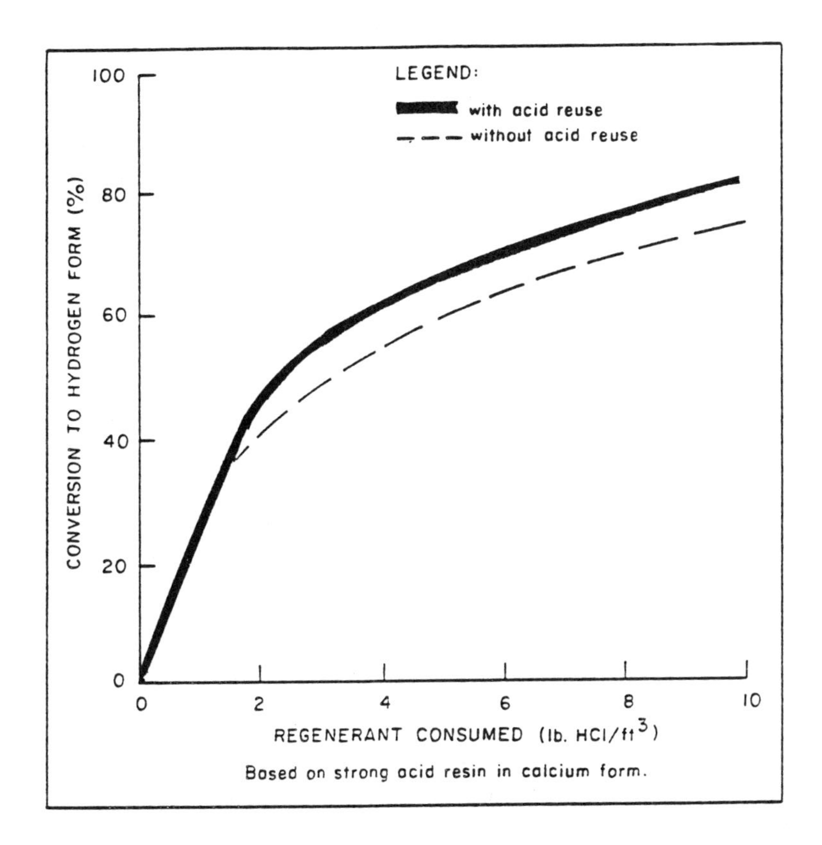

Figure 5.3.2. Effect of acid regeneration on chemical efficiency.

Source: Reference 4.

reused in the next cycle before the resin bed is contacted with fresh HCl, the exchange capacity would increase from 60 to 67 percent at equal chemical doses.[4]

Ion exchange systems generate a waste stream of spent regenerant, which will typically require neutralization and disposal. The use of countercurrent flow between feed and regenerant uses regenerant chemicals more efficiently than is possible with cocurrent systems. Counter current systems also achieve a higher product concentration than is possible with conventional cocurrent flow. A counter current system that is widely used for chemical recovery from plating rinses is the reverse or reciprocating flow ion exchanger (RIFE). An example of an RFIE system used for recovery of chromic acid from a dilute solution is shown in Figure 5.3.3.

Eco-Tech, Ltd. (in Pickering, Ontario) has developed an acid purification system that uses the RFIE technology to efficiently remove high concentrations of metal contaminants from acid baths.[5] Although the acid purification unit (APU) uses countercurrent flow techniques, the bed is not fixed. As shown in Figure 5.3.4, the acid purification process consists of an upstroke adsorption cycle and a downstroke regeneration cycle. During the upstroke cycle, spent acid is forced by air pressure through the resin bed. The acid is adsorbed and the de-acidified metallic salts are collected from the top of the resin bed. During the downstroke regeneration cycle, compressed air is used to force water into the top of the resin bed. The water passes countercurrently through the bed, displacing the adsorbed acid from the resin. The purified acid product is collected from the bottom of the resin bed and recycled into the process operation. The only waste product from the acid purification process is a metallic salt sludge.[1,6,7]

Typical components of the APU include the resin bed, metering tanks, an electrical control panel, plastic piping, and pneumatically operated plastic control valves. The size of the unit varies from 6 to 24 square feet with a height of less than 6 feet.[6]

Although the conventional cocurrent flow ion exchange systems can be used effectively to recover process rinsewaters, RFIE systems have wider application in the recovery of corrosive wastes.

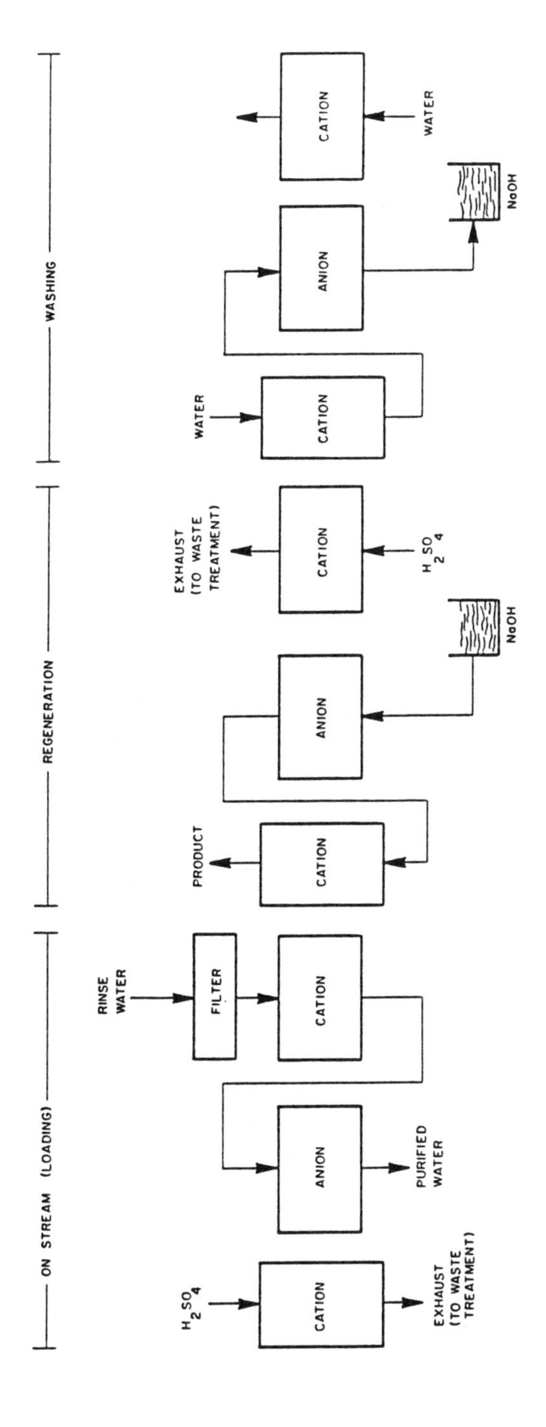

Figure 5.3.3. Schematic of a fixed bed reverse flow ion exchange (RFIE) system for the recovery of chromic acid from a dilute solution.

Source: Reference 4.

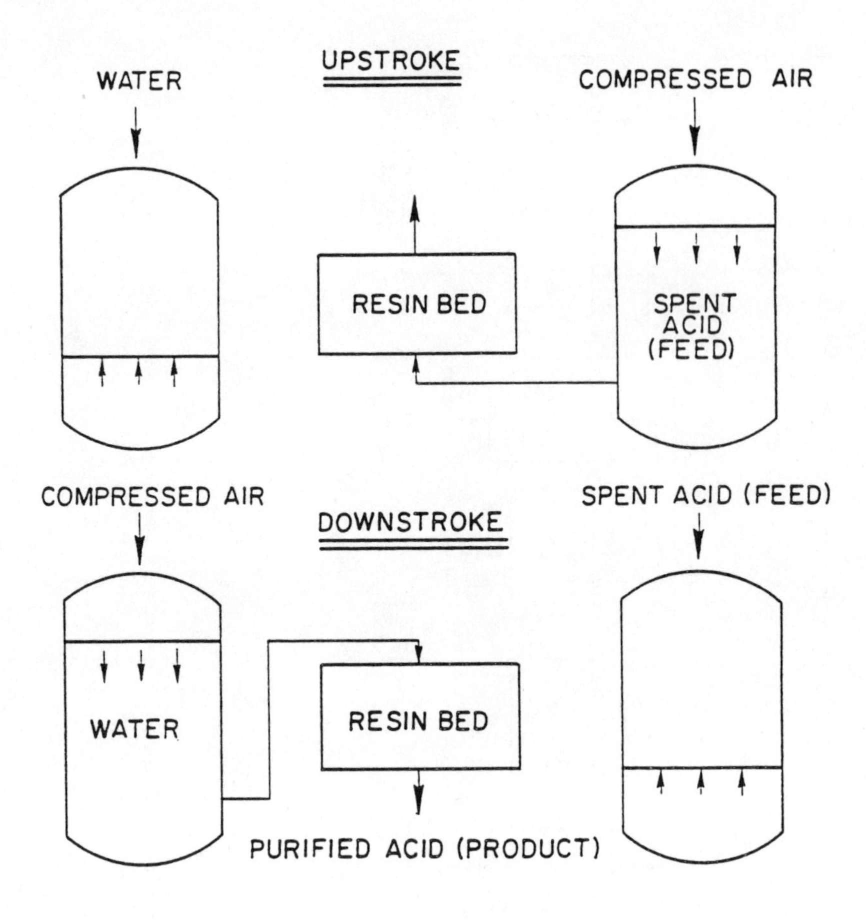

Figure 5.3.4. Basic operation of the acid purification unit (APU) using a continuous bed RFIE system.

Sources: References 7 and 9.

Operating and Design Parameters--

The most significant design parameter in an ion exchange system is the selection of an appropriate resin. Resin selection is based on the type of ion exchanger, the strength of the resin, its exchange capacity and selectivity, and the volume required.

Resins can be classified as acid cation exchangers or base anion exchangers. Cation exchangers have positively charged exchangeable ions, and anion exchangers have negatively charged exchangeable ions. Heavy metal selective chelating resins are weak acid cation resins that exhibit a high affinity for heavy metal cations.

The exchange capacity of a resin is generally expressed as equivalents per liter (eg/L), where an equivalent is equal to the molecular weight of the ion in grams divided by its electrical charge or valence.[4] For example, a resin with an exchange capacity of 1 eq/L could remove 37.5 g of divalent zinc (Zn^{+2}, molecular weight = 65 g) from solution.

The strength of a resin is determined by the degree to which its exchangeable ions dissociate in solution. The exchangeable ions of strong acid and strong base resins are highly dissociated and therefore are readily available for exchange, and are unaffected by solution pH. Conversely, the exchange capacity of weak acid and weak base resins is strongly influenced by solution pH.

Ion exchange reactions are stoichiometric and reversible. A generalized form of an ion exchange reaction can be depicted as follows:

$$R\text{-}A^+ + B^+ \rightleftharpoons R\text{-}B^+ + A^+$$

where R is the resin, A^+ is the ion originally associated with the resin, and B^+ is the ion originally in solution. The degree to which the exchange reaction proceeds is dependent on the preference (or selectivity) of the resin for the exchanged ion. The selectivity coefficient, K, is used to measure the preference of a resin for a particular ion. It expresses the relative distribution of ions when a charged resin is contacted with solutions of different (but similarly charged) ions. For example, in the generalized ion exchange reaction presented above, the selectivity coefficient (K) is defined as follows:

$$K = \frac{[B^+] \text{ in resin}}{[A^+] \text{ in resin}} \times \frac{[A\ ]^+ \text{in solution}}{[B^+] \text{ in solution}}$$

The selectivity coefficient of a resin will vary with changes in solution characteristics and the strength of the resin. Table 5.3.1 shows the selectivites of strong acid and strong base resins for various ionic species.

Operating parameters will vary with the particular application. The following factors will influence the selection of a resin type, pretreatment requirements, flow rates, cycle times, and the sizing of a system for a particular application:

- Types and concentrations of constituents present in the feed
- Rate of metal salt accumulation in the bath
- Flow rate
- Number of hours of operation.

The types and concentrations of constituents present in the spent solution will determine the type of resin selected. For corrosive wastes, these constituents will typically be metal salts. Weak cation exchangers can be used for spent solutions containing low concentrations of metal ions. For solutions containing high concentrations of metal ions, a strong anion exchanger can be used effectively in an APU.

The constituent concentrations will also determine the resin volume needed to treat the stream; larger concentrations will generally require a larger volume of resin to reduce regeneration frequency. The system size will also increase along with the rate of contaminant accumulation in the bath and the volume of solution to be treated per unit time. Smaller systems can generally be used if the process is operated for a longer period of time due to stabilization of the bath. Commericially available systems are able to process wastes at throughput rates ranging from 38 to 6,700 liters per hour.[11] Cycle times for RFIE systems generally range from 5 to 15 minutes.[11,12], whereas for cocurrent systems they can be as much as 1 to 2 hours because of the time needed to regenerate the column.[4] As a result, dual sets of columns are typically used in cocurrent systems to avoid excessive downtime.

TABLE 5.3.1. SELECTIVITIES OF ION EXCHANGE RESINS IN ORDER OF DECREASING PREFERENCES[a]

Strong acid cation exchanger	Strong base anion exchanger	Weak acid cation exchanger	Weak base anion exchanger	Weak acid chelate exchanger
Barium (+2)	Iodide (-1)	Hydrogen (+1)	Hydroxide (-1)	Copper (+2)
Lead (+2)	Nitrate (-1)	Copper (+2)	Sulfate (-2)	Iron (+2)
Mercury (+2)	Bisulfite (-1)	Cobalt (+2)	Chromate (-2)	Nickel (+2)
Copper (+1)	Chloride (-1)	Nickel (+2)	Phosphate (-2)	Lead (+2)
Calcium (+2)	Cyanide (-1)	Calcium (+2)	Chloride (-1)	Manganese (+2)
Nickel (+2)	Bicarbonate (-1)	Magnesium (+2)		Calcium (+2)
Cadmium (+2)	Hydroxide (-1)	Sodium (+2)		Magnesium (+2)
Copper (+2)	Fluoride (-1)			Sodium (+1)
Cobalt (+2)	Sulfate (-2)			
Zinc (+2)				
Cesium (+1)				
Iron (+2)				
Magnesium (+2)				
Potassium (+1)				
Manganese (+2)				
Ammonia (+1)				
Sodium (+1)				
Hydrogen (+1)				
Lithium (+1)				

[a]Valence number is given in parentheses.

Source: References 4 and 10.

Pretreatment Requirements/Restrictive Waste Characteristics--

Pretreatment of the waste stream (usually via filtration) is necessary to remove any constituents which would adversely affect the resin. Certain organics (e.g., aromatics) become irreversibly sorbed by the resin. Oxidants (such as chromic or nitric acid) can also damage the resin. Sodium metabisulfite, which converts hexavalent chrome to its triavalent state, can be added to the solution to prevent damage to the resin.[13] Eco-Tec has stated that resin degradation is not a problem with the RFIE process due to the short duration of contact (approximately. 1.5 mins) between the acid and the resin.[9,14]

Ions exchange units are only able to treat contaminants in solution. High concentrations of suspended solids which can foul the resin bed are typically pretreated through some form of filtering. Examples of filters which may be employed include activated carbon, deep bed, diatomaceous earth precoat, and resins. The filters eventually become clogged with particulates, and are replaced when overall cycle time decreases due to the increased time required for passage of the solution into the ion exchange unit.[7] The filter size and replacement frequency will depend on the quantity of suspended particulates passed through the filter per unit time.

For large volume systems, which require more frequent changing of the filter cartridges, it may be more cost-effective to use a multimedia sand filter with a backwashing system to regenerate the filter. Although initial capital costs for this type of filtration system are higher, savings in filter replacement costs will be realized.[11]

The use of weak acid and weak base exchangers for treating corrosive wastes will require additional pretreatment. The exchange capacity of weak acid exchangers is limited below pH 6.0, and weak base exchangers are not effective above pH 7.[4] Therefore, a pH adjustment system must be incorporated prior to feeding the waste stream through weak ion exchangers. Although weak exchangers cannot be used directly for the treatment of corrosive wastes, they can be used to reduce the costs of neutralization treatment by eliminating the need for over-neutralization (discussed in Section 4) to remove the metals.

Ion exchange using cocurrent flow is not suitable for removal of high concentrations of exchangeable ions (above 2,500 mg/L expressed as calcium carbonate equivalents). The resin material is rapidly exhausted during the exchange process and regeneration becomes prohibitively expensive.[4,15] However, the reverse is true for acid purification units since they are capable of recycling the regenerate. In addition, higher concentrations of metal contaminants present in the solution will improve removal efficiencies (see Section 5.3.2).

Post-Treatment Requirements--

Waste streams from the ion exchange process include: spent regenerant solution, byproduct stream, and filtrate from the pre-filtering system. Spent regenerant, produced when cocurrent flow ion exchange is used, generally requires neutralization and disposal. Optimal neutralization methods will depend upon the waste type (refer to Section 4). The byproduct may require treatment and disposal, or in some cases the recovered metal can be reused or marketed. Filtrate from the pre-filtering system can generally be land disposed without further treatment.[16] The quantities of wastes generated will depend on the types and concentrations of contaminants present in the solution being treated. However, the total volume of wastes from the ion exchange system requiring treatment or disposal will be less than if a recovery system was not employed. Thus, an overall savings in treatment and disposal costs is realized.

5.3.2 Process Performance

The performance of an ion exchange system will be predominantly influenced by the characteristics and quantity of the waste stream being treated. Parameters which need to be considered when evaluating the applicability of the system for a particular waste stream include: types and concentrations of constituents present in the waste stream, acidity of the spent stream versus acidity of the fresh stream, and required quality of the recovered stream.

Additional factors which can be used to evaluate the performance of an ion exchange system include: the quantity of byproduct streams generated, cycle times, product concentration, process modifications required, flow rates, processing capabilities (system size), and costs.

Cocurrent flow generates an additional corrosive waste stream which requires treatment and disposal. Cocurrent flow methods are inefficient for the treatment of corrosive wastes and are generally not used for this purpose. Reciprocating Flow Ion Exchange (RFIE) units are more cost-effective than cocurrent fixed-bed systems for use with corrosive wastes. RFIE units use smaller resin volumes, and therefore have lower capital costs and space requirements. Also, operating costs are lower due to the reuse of regenerant. Automation of the process operation provides additional savings in labor requirements. In addition, these units are able to generate a higher product concentration when used for chemical and acid bath recovery.

Acid purification systems using RFIE have been commercially demonstrated to be effective in the recovery of acids from aluminum anodizing solutions, acid pickling liquors, and rack-stripping solutions. Acid purification systems are most effective in recovering acids which have high concentrations of metal ion contaminants.

An acid purification unit using RFIE technology was installed at the Continuous Colour Coat, Ltd. plant in Rexdale, Ontario, to recover sulfuric acid from a steel pickling process.[17] During pickling, as iron buildup begins to cause poor product quality, acid additions are made to the bath or the tank is drained and a fresh acid solution is prepared. As an alternative, to neutralization and disposal of spent baths, Colour Coat, Ltd. employed an APU to remove the iron so that the solution could be recycled.

Typical operating parameters and results for the APU application are presented in Table 5.3.2. As shown, iron concentration in the reclaimed acid was reduced by 80 percent. Also, acidity and acid volume losses were minimal. Occasional replenishment of the bath was necessary, but draining of the tank (an expensive process) was no longer required. Also, improvements in product quality were noted due to the consistency of the bath concentration that was maintained during the APU operation.

Savings were realized in neutralization and disposal requirements. Also, there were net reductions in labor requirements due to reduced bath maintenance. An economic evaluation of the system (Table 5.3.3.) showed an estimated payback period for the unit of less than 2 years.

TABLE 5.3.2. TYPICAL OPERATING PARAMETERS AND RESULTS FOR THE APU INSTALLED AT CONTINUOUS COLOUR COAT, LTD. IN REXDALE, ONTARIO

Parameter	Feed acid	By product	Return acid
Flow Rate (gal/min)	19	N/A	19
Temperature (°F)	119	N/A	119
%-Sulfuric Acid	15.8	0.5	15.7
%-Iron	3.0	2.6	0.6

Note: N/A = Not applicable.

Source: Reference 11.

TABLE 5.3.3. ECONOMIC EVALUATION OF THE APU INSTALLED AT CONTINUOUS COLOUR COAT, LTD.

Item	Cost
CAPITAL COSTS[a] (includes costs for equipment & installation)	$100,000
OPERATING COSTS	
Resin Replacement (every 4 years at $58/liter)	$3,770/year
Utilities (0.5 KW x 16 hrs/day x 250 days/yr x $0.055/KWH)	$ 110/year
Taxes and Insurance (1% of TIC)	$1,000/year
TOTAL OPERATING COSTS	$4,880/year
COST SAVINGS	
Reduction in Acid Purchase @ $92.40/ton	$25,875/year
Reduction in Neutralization Costs (Lime) @ $80/ton	$18,000/year
Reduction in Sludge Disposal Costs	$20,000/year
TOTAL COST SAVINGS	$63,875/year
NET COST SAVINGS (Gross Savings - Operating Costs)	$58,995/year
PAYBACK PERIOD (Capital Costs + Net Savings)	1.7 years (approx. 20 months)

[a]Capital Equipment included APU Model No. AP30-24, multimedia sand filter, water supply tank, and piping.

Source: Reference 11 (As Ecotech cost quote, August 1986).

Another common application in which the APU works effectively is the recovery of acids from aluminum anodizing solutions. An APU system was installed at the Modine Manufacturing Company in Racine, Wisconsin to recover nitric acid from an aluminum etching process.[7] The APU was connected directly on-line, which allowed for continuous process operation. The APU generated a more concentrated solution of recovered nitric acid than it was fed; thus a slightly lesser volume of acid was returned to the tank. After a certain period of time, the total acid volume in the tank was reduced enough so that additional concentrated acid could be added without draining the tank. Table 5.3.4 presents a summary of the results and operating parameters of the APU at the Modine plant. Improvements in product quality and savings in neutralization, disposal, and fresh acid makeup were noted by Modine personnel.[7] An economic evaluation of this system (Table 5.3.5.) shows nitric acid recovery with an APU to be very cost-effective.

A full-scale demonstration of the APU in recovering sulfuric acid from an aluminum anodizing solution was performed at Springfield Machine and Stamping, Inc. of Warren, Michigan.[6] Typical operating parameters and results during the 6 month testing period are summarized in Table 5.3.6. The system proved to be cost-effective due to the high aluminum removal efficiency, retention of acid strength, and reductions in raw material purchase, disposal and labor costs.

A pilot scale unit for recovering hydrochloric acid from an electroplating pickling liquor was tested at Electroplating Engineering, Inc. in St. Paul, Minnesota.[13] The results were not as successful as with the other cases presented. Reduced metal removal efficiencies and high acidity losses were experienced.

Typical concentrations of metals in the new, intermediate, and spent pickling solution at this plant are presented in Table 5.3.7. As shown, zinc and iron are the primary constituents of concern. A different system configuration was required for this application because the zinc was present in the form of zinc chloride complexes. As described in Section 5.3.1, the resin used in an APU shows a preferential affinity for acid anions as opposed to metal cations, which causes metals to pass through the resin while the acid is retained. However, instead of passing through the resin, zinc chloride complexes will also be retained by the anion resin.

TABLE 5.3.4. TYPICAL OPERATING PARAMETERS AND RESULTS DURING TESTING OF THE APU FOR RECOVERY OF NITRIC ACID AT MODINE MANUFACTURING COMPANY IN RACINE, WISCONSIN

Parameter	Result
Feed to APU from etch tank	6.2 N
Product returned to etch tank	6.5 N
By-product going to waste treatment	0.6 N
Level of aluminum contamination:	
Coming into APU from etch tank	793 mg/L
Returning to the etch tank	231 mg/L
Average cycle time	12.7 min
Volume of water removed from etch tank/APU cycle	0.89 gal
Mass balance	
Equivalents of nitric acid into APU from etch tank	251
Equivalents returned to etch tank and waste	257

Source: References 6 and 7.

TABLE 5.3.5. ECONOMIC EVALUATION OF THE APU INSTALLED AT MODINE MANUFACTURING CO. IN RACINE, WISCONSIN FOR THE RECOVERY OF NITRIC ACID

Item	Cost
CAPITAL COSTS (includes costs for equipment and installation)	$37,234
COST SAVINGS	
Reduction in nitric acid purchase	$20,064/year
Reduction in neutralization costs	$ 6,276/year
Reduction in disposal costs	$ 7,236/year
Reduction in labor	$ 2,400/year
TOTAL COST SAVINGS	$35,976/year
OPERATING COSTS	
Resin replacement (every 4 years at $58/liter)	$ 1,305/year
Utilities (0.5 KW x 16 hrs/day x 350 days/yr x $0.055/KWH)	$ 132/year
Taxes and insurance (1% of TIC)	$ 372/year
TOTAL OPERATING COSTS	$ 1,809/year
NET COST SAVINGS (Gross savings - Operating costs)	$34,167/year
PAYBACK PERIOD (Capital cost ÷ Net savings)	1.1 years (approx. 13 months)

Source: References 7 and 11 using August 1986 cost data.

TABLE 5.3.6. TYPICAL OPERATING PARAMETERS AND RESULTS FOR THE APU INSTALLED AT SPRINGFIELD MACHINE & STAMPING, INC. IN WARREN, MICHIGAN FOR SULFURIC ACID RECOVERY

Parameter	Feed	Product	By-product
Flow rate (liters/hr)	298	296	175
Sulfuric acid concentration (g/L)	183.8	175.0	13.0
Aluminum concentration (g/L)	12.2	4.2	12.0

Source: Reference 8.

TABLE 5.3.7 AVERAGE CONSTITUENT CONCENTRATIONS IN THE NEW, INTERMEDIATE, AND SPENT BATH SOLUTIONS

Parameter	New solution	Intermediate solution	Spent solution
Acidity, mg/L $CaCO_3$	246,333	168,667	112,833
Iron concentration (mg/L)	219	1,333	1,650
Zinc concentration (mg/L)	1,052	3,693	4,283
Nickel concentration (mg/L)	ND	ND	1.3
Copper concentration (mg/L)	ND	ND	0.6
Chrome concentration (mg/L)	ND	ND	6.3

Source: Reference 13.

In order to remove both the zinc chloride complex and the iron contaminants, it was necessary to operate the system in two stages. Initially, the spent solution was passed through one resin to remove zinc (termed the inverse mode since the acid ions are not retarded). Then the acid ions which have passed through the first resin are passed over a second resin in the normal mode of operation, retaining the acid while allowing the iron ions to pass through the resin. As with typical APU processes, the acid is recovered during the regeneration cycle. Figures 5.3.5 and 5.3.6 illustrate these two modes of operation.

Pickling solutions from three different HCl pickling liquors were used to test the performance of the acid purification system. Analysis of these spent solutions yielded the following:

Parameter	Range of Concentrations
Acidity	77,000 - 284,000 mg/L (expressed as $CaCO_3$)
Zinc Content	640 - 52,000 mg/L
Iron Content	1,100 - 7,000 mg/L

Several test runs were performed using the acid purification unit as summarized in Table 5.3.8. The results show that good zinc removal efficiencies (99.3 percent) were achieved during the inverse mode of operation with minimal losses in acidity (3.5 percent). However, during the normal mode of operation, only an average of 60 percent iron removal was achieved and acidity losses were quite high (averaging 38 percent). The results showed that increased iron removal could only be achieved at the expense of greater reductions in acidity. It was determined that the ratio of iron to acidity in the feed has to approach 1:15 in order to achieve effective performance. The iron to acidity ratio for the feed used during these tests was 1:67, which contributed to the poor performance results.

It is possible that the intermediate byproduct solution generated after the inverse mode may be of sufficient quality to be returned to the pickling bath.[13] Iron content of the byproduct solution was comparable to the iron concentrations measured in the bath during its intermediate solution age (Table 5.3.7). Additional testing would be required in order to determine whether bath quality would be acceptable under these conditions.

STEP ONE - WATER DISPLACEMENT

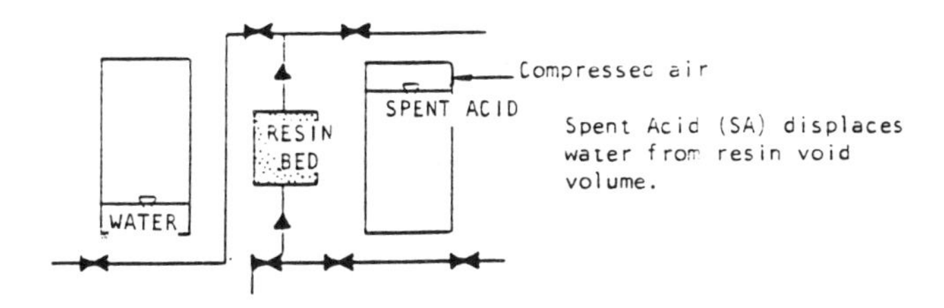

STEP TWO - INTERMEDIATE BYPRODUCT GENERATION

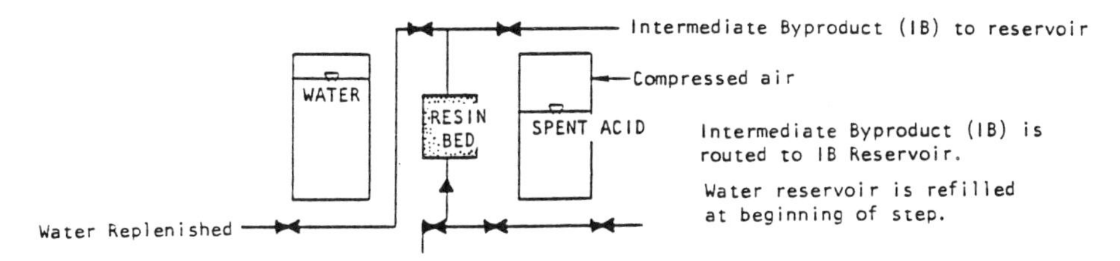

STEP THREE - SPENT ACID DISPLACEMENT

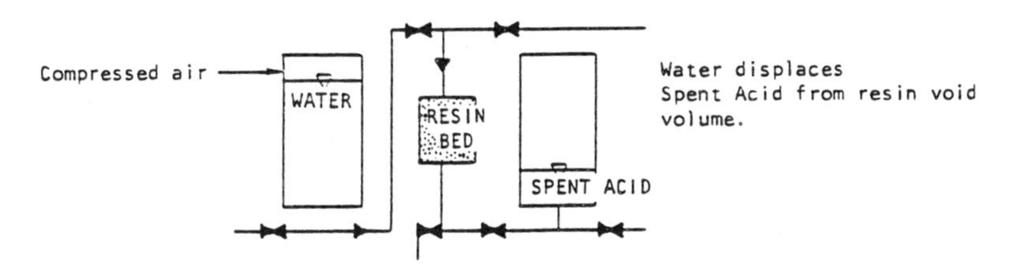

STEP FOUR - INTERMEDIATE WASTE PRODUCT GENERATION

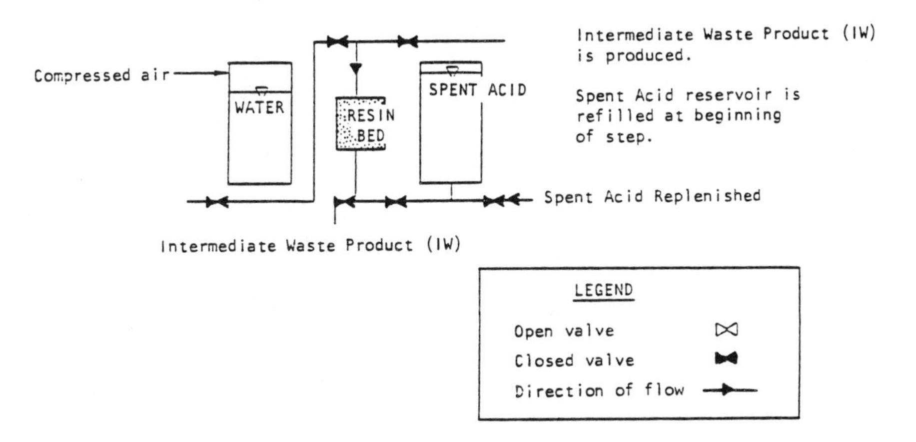

Figure 5.3.5. Schematic of inverse mode of operation.

Source: Reference 13.

STEP ONE - WATER DISPLACEMENT

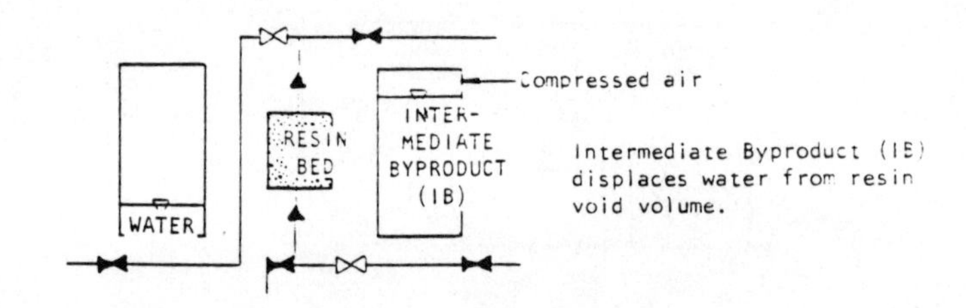

STEP TWO - FINAL WASTE GENERATION

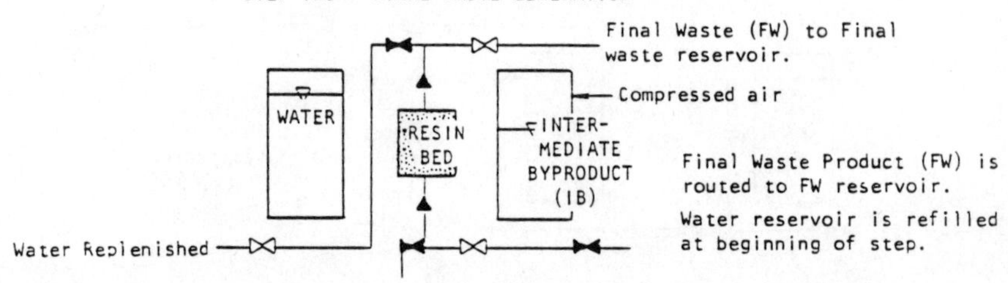

STEP THREE - INTERMEDIATE BYPRODUCT DISPLACEMENT

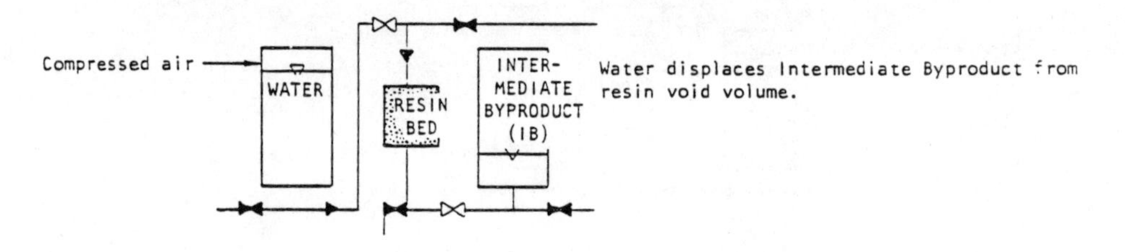

STEP FOUR - RECLAIMED ACID GENERATION

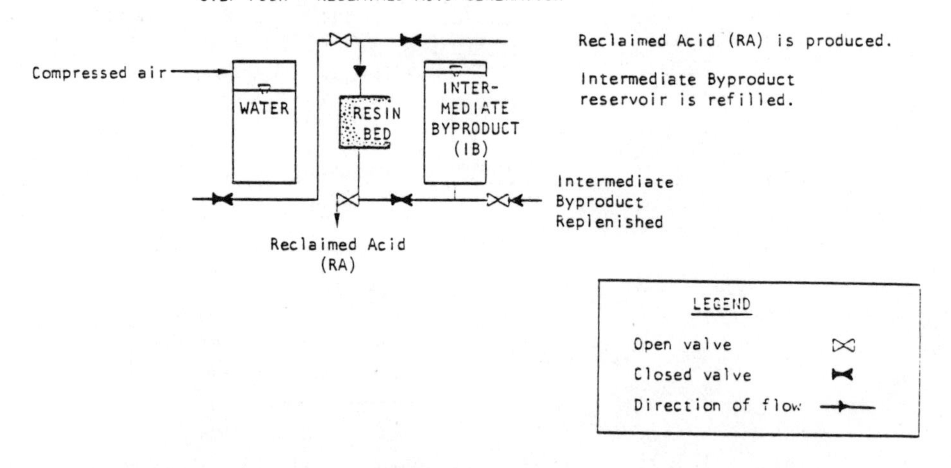

Figure 5.3.6. Schematic of normal mode of operation.

Source: Reference 13.

TABLE 5.3.8. SUMMARY OF RESULTS OF RESEARCH PERFORMED AT ELECTROPLATING ENGINEERING, INC.[a]

Parameter	Preliminary runs (No.2,3,4)[b]	Preliminary runs (No.5A,5B,6)[c]	Final runs (No. 7A-7P)[d]
VOLUME TREATED/GENERATED (LITERS):			
Spent acid	98	174	189
Intermediate by-product	86	174	189
Reclaimed acid	70	162	4.6
Intermediate waste product	64	96	106
Final waste product	67	121	3.3
INVERSE MODE LOADINGS TO RESIN:			
Zinc (grams/cycle)	34.2	4.4	2.6
Volume (bed volumes/cycle)	4.05	4.05	4.05
Feed rate (liters/hour)	11.4	11.4	11.4
NORMAL MODE LOADINGS TO RESIN:			
Acidity (grams $CaCO_3$/cycle)	43.4	35.2	14.6
Volume (bed volumes/cycle)	0.40	0.40	0.34
Feed rate (liters/hour)	6.0	6.0	5.5
STREAM CONCENTRATIONS OF ACIDITY: (expressed as g/L $CaCO_3$ equivalents)			
Spent acid	242	156	77
Intermediate by-product	217	151	74
Reclaimed acid	116	94	46
Intermediate waste product	39	15	5.9
Final waste product	90	57	30

(continued)

TABLE 5.3.8 (Continued)

Parameter	Preliminary runs (No.2,3,4)[b]	Preliminary runs (No.5A,5B,6)[c]	Final runs (No. 7A-7P)[d]
STREAM CONCENTRATIONS OF ZINC (mg/L):			
Spent acid	40,333	1,100	640
Intermediate by-product	34,667	8	1.4
Reclaimed acid	19,233	18	5.47
Intermediate waste product	13,000	2,000	1,200
Final waste product	7,366	0.61	0.25
STREAM CONCENTRATIONS OF IRON (mg/L):			
Spent acid	4,700	2,600	1,100
Intermediate by-product	4,400	2,433	1,100
Reclaimed acid	1,450	920	439
Intermediate waste product	1,277	357	120
Final waste product	3,367	1,733	728
STREAM CONCENTRATIONS OF CHROME (mg/L):			
Spent acid	43	2.7	NA
Intermediate by-product	42	2.6	NA
Reclaimed Acid	12	1.6	NA
Intermediate waste product	4.6	0.51	NA
Final waste product	28	3.4	NA

[a]The results of Run 1 were discarded due to improper installation.

[b]The zinc loadings for Runs 2, 3, and 4 were above the recommended 18 grams/cycle maximum loading recommended for the system.

[c]The objective for Runs 5A, 5B, and 6 were to process a sufficient quantity of acid for reuse and to optimize loadings.

[d]The objective for Runs 7A through 7P were to optimize loadings for the normal mode of operation.

Source: Reference 13.

Based on the results of these pilot-scale tests, Pace Laboratories (1986) conducted an economic evaluation of the APU system for this application. They determined that the APU could only be cost-effective for this application if a large volume of spent solution is processed.

In summary, available performance data suggest that the technical and economic feasibility of acid purification systems will mainly depend on the types and concentrations of metal ions present. These systems work well in recovering solutions with highly positively charged contaminant ions (e.g., aluminum, iron) because these ions pass rapidly through the strong base anion exchanger resin. Solutions containing low concentrations of contaminant ions are not efficiently recovered using the APU. Recommended minimum concentrations for efficient results are presented in Table 5.3.9. Although lower concentrations may be treated, removal efficiencies will be lower unless larger systems are employed. However, cost-effectiveness is compromised due to the increased capital costs for larger systems. Generally, for low concentrations of metal ions, ion exchange methods using weak cation-exchangers to retard the metals are recommended.

5.3.3 Process Costs

An economic evaluation of countercurrent (RFIE) systems is presented in this section. Cocurrent flow methods are generally not technically or economically feasible for the treatment of corrosive wastes and will therefore not be discussed.

Costs for RFIE systems vary with the specific application. Factors that affect the costs include: quantity and quality of constituents recovered, production rates, volume of spent solution to be treated, concentration of metal salts present in the spent solution, rate of build-up of metal ions in the bath, concentration of the bath, and number of hours of process operation.

Capital costs, which include equipment, installation, and peripheral costs, increase with system size. These costs are offset by savings which are realized through reduced volumes of wastes requiring treatment (e.g., neutralization) and disposal, and reduced purchase requirements for bath reagents. Operating costs will include replacement of filter cartridges, resin replacement (approximately every 5 years), and utilities.

TABLE 5.3.9. RECOMMENDED MINIMUM CONCENTRATIONS (g/L) FOR EFFICIENT METALS REMOVAL USING THE ECO-TECH APU

Solution	Iron	Zinc	Aluminum	Copper	Total metals
Hydrochloric acid	30-50	130-150	-	-	-
Sulfuric acid	30-50	-	5	20	-
Nitric/hydrofluoric acid	-	-	-	-	30
Nitric acid rack stripping	-	-	-	-	75-100

Note: The APU can be used for solutions with lower concentrations of these metals, but the metal removal efficiencies will be lower unless a larger unit is used. Metal removal efficiencies average 55% for typical systems.

Source: Reference 16.

Capital costs for acid purification systems typically range from $15,000 to $100,000 depending on the system size (Fontana, 1986). Typical equipment costs for systems with various throughput rates are presented in Table 5.3.10. These costs include installation, equipment and peripherals, and a pre-filter system. Costs presented in this table are for the recovery of sulfuric acid from aluminum anodizing solutions. Costs may be slightly higher for other applications (Fontana, 1986).

Typical operating costs are presented in Table 5.3.11. Savings will depend on the specific application. Table 5.3.12 presents an economic evaluation of several hypothetical systems.

5.3.4 Process Status

Cocurrent ion exchange systems are generally not employed for direct treatment of corrosive wastes. Cocurrent systems using weak exchangers have inefficient exchange capacities in the corrosive pH ranges, and are generally only used as polishing systems following other treatment operations (e.g., neutralization). Cocurrent systems using strong exchangers are technically feasible for the treatment of corrosive wastes, but they are not cost-effective because of the high costs for column regeneration.

Ion exchange systems, using the reverse or reciprocating flow mode (countercurrent), have been shown to be effective in the treatment of corrosive wastes. The process has been demonstrated commercially for chemical recovery from acid copper, acid zinc, nickel, cobalt, tin, and chromium plating baths as well as for purification of spent acid solutions.

Chemical recovery systems using fixed bed RFIE have been used to recover chromic acid and metal salts. It has also been used to deionize mixed-metal rinse solutions for recovering process water and concentrating the metals for subsequent treatment.[4] Commercial units are available from several vendors.

Acid purification systems using continuous RFIE have been used to remove aluminum salts from sulfuric acid anodizing solutions, to remove metals from nitric and rack-stripping solutions, and to remove metals from sulfuric and hydrochloric acid pickling solutions.[4] The APU is primarily used for recovering aluminum anodizing solutions.[4] Acid purification systems are more cost effective for removing high concentrations of contaminants than other ion exchange systems. Demonstrated applications are listed in Table 5.3.13.

TABLE 5.3.10. TYPICAL CAPITAL COSTS FOR ECO-TECH APU

Item	Small unit	Medium unit	Medium unit	Large unit
APU Model No.	AP-6	AP-24	AP-54	AP-72
Flow rate	38 L/hr	500 L/hr	800 L/hr	6700 L/hr
Capital cost	$14,000	$37,000	$116,000	$184,000

Notes: Capital Costs include equipment, installation, peripherals, and cartridge-type prefilter system.

Costs presented in this table are for application to recovery of a sulfuric acid anodizing solution. Costs for other applications may be slightly higher.

Twelve different size units are available from Eco-Tech, Ltd. The model numbers, which indicate bed diameters, for these units are: AP-6, AP-12, AP-18, AP-24, AP-30, AP-36, AP-42, AP-48, AP-54, AP-60, AP-66, and AP-72.

Source: References 16 and 18 (Ecotech quote July 1986).

TABLE 5.3.11. TYPICAL OPERATING COSTS FOR ACID PURIFICATION USING CONTINUOUS COUNTERCURRENT ION EXCHANGE (RFIE)

Item	Cost
Filter cartridges for prefilter system	\$10.00/month
Utilities:	
(0.5 KW x 16 hrs/day x 20 days/month x 0.055 \$/KWH)	\$8.80/month
Resin replacement (specific cost depends on system size)	\$58/liter every 4 years

Source: References 11 and 13 (Based on August 1986 cost data).

TABLE 5.3.12. ECONOMIC EVALUATION OF ACID PURIFICATION PROCESS

Description	30,000 gpy throughput	100,000 gpy throughput	500,000 gpy throughput
Case 1 - Purification of Sulfuric Acid Anodizing Solution: Previous approach used caustic acid neutralization: New approach uses APU with caustic neutralization.			
Approx. APU Cost	$ 6,000	$11,000	$ 25,000
Previous treatment cost	$ 9,690	$32,300	$161,500
Previous acid cost	$ 2,349	$ 7,830	$ 39,150
Annual savings	$ 8,427	$28,891	$140,455
Payback (months)	9	5	2
Case 2 - Purification of Sulfuric Acid Anodizing Solution: Previous approach used lime neutralization: New approach uses APU with lime neutralization.			
Approx. APU cost	$ 6,000	$11,000	$ 25,000
Previous treatment cost	$ 2,250	$ 8,500	$ 38,500
Previous acid cost	$ 2,349	$ 7,830	$ 39,150
Annual savings	$ 3,216	$10,731	$ 53,655
Payback (months)	22	12	6
Case 3 - Purification of Sulfuric Acid Anodizing Solution: Previous approach used waste haulage: New approach uses APU with caustic neutralization.			
Approx. APU cost	$ 6,000	$11,000	$ 25,000
Previous treatment cost	$ 3,000	$10,000	$ 50,000
Previous acid cost	$ 2,349	$ 7,830	$ 39,150
Present treatment cost	$ 2,907	$ 9,690	$ 48,450
Annual savings	$ 1,737	$ 5,791	$ 28,955
Payback (months)	41	23	10

(continued)

TABLE 5.3.12 (Continued)

Description	30,000 gpy throughput	100,000 gpy throughput	500,000 gpy throughput
Case 4 - Purification of Sulfuric Acid Anodizing Solution: Previous approach used waste haulage: New approach uses APU with lime neutralization.			
Approx. APU cost	$6,000	$11,000	$ 25,000
Previous treatment cost	$3,000	$10,000	$ 50,000
Previous acid cost	$2,349	$ 7,830	$ 39,150
Present treatment cost	$ 675	$ 2,250	$ 11,250
Annual savings	$3,969	$13,245	$ 66,155
Payback (months)	18	10	5
Case 5 - Nitric Acid Recovery: Previous approach used caustic neutralization: New approach uses APU with caustic neutralization.			
Approx. APU cost	$9,400	$11,300	$ 18,400
Previous treatment cost	$7,575	$30,300	$ 50,500
Previous acid cost	$8,775	$35,100	$ 58,500
Total previous cost	$16,350	$65,400	$109,000
Annual savings	$ 9,810	$39,240	$ 65,400
Payback (months)	11	3	3

Source: References 6, 8, 9, and 14. (Based on July 1986 Ecotech cost data)

TABLE 5.3.13. DEMONSTRATED APPLICATIONS OF ECO-TECH ACID PURIFICATION UNIT USING RFIE

Application/ bath components	Typical bath concentration (g/L)	Typical product concentration (g/L)	Typical by-product concentration (g/L)
Sulfuric acid	190	182	13
Aluminum	10	5.5	6
Sulfuric acid	127	116	10
Iron	36	10.5	21
Nitric acid	514	581	10
Nickel and copper	99	47.5	70.8
Sulfuric acid	128	113	18
Hydrogen peroxide	41	35	7
Copper	13.3	5.9	9.2
Hydrochloric acid	146	146	10
Iron	34	25	15
Nitric acid	150	139	4.5
Hydrofluoric acid	36	28.8	7.2
Iron	29	8.7	20.3
Nickel	7.02	2.1	4.9
Chrome	7.33	2.2	5.1
Sulfuric acid	61.3	54.9	5.88
Sodium	7.8	0.8	5.56

Source: Reference 9 (Based on July 1986 Ecotech cost data).

Although the use of ion exchange for acid purification is currently under investigation by several ion exchange vendors (e.g., Alpha Process Systems; Illinois Water Treatment Company; Ionics, Inc.; etc.), Eco-Tec, Ltd. is the only vendor with commercial units currently in operation.[13,19,20]

Table 5.3.14 compares the advantages and disadvantages of the various ion exchange alternatives. Cocurrent fixed beds are generally used in wastewater treatment; they are not cost-effective for treatment of corrosives due to the high costs associated with regenerant purchase and treatment. The countercurrent (RFIE) continuous mode of operation, which is utilized by the APU, is generally more efficient in the treatment of solutions containing high concentrations of metal ion contaminants. The countercurrent (RFIE) fixed mode is most effective in the recovery of dilute solutions. Overall, RFIE systems may be more cost-effective for the treatment of corrosives than conventional neutralization, particularly when enactment of land disposal restrictions increase costs associated with land disposal of residuals.

TABLE 5.3.14. COMPARISON OF ION EXCHANGE OPERATING MODES

Criteria	Cocurrent fixed bed	Countercurrent (RFIE) fixed bed	Countercurrent (RFIE) continuous bed
Capacity for high feed flow and concentration	Least	Middle	Highest
Effluent quality	Fluctuates with bed exhaustion	High, minor fluctuations	High
Regenerant and rinse requirements	Highest	Somewhat less than cocurrent	Least, yields most concentration regenerant waste
Equipment complexity	Simplest; can use manual operation	More complex; automatic controls for regeneration	Most complex; completely automated
Equipment for continuous operation	Multiple beds, single regeneration equipment	Multiple beds, single regeneration equipment	Provides continuous service
Relative costs (per unit volume)			
Investment	Least	Middle	Highest
Operating	Highest chemicals and labor; highest resin inventory	Less chemicals, water and labor than cocurrent	Least chemical and labor; lowest resin inventory
Application for corrosive wastes	Useful only as a polishing step for neutralization treatment	Effective in the treatment of more dilute solutions (i.e., corrosive plating rinses)	Effective in the treatment of more concentrated solutions (i.e., direct treatment of acid baths).

Source: References 16 and 21.

REFERENCES

1. Camp Dresser & McKee, Inc. (CDM). Technical Assessment of Treatment Alternatives for Wastes Containing Corrosives. Prepared for the U.S. EPA-OSW, Waste Treatment Branch, Washington, D.C., under EPA Contract No. 68-01-6403, Work Assignment No. 39. September 1984.

2. Higgins, T.E., CH2MHILL. Industrial Processes to Reduce Generation of Hazardous Waste at DOD Facilities - Phase 2 Report, Evaluation of 18 Case Studies. Prepared for the DOD Environmental Leadership Project and the U.S. Army Corps of Engineers. July 1985.

3. GCA Technology. Industrial Waste Management Alternatives And Their Associated Technologies/Processes. Prepared for the Illinois Environmental Protection Agency, Division of Land Pollution Control, Springfield, Illinois. GCA Contract No. 2-053-011 and 2-053-012. GCA-TR-80-80-G. February 1981.

4. U.S. EPA, Industrial Environmental Research Laboratory, Cincinnati, Ohio. Summary Report: Control and Treatment Technology for the Metal Finishing Industry - Ion Exchange. EPA-625-8-81-007. June 1981.

5. Hatch, M.J., and J.A. Dillon. Acid Retardation: A Simple Physical Method for Separation of Strong Acids from Their Salts. I&EC Process Design and Development, 2(4):253-263. October 1963.

6. Brown, C.J., Davy, D., and P.J. Simmons. Recovery of Nitric Acid from Solutions Used for Treating Metal Surfaces. Plating and Surface Finishing. February 1980.

7. Robertson, W.M., James, C.E., and J.Y.C. Huang. Recovery and Reuse of Waste Nitric Acid From An Aluminum Etch Process. In: Proceedings of the 35th Industrial Waste Conference at Purdue University. May 13-15, 1980.

8. Brown, C.J., Davy, D., and P.J. Simmons. Purification of Sulfuric Acid Anodizing Solutions. Plating and Surface Finishing. January 1979.

9. Eco-Tec, Ltd. Product Literature: Acid Purification Unit (APU). Bulletin No. ET-4-84-5M. Received July 1986.

10. U.S. EPA, Industrial Environmental Research Laboratory, Cincinnati, Ohio. Sources and Treatment of Wastewater in the Nonferrous Metals Industry. EPA-600/2-80-074. April 1980.

11. Fontana, C., Eco-Tech, Ltd. Telephone Conversation with L. Wilk, GCA Technology Division, Inc. Re: Acid Purification Unit. August 21, 1986.

12. Dejak, M. Acid Recovery Proves Viable in Steel Pickling. Finishing, 10(1): 24-27. January 1986.

13. Pace Laboratories, Inc. Final Report: Reclamation and Reuse of Spent Hydrochloric Acid, Hazardous Waste Reduction Grant. Prepared for the Minnesota Waste Management Board on behalf of Electro-Plating Engineering Company, Inc. February 14, 1986.

14. Eco-Tec, Ltd. Product Literature: Ion Exchange Systems. Bulletin No. ET-11-83-3M. Received July 1986.

15. GCA Technology. Corrective Measures for Releases to Ground Water from Solid Waste Management Units. Prepared for U.S. EPA-OSW Land Disposal Branch, under EPA Contract No. 68-01-6871, Work Assignment No. 51. GCA-TR-85-69-G. August 1985.

16. Fontana, C., Eco-Tech, Ltd. Telephone Conversation with L. Wilk, GCA Technology Division, Inc. Re: Acid Purification Unit. August 26, 1986.

17. Chemical Processing Staff. Spotlight: Pickling Acid Recovery Unit Saves $40,000/year, Purifies Spent Sulfuric Acid. Chemical Processing, 49(3): 36-38. March 1986.

18. Fontana, C., Eco-Tech, Ltd. Telephone Conversation with L. Wilk, GCA Technology Division, Inc. Re: Acid Purification Unit. July 7, 1986.

19. Parcy, E., Ionics, Inc. Telephone conversation with J. Spielman, GCA Technology, Inc. August 14, 1986.

20. Jain, S.M., Ionics, Inc. Telephone conversation with J. Spielman, GCA Technology Division, Inc. August 12, 1986.

21. U.S. EPA, Office of Research and Development, Washington, D.C. Treatability Manual, Volume III: Technologies for Control/Removal of Pollutants. EPA-600-8-80-042c. July 1980.

5.4 ELECTRODIALYSIS

5.4.1 Process Description

Electrodialysis (ED) uses an electric field and a semipermeable ion-selective membrane to concentrate or separate ionic species in an aqueous solution.[1,2] Its primary application in the treatment of corrosive wastes is for use in the recovery of acids from plating solutions, pickling solutions, and etchants. Three types of configurations may be used in the design of ED units: concentrating-diluting, ion-substituting, and electrolytic.

The concentrating-diluting configuration, shown in Figure 5.4.1, typically contains a series of alternating cation and anion membranes arranged in parallel between two electrodes (cathode and anode) to form an ED "multicell" or membrane "stack."[2,4] Cationic membranes are only permeable to positive ions and anionic membranes are only permeable to negative ions.[1,4] Spacers (gaskets) are used to separate the adjacent membranes into leak-tight compartments.[4]

An electrical potential is applied across the ion exchange membrane which causes the migration of cations (e.g., copper, nickel, and zinc) toward the negative electrode and anions (e.g., sulfates, chlorides, and cyanides) toward the positive electrode. The cathode is typically comprised of stainless steel, and the anode usually consists of platinized titanium.[4] Thus, under the influence of an electric potential applied across the membranes, alternating cells of concentrated and dilute solutions are formed between the cationic and anionic membranes.[1]

Electrodialysis units are generally characterized by the number of "cell-pairs" comprising a multicell unit. One concentrating and one diluting cell comprise a cell pair. A cell-pair consists of a cation-selective membrane, a diluting spacer, an anion-selective membrane, and a concentrating spacer.[4] A stack or multicell refers to the entire arrangement of repeating cell-pairs between two electrodes.[3]. Industrial stacks typically have 50 to 300 cell-pairs.[4]

The ion-substituting configuration is similar to the concentrating-diluting configuration, but it involves a transfer of ions. As illustrated in Figure 5.4.2, the waste feed is passed between two membranes of like charge.

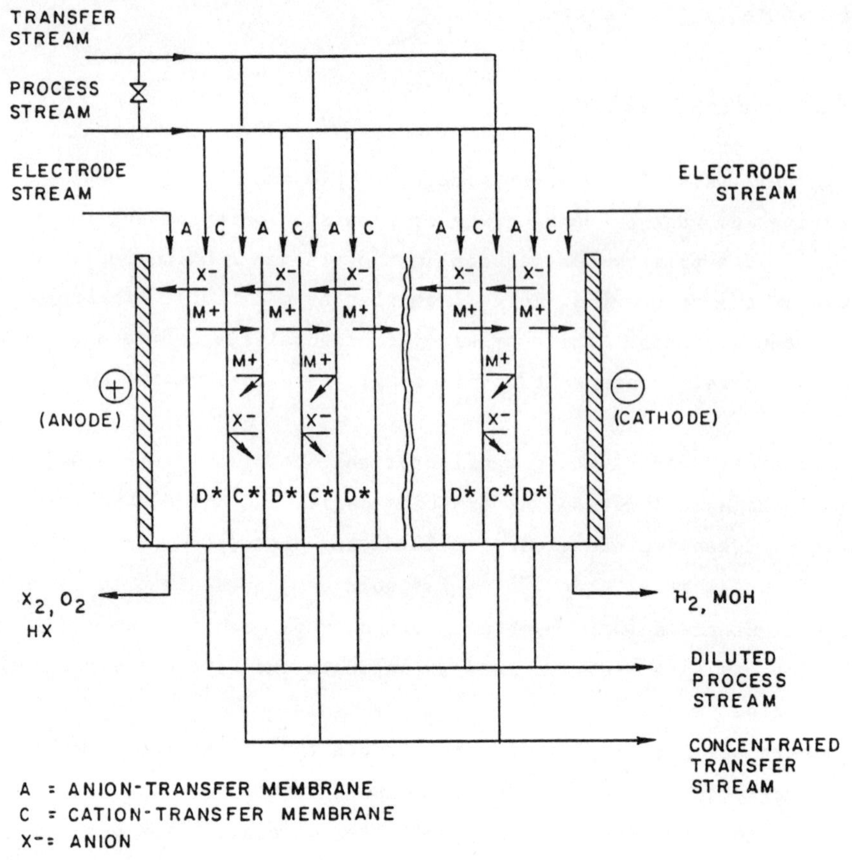

Figure 5.4.1. Schematic of a concentrating-diluting electrodialysis process.

Source: Reference 3.

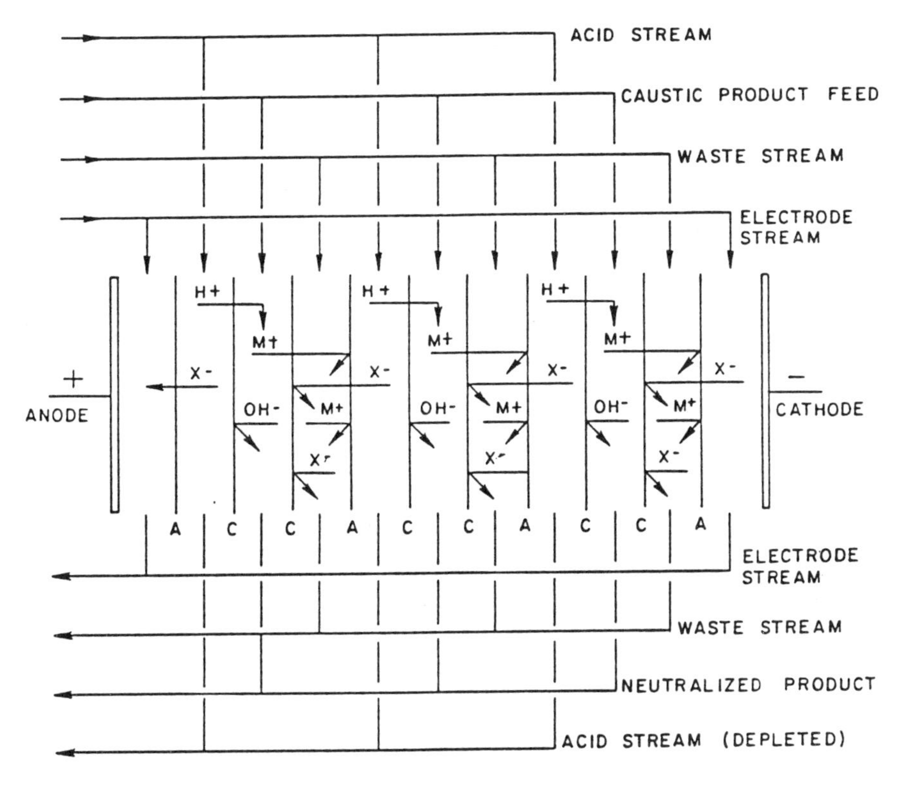

Figure 5.4.2. Schematic of an ion transfer/ion substituting electrodialysis process.

Source: Reference 3.

A solution containing the ions which are to be substituted is passed through an adjacent compartment. Upon application of an electric potential, an ionic substitution occurs across the membranes.

Electrolytic ED units are similar to conventional electrolysis cells, except that a membrane is placed between the electrodes. The membrane acts to separate the products of the electrode reactions.[3] In typical applications for corrosives, the diaphragm cell is fitted with cation selective membranes.[5] Figure 5.4.3 shows a schematic representation of a diaphragm cell containing one anode chamber and two cathode chambers. Spent etchant is fed into the anode chamber. Upon application of an electric potential, metal contaminants migrate to the cathode chambers. The anolyte (etchant solution) is regenerated and reused, and the metals can be recovered at the cathode.

Operating Parameters--

Important design factors which affect the operation of an ED unit include membrane characteristics, anode and cathode materials, voltage density, current density, temperature, and the number of cells. Parameters used to select an appropriate membrane for a specific application include transport number for the appropriate ions and electrical, physical and chemical resistance of the membrane.

Higher current densities will improve the product concentration. Also, increased solution concentrations will require higher voltages to maintain current densities. However, increases in the current density also cause increases in temperature. Operating temperature is a limiting factor in that high temperatures (above 100°F) can cause damage to the ED unit.[4] A cooling system may be required to prevent heat damage.

The number of cells determine the maximum throughput rate of the system. Selection of the appropriate number of cells is based on spent bath ion concentration and required reduction, rate of metal salt accumulation in the bath, volume of spent bath to be treated per unit time, and the number of hours of process operation.

Pre-Treatment Requirements/Restrictive Waste Characteristics--

Spent finishing and/or plating solutions often contain small metal parts and extraneous metal chips. Buildup of these particulates can foul the membranes and lower the cell resistance.[5] High organic levels can also foul

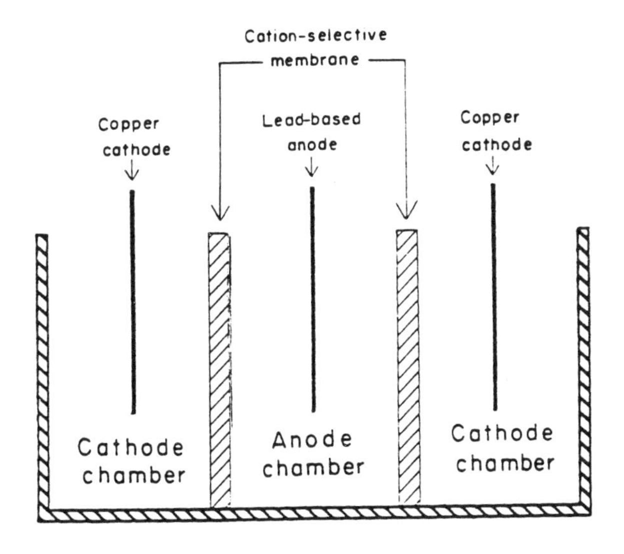

Figure 5.4.3. Diagram of an electrolytic electrodialysis cell.

Source: Reference 6.

the membrane. Thus, filtration generally needs to be employed as a pretreatment step to avoid plugging of the membrane.[5] In addition, oxidizing substances will attack the membranes and must be chemically removed.[7]

Post-Treatment Requirements--

Post-treatment requirements for electrodialysis systems are minimal because it is a closed-loop operation. Electrodialysis generally creates a concentrated solution such that an evaporation/concentration unit is not required.[1] ED units do not need to be regenerated and are able to operate continuously.[1]

Electrodialysis removes ionic species nonselectively. This may cause ionic impurities to be returned to the plating baths in cases where there is more than one contaminant ion present in the solution.[1,7] Therefore, periodic treatment of the plating baths using an ion exchange technique may be required to remove impurities.[1]

As discussed in Section 5.4.2, the applicability of the electrolytic ED process may be limited if the contaminant metal ions do not electrodeposit homogeneously on the cathodes.[8,9] Difficulty in removing the metal from the cathode may make the process economically inefficient.

5.4.2 Process Performance

Factors to consider when evaluating the applicability of an ED system for a particular use include types and concentrations of constituents present in the spent solution, required product concentration, required throughput rates, and spent bath temperature. Factors which affect the performance of a system for a particular application include: the chemical, physical, and electrical resistance of the membrane for the spent solution; the number, type, and concentrations of metal ion contaminants present in the spent solution; the ability of the contaminant ions to electrodeposit on the cathode; the ease of removal of the metal contaminants from the cathode; and the ability of the system to maintain satisfactory product concentration.

Research was conducted using a concentrating-diluting electrodialysis unit to recover chromic acid from chrome plating rinses.[4] Preliminary testing was performed using a laboratory-scale (five cell-pairs) ED unit to determine the chemical resistance of the membrane and to optimize current densities for maximum product concentration.[4]

Chemical resistance was tested by immersing the membranes at room temperature in a chromic acid plate solution and removing sections after 7, 14, 42, and 69 days.[4] The membranes appeared to be unaffected by the acid, with the exception of a slight roughening of the surface. The membranes did not exhibit a loss in physical strength or exchange capacity and remained leak-tight.

Electrodialysis was performed on a simulated chromic acid rinse solution (chromium trioxide dissolved in tap water) over a range of current densities.[4] Samples were collected after several hours operation at each operating condition. The results are presented in Table 5.4.1. The data indicate that chromic acid from the rinsewater could be concentrated to 70 percent original strength. The maximum chromic acid concentration achieved was approximately 175 g/L.[4] As expected, the product/feed ratio decreased with increasing feed concentration.

Following the laboratory testing, a 50 cell-pair demonstration module was constructed for a chrome plating line at the Seaboard Metal Finishing Company in West Haven, Connecticut.[4] Figure 5.4.4 presents a flow diagram for the system. Operating parameters for this series of tests are presented in Table 5.4.2. Objectives of the demonstration were to test the chemical resistance of the membranes, determine proper operating parameters for optimum chrome recovery, and to familiarize company personnel with its operation.[4]

Results similar to those obtained in the laboratory-scale tests were obtained. However, problems were encountered with high rinse temperatures due to heat generated from the electrical current and pumps.[4] During much of the operation, rinse temperatures were higher than 100°F, the maximum recommended operating temperature. The high temperatures caused extensive slippage of the stack spacers and membranes in the ED cells which led to external and internal leakage problems. A significant drop in product concentration occurred as a result of the leakage.

TABLE 5.4.1. LABORATORY-SCALE ELECTRODIALYSIS OF SIMULATED CHROMIC ACID RINSES OVER A RANGE OF CURRENT DENSITIES

Run	Current density (ma/cm^2)	Feed concentration g/L CrO_3	Product concentration g/L CrO_3	Product/ rinse ratio
1	10	0.37	61	165
2	12	0.32	106	331
3	14	1.24	143	115
4	16	0.98	167	170
5	18	0.70	174	249

Source: Reference 4.

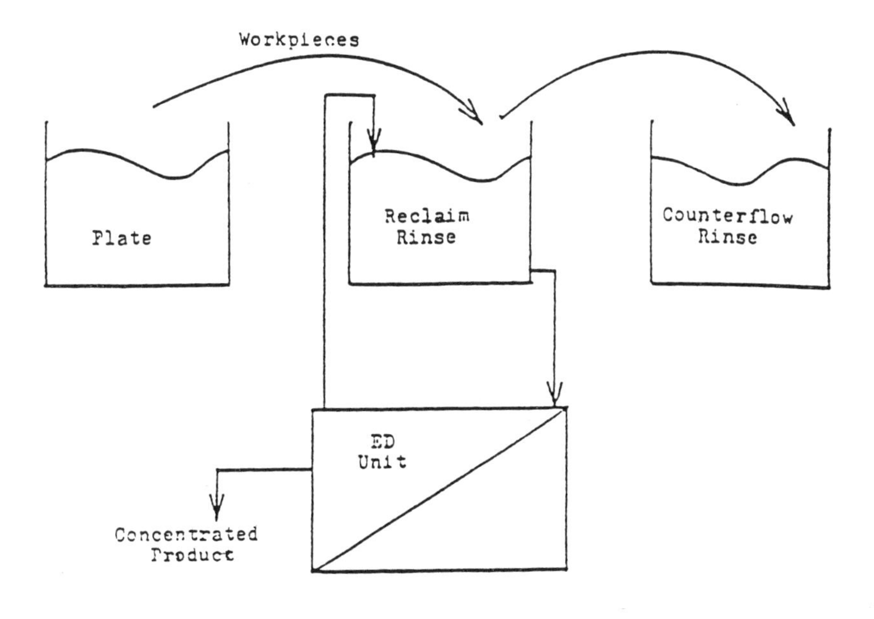

Figure 5.4.4. Flow diagram for electrodialysis treatment of chrome line at Seaboard Metal Finishing Company.

Source: Reference 4.

TABLE 5.4.2. OPERATING PARAMETERS FOR ELECTRODIALYSIS DEMONSTRATION UNIT FOR RECOVERY OF CHROMIC ACID AT SEABOARD METAL FINISHING COMPANY

Parameter	Measurement
Operating time	250 hrs over > 2 month period (day shift)
Maximum current	20 to 21 amps
Product concentration	160 to 212 g/L chromic acid
Rinse concentration	50 to 70 g/L CrO_3
Product/feed concentration ratios	2 to 4
Maximum observed rinse temperature	118°F
Maximum recommended rinse temperature	100°F

Source: Reference 4.

Figure 5.4.5 shows the product concentrations achieved during operation of the ED unit as a function of operating current. As expected, product concentrations increased with increasing current densities. However, as shown in Figure 5.4.6, the increasing rinse temperatures and the resulting damage to the ED cells caused a sharp drop in product concentrations as the testing progressed.[4]

As a result of the leakage problems, operations were temporarily ceased to evaluate the system and to make improvements. It was determined that product concentrations could be increased by: (1) running the ED unit concurrently with plating shifts; (2) increasing the membrane area to increase recovery capacity of the stack; and (3) adding a cooler to the rinse so that higher current densities could be explored.[4] However, results from the second phase of tests are not available because the project was never completed.[10]

An electrodialytic process has recently been developed by Aquatech Systems (Bethel, New Jersey) for the recovery of hydrofluoric/nitric acid pickling liquors.[11,12] With this process, electrodialysis is used following neutralization of the waste acid with potassium hydroxide. Thus, membrane degradation is minimized. The potassium hydroxide neutralization forms a potassium fluoride/nitrate solution which can be filtered with diatomaceous earth to remove metal hydroxides.[11] The filtered solution is then directed to the electrodialysis stack where, upon application of an electric potential, a 3 M stream of mixed hydrofluoric/nitric acid and a 2 M potassium hydroxide solution are produced.[11,13] The HF/HNO_3 can be recycled back to the pickling bath and the KOH solution can be reused in the neutralization step. Typical operating parameters for this process are summarized in Table 5.4.3.

Innova Technology, Inc. has developed ion transfer electrodialysis units for commercial application in the recovery of chromic acid from plating rinse baths. Typical results are presented in Table 5.4.4. Although more concentrated solutions (such as the actual plating bath) can be fed to the unit without causing degradation of the resin, much larger systems would be required to achieve satisfactory product concentration.[14] At the present time, these systems would not be cost-effective.[14]

The Bureau of Mines has conducted research using an electrolytic membrane cell system to recover chromic acid and sulfuric acid from spent brass

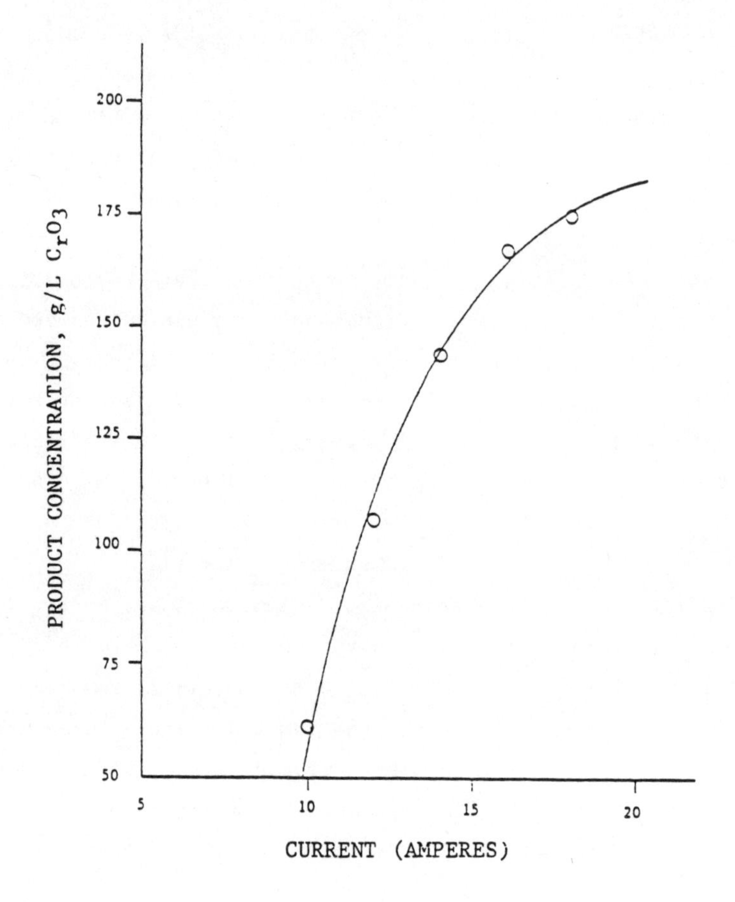

Figure 5.4.5. Relationship between current and product concentration during operation of ED unit for chromic acid recovery.

Source: Reference 4.

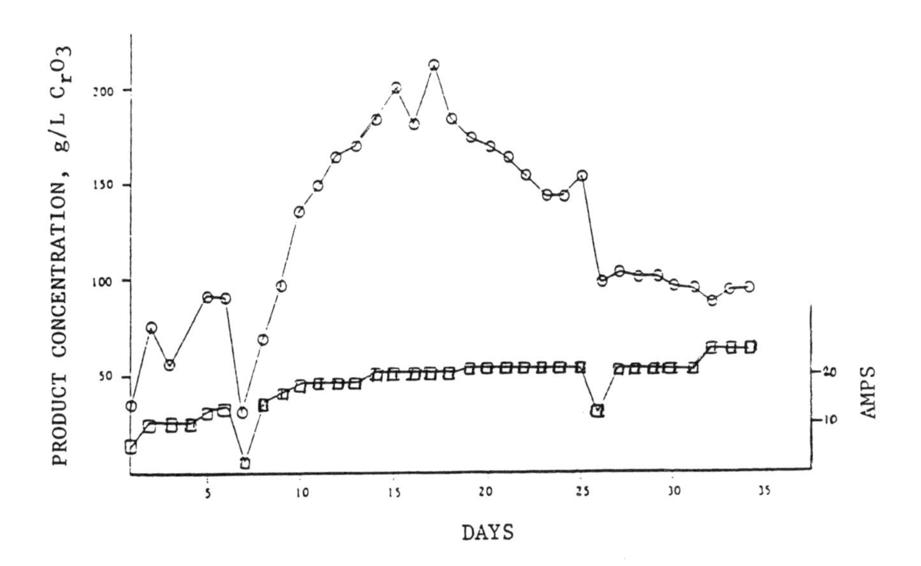

Figure 5.4.6. Production concentrations achieved during operation of the pilot scale ED unit at Seaboard Metal Finishing Company.

Source: Reference 4.

TABLE 5.4.3. TYPICAL OPERATING PARAMETERS AND RESULTS FOR THE AQUATECH ED PROCESS

Parameter	Result
Operating temperature	80 to 100°F
Current density	70 to 100 amp/ft^2
Pressure drop	5 psi
Voltage	2.0 to 2.5 volts/cell
Feed solution	
Conductivity	>100,000 mohms
Maximum particle size	<10 microns
Maximum metal concentrations	
Chromium	<1 to 2 ppm
Nickel	<1 to 2 ppm
Iron	<1 to 2 ppm
Calcium	<0.1 ppm
Magnesium	<0.1 ppm
Molybdenum	<70 ppm
Product concentrations	
Hydrofluoric/nitric acid	3 M
Potassium hydroxide	2 M

Source: Reference 13.

TABLE 5.4.4. TYPICAL OPERATING PARAMETERS FOR ION TRANSFER SYSTEM DEVELOPED BY INNOVA TECHNOLOGIES, INC.

Parameter	Result
Cell size	18 in. x 27.6 in. x 2.5 in. (46 cm x 69 cm x 6.4 cm)
Voltage	25 Vdc
Product concentration	15%
Bath temperature	112°F (45°C)
Rinse temperature	80 to 85°F (27 to 30°C)

Source: References 14 and 15.

etchants.[5] Their pilot-scale unit, termed the Process Research Unit (PRU), is designed for online regeneration. The PRU is able to use one to five anode membrane cells in single catholyte tank.

Spent etchant solution is pumped from the bottom of the etchant tank through a filter into a holding tank. Etchant from the holding tank is pumped into the anode chambers of the PRU. An electric potential is applied, causing migration of copper and zinc through the cation-selective membranes and into the catholyte (180 g/L sulfuric acid). Trivalent chromium (Cr^{+3}) is oxidized to hexavalent chromium (Cr^{+6}) in the anode chamber. Metal contaminants are recovered at the cathode, and the etchant solution is regenerated in the anode chamber to a level comparable to fresh etchant. The regenerated etchant is then pumped directly back to the etchant tank.

The condition of the PRU is monitored by a control panel, which includes an ampere-hour meter, a voltage meter, and an ampere meter. A fume hood located directly above the diaphragm cell is used to remove hydrogen, oxygen, and chromic acid mist.

Preliminary testing of the PRU was conducted with samples from brass etchant solutions.[5,16] Two anode chambers were used for Tests 1 and 2, and five anode chambers were used for Test 3. The test results, presented in Table 5.4.5, indicate that sodium dichromate production increases with increasing flow rate.

Following the preliminary tests, online testing of the Bureau of Mines PRU was performed at the Valve and Fitting Division of Gould Manufacturing, Inc. (now known as Impervial Clevite, Inc.) in Niles, Illinois.[17,18,19,20] The PRU was operated over a 17-day period, which included 11 days when two 10-hour shifts were operated and three weekends when the plant was shut down. Operating parameters recorded during the testing period and test results are summarized in Table 5.4.6. The PRU proved to be technically and economically effective. The PRU regenerated and recycled the chromic acid/sulfuric acid etching solution at an acceptable level of performance and reduced sodium dichromate consumption, disposal volumes, and process operating costs.

The Bureau of Mines research led to the development of a commercial-scale electrolytic ED unit, which is currently available from Scientific Control, Inc. in Chicago, Illinois.[21,22] The unit is applicable to the recovery of chromic acid/sulfuric acid etching solutions in which copper is

TABLE 5.4.5. SUMMARY OF RESULTS OF PRELIMINARY TESTING OF PRU

Parameter	Test 1	Test 2	Test 3
Number of anode chambers used	2	2	5
Flow rate (liters/hr)	2.6	5.7	7.3
Cr^{+3} oxidation (percent)	>96	88	92
Cu removal (percent)	41	20	31
Zn removal (percent)	41	23	28
Cr lost to catholyte (percent)	11	4	5
Energy consumption (KWH/kg sodium dichromate)	8.6	5.7	8.0
Sodium dichromate produced (kg/hr)	0.43	0.70	0.48
Anode efficiency (percent)	19	31	21
Duration of run (days)	3	3	7

Source: Reference 5.

TABLE 5.4.6. SUMMARY OF OPERATING PARAMETERS AND RESULTS OF BUREAU OF MINES TESTS USING ELECTROLYTIC ED TO TREAT SPENT CHROMIC ACID-SULFURIC ACID ETCHANTS

Parameter	Result
Total operating period	17 days (220 hrs)
Flow rate	
First 5 days	26.5 L/hr (7 gal/hr)
Remaining 12 days	41.6 L/hr (11 gal/hr)
Etchant (working) volume	378.5 L (100 gal)
Residence time	
First 5 days	15 hours
Remaining 12 days	9 hours
Cathode current density	215 amp/sq meter (20 amp/sq ft)
Average applied current	1,181 amps
Average applied potential	3.77 volts
Total electric power consumption	1,518 KW-hr
Products generated	
Sodium dichromate	40.0 kg (88.3 lb)
Copper	19.8 kg (43.8 lb)
Zinc	8.6 kg (18.9 lb)
Percent removal	
Copper	40.9 percent
Zinc	21.1 percent
Percent conversion of trivalent chromium to hexavalent chromium	81.2 percent
Overall current efficiency	11.2 percent

Source: Reference 17.

the primary metal contaminant. Companies using the systems have reported product quality equivalent to that of fresh make-up solution.[21] Typical operating parameters for the system are listed in Table 5.4.7. The system operates most efficiently with acid baths containing copper concentrations of 2 to 4 oz/gal.[21]

Additional studies have been performed using the Electrolytic ED unit to recover acids contaminated with iron.[8,9,23] At present, difficulties in removing the iron metal from the cathode have made the process economically infeasible. However, research is currently being performed at the U.S. Bureau of Mines to address this problem.[8,23] Preliminary results are expected in late 1986.[8]

In summary, at the present time the concentrating-diluting and the ion transfer types of electrodialysis units are only technically feasible for treatment of dilute solutions; e.g., plating rinse waters. A recently developed ED process is able to recover HNO_3 and HF from pickling liquors that have been neutralized with KOH. However, limited performance data is available for this process. Electrolytic ED units are capable of treating concentrated solutions. However, the application of electrolytic ED units is currently limited to the recovery of solutions which contain one predominant metal ion contaminant that can be homogeneously electroplated and easily removed from the cathode. Performance data have demonstrated electrolytic ED units to be effective in the removal of copper ions from acid baths. However, difficulties have been encountered with the removal of iron metal from the cathode.

TABLE 5.4.7. TYPICAL OPERATING PARAMETERS FOR SCIENTIFIC CONTROL, INC. ELECTROLYTIC ED UNIT FOR RECOVERY OF BRASS ETCHANTS

Parameter	Typical value
Temperature	70°F
Voltage requirements	4 to 6 volts
Current density	30 amps/ft^2
Optimum copper concentration	2 to 4 oz/gal

Source: Reference 21.

5.4.3 Costs

Typical costs for electrodialysis systems to treat plating rinsewaters range from $30,000 to $45,000, depending on the application.[1] Capital costs for the "Chrome Napper" system available from Innova Technology, Inc. in Clearwater, Florida, range from $9,900 to $30,000, including installation and power supply.[14] Systems are sized according to bath temperature, dragout concentrations, number of rinse tanks, concentration of the bath, and the volume of spent solution to be treated per unit time.

The ED system developed by Aquatech Systems is generally only cost-effective for large quantity generators.[13] An economic evaluation for a typical application is presented in Table 5.4.8. Systems are also available for leasing at the costs listed in Table 5.4.9.

Scientific Control, Inc. sells an electrolytic ED unit to recover chromic/sulfuric acid brass etchants.[21] System sizes are based on the amount of copper the system is capable of removing per unit of time. Available unit sizes range from 0.05 to 0.5 lb copper removal/hour. Capital equipment costs for these units range from $24,000 to $80,000.[21] These costs do not include installation which would include a hoist, plumbing, and a ventilation/exhaust system. Additional costs for the exhaust system cost could range from $5,000 to $15,000, depending on the size required. Operating and maintenance costs are relatively low. Membranes will need to be replaced approximately every 9 months, depending on usage, at a replacement cost of approximately 10 to 15 percent of the original equipment costs. Additional maintenance costs will include approximately $10/month for replacement of filter cartridges (a pre-filter system is incorporated into the unit). The estimated payback period for the system is approximately 2 years, based on savings in treatment and disposal costs.[21]

The Bureau of Mines conducted an economic analysis of electrolysis versus conventional neutralization/disposal for treatment of spent chromic/sulfuric acid etching solutions.[24] Capital costs will depend upon the size of the system employed. Table 5.4.10 presents fixed capital and installation labor costs for 500 and 1,000 gallon catholyte holding tank capacities.

TABLE 5.4.8. ECONOMIC EVALUATION OF AQUATECH ED SYSTEM TO REGENERATE 1.5 MILLION GALLONS OF HF/HNO_3 PICKLING LIQUOR PER YEAR

Item	Cost ($)
Capital costs (installed system)	1.5 to 2.0 million
Operating costs (includes electricity, chemical make-up, and cell maintenance)	415,000/year
Cost savings (includes HF and HNO_3 purchase costs, and waste disposal savings)	808,000/year
Net savings	393,000/year
Payback period	3.8 to 5.1 years

Source: Reference 13 (Aquatech cost data August 1986).

TABLE 5.4.9. COSTS FOR LEASING AQUATECH SYSTEM

Description	Size	Minimum lease period	Cost ($)
Bench-scale cell stack	6 in. x 8 in., 8 cells, effective area = 1 sq ft	3 months	1,000/month
Pilot-scale cell stack	1 ft x 2 ft, 50 cells, effective area = 50 sq ft	6 months	5,000/month
Pilot plant	Full scale, effective area = 50 sq ft	4 months	10,000/month

Source: Reference 25 (Aquatech cost data January 1986).

TABLE 5.4.10. ESTIMATED FIXED CAPITAL COSTS FOR ELECTROLYTIC RECOVERY UNIT

Item	1,000 gal cell ($)	500 gal cell ($)
Electrolytic cell	65,700	42,000
Hoist	3,400	3,400
Blower	1,000	1,000
Ductwork	100	100
Electrical	1,800	1,100
Piping	200	200
Total:	72,200	47,800

Source: Reference 24 (Based on 1983 cost data).

Estimated annual operating costs for the Bureau of Mines electrolytic process are presented in Table 5.4.11. These costs are based on an average of 350 days of operation per year over the life of the equipment. It is assumed that maintenance activities, such as copper removal and occasional system checks, will require a minimal amount of time and can be performed by existing employees. Equipment depreciation is based on a straight-line 20-year period. The following cost savings were noted in the Bureau of Mines economic evaluation:[24]

- better product performance,
- reduced waste solution treatment and disposal costs,
- reduced sodium dichromate consumption,
- reduced drag-out losses, and
- copper recovery.

Table 5.4.12 presents quantitative estimates of the cost savings for a 1,000 gallon electrolytic unit operating 350 days per year. Figures 5.4.7 and 5.4.8 show the effects of variable costs on the operating costs for 500 and 1,000 gallon systems. The cost savings presented in Table 5.4.12 do not include the value of recovered copper, which may be marketed, or benefits resulting from improved product quality. Inclusion of these factors would decrease the payback period slightly.[24]

5.4.4 Process Status

Currently, electrodialysis has a limited area of application in the recovery/reuse of corrosive wastes. Concentrating-diluting and ion transfer ED units have been successful in the recovery of chromic acids from dilute solutions. The electrolytic ED unit has shown success in removing copper ions from spent chromic/sulfuric acid brass etchants and bright dipping solutions. Another recently developed application uses electrodialysis in conjunction with neutralization to recover spent hydrofluoric/nitric acid pickling liquors.

TABLE 5.4.11. ESTIMATED ANNUAL OPERATING COSTS FOR ELECTROLYTIC RECOVERY UNIT

Item	1,000 gal cell ($)	500 gal cell ($)
DIRECT COST		
Utilities		
Electric power (@ 0.045 $/KW-h)	5,500	2,800
Total Utilities:	5,500	2,800
Plant Maintenance		
Labor	600	300
Materials	600	400
Total plant maintenance:	1,200	700
Payroll overhead (33% of payroll)	200	100
TOTAL DIRECT COST	6,900	3,600
FIXED COST		
Taxes (1% of total capital cost)	700	500
Insurance (1% of total capital cost)	700	500
Depreciation (20-year life)	3,600	2,400
TOTAL OPERATING COST	11,900	7,000

Source: Reference 24 (Based on 1983 cost data).

TABLE 5.4.12. COST COMPARISON OF ELECTROLYTIC UNIT VS. CONVENTIONAL TREATMENT

Item	Cost
Electrical consumption/lb of sodium dichromate regenerated	3.6 KW-hr
Sludge reduction/lb of sodium dichromate regenerated	9.4 gal
Reduction in sodium dichromate consumption for a 1,000 gal system	100 lb/day
Sludge reduction for a 1,000 gal system	940 gal/day
Cost for sludge treatment and disposal	0.20 $/gal
Cost for sodium dichromate purchase	0.68 $/lb
Total cost savings	256 $/day
Electric power cost for electrolytic unit	0.045 $/Kw-hr
Electric power cost for 1,000 gal system (not including taxes and depreciation)	20 $/day
Net savings	236 $/day
Payback period for 1,000 gal system	0.84 years (10 months)

Source: Reference 24 (Based on 1983 cost data).

The primary application of electrodialysis is in the treatment of metal wastes. The only commercially available electrodialysis system for the direct recovery/reuse of corrosive wastes is an electrolytic ED unit sold by Scientific Control, Inc. in Chicago, Illinois.[21] Development of the unit was based on research performed by the U.S. Bureau of Mines. It is only capable of treating brass etchants or bright dipping solutions where copper is the primary metal contaminant. As a result of its limited proven area of application, only four units are currently in operation. The companies using the system have reported satisfactory results.

The Aquatech ED process, developed for the recovery of hydrofluoric/ nitric acid pickling liquors that have been neutralized with potassium hydroxide, has recently been made available for purchase or lease. However, no commercial systems have been installed to date. Additionally, the system is only cost-effective for large quantity generators.[12,13]

Innova Technology, Inc. in Clearwater, Florida sells an ED system which uses ion transfer technology to recover chromic acid from chromic acid plating rinsewaters.[14] The system, called the "Chrome Napper", has been employed cost-effectively by a number of companies to recover chromic acid plating bath of rinse water. However, this system is not cost-effective for direct treatment of the plating bath.

Current research in the application of electrodialysis to the treatment of corrosive wastes is directed at using the electrolytic ED configuration. The U.S. Bureau of Mines in Rolla, Missouri is experimenting with techniques to recover other types of acid baths. They are currently bench-testing various techniques for the removal of iron from nitric acid/hydrofluoric acid pickling liquors.[8,20]

In addition, Ionics, Inc. in Watertown, Massachusetts is currently developing improved ED membranes for use with corrosive wastes. However, their research is only at the bench scale at this time, and preliminary results have not yet been released.

In summary, the predominant use of ED units is in the recovery/reuse of rinsewaters rather than concentrated wastes.[2] Electrodialysis can be used to concentrate acids for recovery from acidic process wastewaters to reduce neutralization and disposal costs. Research and commercial demonstrations have shown the technical and economic feasibility of regenerating and recycling spent brass etchants using electrolytic ED units.[5] This research has also indicated that ED units show potential applications in the recovery of other corrosive wastes.

REFERENCES

1. Higgins, T. C. CH2MHILL. Industrial Processes to Reduce Generation of Hazardous Waste at DOD Facilities - Phase 2 Report, Evaluation of 18 Case Studies. Prepared for the DOD Environmental Leadership Project and the U.S. Army Corps of Engineers. July 1985.

2. Radimsky, J., Daniels, D. I., Eriksson, M. R., and R. Piacentini. California Department of Health Services. Recycling and/or Treatment Capacity for Hazardous Wastes Containing Dissolved Metals and Strong Acids. October 1983.

3. Jain, S. M. Ionics, Inc. Electrodialysis and Its Applications. American Laboratory. October 1979.

4. Eisenmann, J. L. Membrane Processes for Metal Recovery From Electroplating Rinse Water. In: Proceedings of the 2nd Conference on Advanced Pollution Control for the Metal Finishing Industry, Co-sponsored by the American Electroplaters Society and the U.S. EPA, Kissimmee, Florida, February 5-7, 1979. EPA-600/8-79-014. May 1979.

5. George, L. C., Soboroff, D. M., and A. A. Cochran. Regeneration of Waste Chromic Acid Etching Solutions in an Industrial Scale Research Unit. In: Third Conference on Advanced Pollution Control for the Metal Finishing Industry, sponsored by the U.S. EPA, Cincinnati, Ohio. EPA-600/2-81-028. 1981.

6. Soboroff, D. M., Troyer, J. D., and A. A. Cochran. Regeneration and Recycling of Waste Chromic Acid-Sulfuric Acid Etchants. Bureau of Mines Report of Investigations No. 8377. 1979.

7. Laughlin, R.G.W., Forrestall, B., and M. McKim. Ontario Research Foundation. Technical Manual: Waste Abatement, Reuse, Recycle, and Reduction Opportunities in the Metal Finishing Industry. Environment Canada, Toronto, Ontario. January 1984.

8. Horter, G. L. U.S. Bureau of Mines, Rolla, Missouri. Telephone conversation with Lisa Wilk, GCA Technology Division, Inc. August 29, 1986.

9. Southern Research Institute, Birmingham, Alabama. Electromembrane Process for Regenerating Acid from Spent Pickle Liquor. Project No. 12010-EOF, submitted to the U.S. EPA Water Quality Office. March 1971.

10. Wheeler, C. Seaboard Metal Finishing Company, West Haven, Connecticut. Telephone conversation with Jon Spielman, GCA Technology Division, Inc. August 1986.

11. Basta, N. Use Electrodialytic Membranes for Waste Recovery. Chemical Engineering. March 3, 1986.

12. Rodgers, B. Aquatech Systems, Bethel, New Jersey. Telephone conversation with Jon Spielman, GCA Technology Division, Inc. August 11, 1986.

13. Aquatech Systems. Bethel, New Jersey. Product Literature. Received August 1986.

14. Pouli, D. Innova Technology, Clearwater, Florida. Telephone conversation with Lisa Wilk, GCA Technology, Inc. August 26, 1986.

15. Industrial Finishing Staff. Ion Transfer Recovers Chrome. Industrial Finishing. March 1981.

16. McDonald, H. O., and L. C. George. Recovery of Chromium From Surface-Finishing Wastes. Bureau of Mines Report of Investigations No. 8760. 1983.

17. Horter, G. L., and L. C. George. Demonstration of Technology to Recycle Chromic Acid Etchants at Gould, Inc. In: Proceedings of the 4th Recycling World Congress and Exposition. 1982.

18. Soboroff, D. M., Troyer, J. D., and A. A. Cochran. U.S. Bureau of Mines, Rolla, Missouri. Regeneration of Waste Metallurgical Process Liquor. U.S. Patent No. 4, 337, 129. June 29, 1982.

19. Herdrich, W. J. Recovery of Acid Etchants at Impervial Clevite, Inc. In: Fourth Conference on Advanced Pollution Control for the Metal Finishing Industry, sponsored by the U.S. EPA. EPA-600/9-82-022. 1982.

20. George, L. C., Rogers, N. L., and G. L. Horter. Recovery and Recycling of Chromium-Bearing Solutions. In: Proceedings of the 29th National SAMPE Symposium. April 3-5, 1984.

21. Gary, S. Scientific Control, Inc., Chicago, Illinois. Telephone conversation with Lisa Wilk, GCA Technology Division, Inc. August 29, 1986.

22. Altmayer, F. Introducing the Cops: One-Step Regeneration and Purification of Chromic Acid Pickling/Stripping Solutions. Plating and Surface Finishing. March 1983.

23. Horter, G. L., Stephenson, J. B., and W. M. Dressel. Permselective Membrane Research for Stainless Steel Pickle Liquors. In: Proceedings of the International Symposium on Recycle and Secondary Recovery of Metals; sponsored by the Metallurgical Society of AIME, Warrendale, Pennsylvania; held in Fort Lauderdale, Florida. December 1-4, 1985.

24. Spotts, D. A. Economic Evaluation of a Method to Regenerate Waste Chromic Acid-Sulfuric Acid Etchants. Bureau of Mines Information Circular No. 8931. 1983.

25. Aquatech Systems. Bethel, New Jersey. Information on Leasing Program. January 21, 1986.

5.5 REVERSE OSMOSIS

5.5.1 Process Description

Reverse osmosis (RO) is a treatment technique used to remove dissolved organic and inorganic materials, and to control amounts of soluble metals, TDS, and TOC in wastewater streams.[1,2] The technology has been applied in the metal finishing industry to recover plating chemicals from rinse water, such that both plating chemicals and rinsewaters can be reused.

RO involves passing the wastewater through a semipermeable membrane at a pressure greater than the osmostic pressure caused by the dissolved materials in the solvent.[2,3,4] Thus, the osmotic flow, defined as the flow from a concentrated solution to a dilute solution, is reversed due to the increase in pressure applied to the system.

The application of RO to treatment of corrosive wastes is generally limited by the pH range in which the membrane can operate. Only a limited number of membranes are applicable for recovery of corrosive plating and other solutions. Table 5.5.1 summarizes the characteristics of these commercially available membranes.

The semipermeable membranes can be fabricated either in the form of a sheet or tube, which is then assembled into modules.[3,9] Figure 5.5.1 shows the three basic module designs, which include:[10]

- Tubular - a porous tubular support with the membrane cast in place, or inserted into the tube. Feed is pumped through the tube; concentrate is removed downstream; and the permeate passes through the membrane/porous support composite.
- Spiral Wound - large porous sheet(s) wound around a central permeate collector tube. Feed is passed over one side of the sheet, and the permeate is withdrawn from the other.
- Hollow Fiber - thousands of fine hollow fiber membranes (40 to 80 μm diameter) arranged in a bundle around a central porous tube. Feed enters the tube, passes over the outside of the fibers, and is removed as concentrate. Water permeates to the inside of the fibers and is collected at one end of the unit.

TABLE 5.5.1. COMMERCIALLY AVAILABLE RO MEMBRANES APPLICABLE TO THE TREATMENT OF CORROSIVE WASTES

Membrane type	Membrane description	Source	Optimal pH range[a]	Operating pressure (psig)	Module replacement cost ($)
RC-100	Flat Sheet Composite Membrane of Polyether/Amide on Polysulfone, rolled into Cartridge	Fluid Systems Division of UOP, Inc. (San Diego, CA)	1 to 12	400 to 800	1,000
PA-300 (TFC-PA)	Polyether/Amide Spiral-Wound	Fluid Systems Division of UOP, Inc. (San Diego, CA)	1 to 7	400 to 800	--
-	Thin Film Spiral-Wound Polyamide	Desalination Systems, Inc. (San Diego, CA)	1 to 7	400 to 800	560
-	Thin Film Spiral-Wound Polyamide	Desalination Systems, Inc. (San Diego, CA)	4 to 11	400 to 800	320
-	Spiral-Wound Polyamide	Osmonics, Inc. (Minnetanka, MN)	1 to 12	200 to 800	--

[a]Treatment of solution with pH outside of this range is possible, but will result in more rapid deterioration of the membrane.

Source: References 5,6,7, and 8.

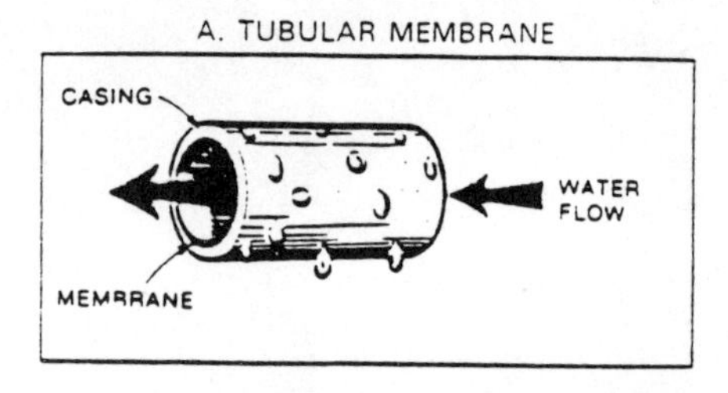

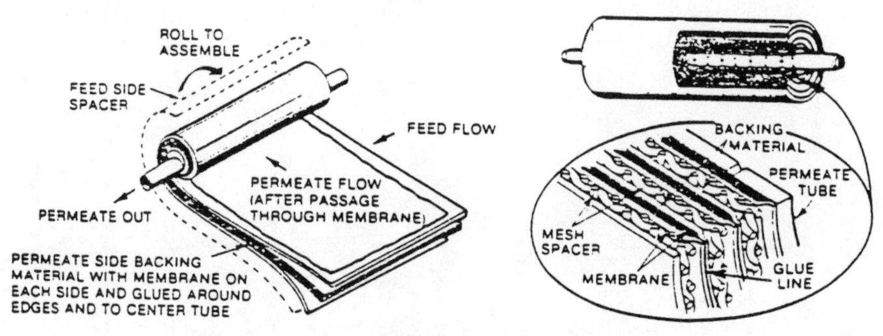

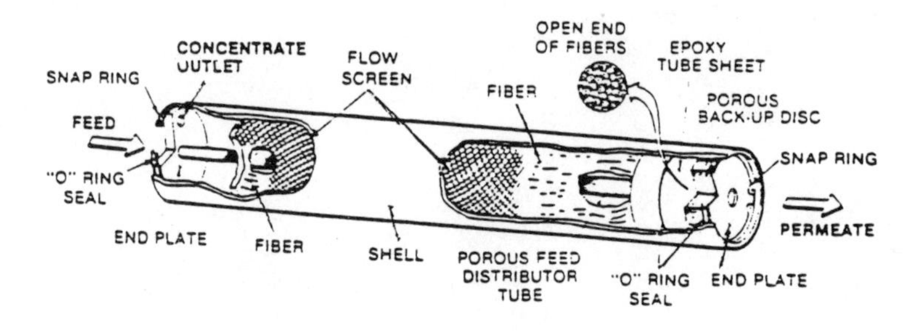

Figure 5.5.1. Reverse osmosis membrane module configurations.
Source: Reference 3.

Although they are able to operate at higher pressures, tubular modules are not applicable for most industrial applications because of large floor space requirements and high capital costs as compared with other modules. Hollow fiber and spiral wound modules have similar capital costs. Hollow fiber modules require less floor space, but spiral wound modules are not as susceptible to plugging by suspended solids.[11,12]

Reverse osmosis systems typically consist of a number of modules connected in series or parallel, or a combination of both arrangements. In a series arrangement, the reject stream from one module is fed directly to another module, such that greater product concentration is achieved.[4] Alternatively, the reject stream may be recycled to the feed stream of the same RO unit. Series treatment may be limited in some cases by the ability of the membrane to withstand concentrated contaminants.[4] The system capacity can be increased through the use of a parallel arrangement of modules; however, product quality will not be enhanced.[4] Schematic flow diagrams of series and a parallel systems are shown in Figure 5.5.2.

Operating Parameters--

The operation of a reverse osmosis system is affected primarily by the feed characteristics, operating pressure, and membrane type. These factors will affect the flux and percent rejection, which in turn, define system size requirements and effluent quality, respectively.

Flux determines the system size for a given waste flow rate; i.e., higher flux permits the use of smaller RO systems. Flux is the volume flow of permeate per unit membrane area. It is proportional to the effective pressure driving force, according to the following relationship:[10]

$$J = K\ (\Delta P - \Delta D)$$

where J=flux, K=membrane constant; ΔP=difference in applied pressure across the membrane; and ΔD=difference in osmotic pressure across the membrane.

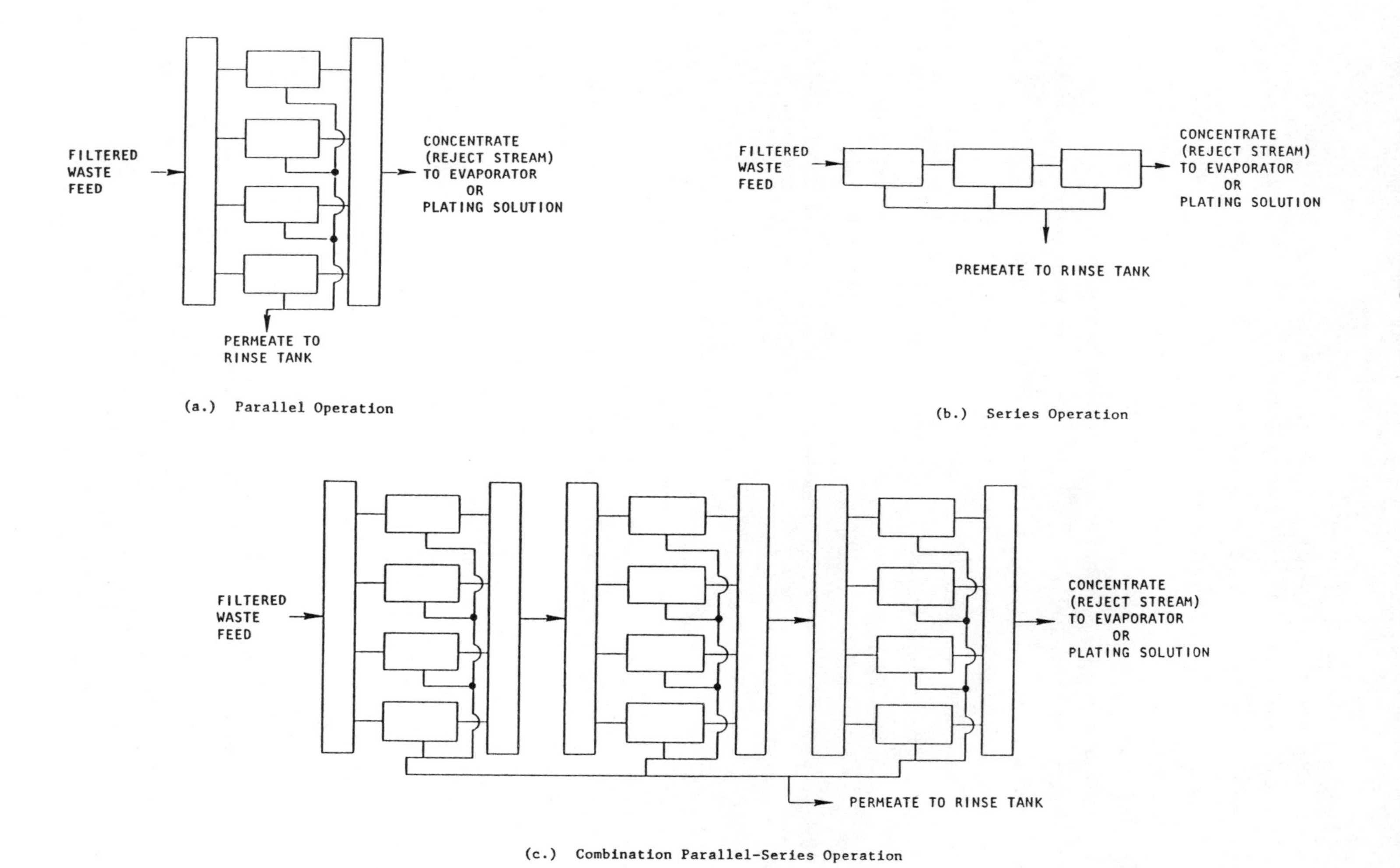

(a.) Parallel Operation

(b.) Series Operation

(c.) Combination Parallel-Series Operation

Figure 5.5.2. Reverse Osmosis Module Arrangements.
Reference: No. 4

Since osmotic pressure is approximately proportional to molar feed concentration, flux increases with increasing operating pressure, and it decreases with increasing feed concentration. Thus, chemicals which form high-molecular weight complexes will have higher flux for a given weight percent in solution.

More concentrated solutions can be achieved by utilizing a large effective driving pressure. Increases in temperature of the waste feed will also increase the flux by lowering viscosity.[13] However, although increased operating temperatures will improve the performance of the system in the short-term, the lifetime of the membrane will be shortened.

Percent rejection is defined as follows:[10]

$$\% - \text{Rejection} = \frac{(\text{feed concentration}) - (\text{permeate concentration})}{\text{permeate concentration}} \times 100\%$$

Higher percent rejections will result in better quality (higher purity) of the permeate and concentrated streams. Percent rejection is primarily affected by the membrane type although rejection will decrease with increasing feed concentration.[10]

Pretreatment Requirements/Restrictive Waste Characteristics--

Colloidal and organic matter can clog the membrane surface, thus reducing the available surface area for permeate flow. Also, low-solubility salts will precipitate on the membrane and similarly reduce membrane efficiency. Pretreatment techniques such as activated carbon absorption, chemical precipitation or filtration may be required to ensure extended service life. Operating costs for membrane systems are a direct function of the concentration of the impurity to be removed, due in part to increased maintenance and membrane replacement costs.

Multi-charged cations and anions are easily removed from the wastewater with this technique. However, most low molecular weight, dissolved organics are, at best, only partially removed with this method. The use of reverse osmosis for recovery/reuse of corrosive process wastes is currently somewhat limited because RO membranes are attacked by solutions with a high oxidation potential (e.g., chromic acid) or corrosive pH levels.[5] However, future development of membranes which are able to withstand corrosive and oxidizing solutions is expected.[5]

Post-Treatment--

Reverse osmosis applied to plating bath wastes is usually supplemented with an evaporation system in order to adequately concentrate constituents for reuse.[5] Increases in concentration of corrosive constituents will tend to degrade the membrane. Thus, the amount of feed concentration permitted in an RO unit is limited by the membrane characteristics. RO units can concentrate most divalent metals (nickel, copper, cadmium, zinc, etc.) from rinse waters to 10 to 20 percent of the bath solution.[5] Further concentration must be achieved through the use of a small evaporator. Evaporators are especially necessary for systems operated at ambient temperatures where atmospheric evaporative losses are minimal.[5]

5.5.2 Process Performance

Reverse osmosis applications in recycling corrosive wastes is somewhat limited due to membrane degradation in the extreme pH regions. However, research has been conducted using RO to recover both acidic and basic plating rinsewaters, which has led to the development of more chemically resistant membranes.

The Walden Division of Abcor, Inc. (Wilmington, MA) conducted a number of studies of RO systems for recovery of plating rinsewaters. Initially, studies were conducted to test the applicability of membrane types to various plating rinses. Test samples were prepared by diluting actual plating bath solutions with deionized water.[10] Bath properties (total dissolved solids, pH) and test solution properties (concentration, pH) for the corrosive wastes* tested is presented in Table 5.5.2.

*The RO performance was evaluated with other wastes during this testing program; however, only the data for the corrosive wastes tested is presented in this section.

TABLE 5.5.2. SUMMARY OF THE CHARACTERISTICS OF THE CORROSIVE WASTES TREATED BY REVERSE OSMOSIS DURING PRELIMINARY TESTING CONDUCTED BY THE WALDEN DIVISION OF ABCOR, INC.

Plating bath	Source of bath	Bath properties		Test soln properties		Membrane modules tested[a]
		% - TDS	pH	Concentration range (%-TDS)	pH range	
Chromic Acid	Whyco Chromium Co.	27.5	0.53	0.4-9.0	0.9-1.9	A,B,C
Cadmium Cyanide	American Electroplating Co.	26.3	13.1	0.3-10.0	11.4-12.9	A
Zinc Cyanide	American Electroplating Co.	11.4	13.9	0.5-4.0	12.3-13.7	A
Copper Cyanide	American Electroplating Co.	37.0	13.3	0.6-8.0	11.8-12.9	A

Notes: [a]A = Dupont B-9 permeator, polyamide hollow-fiber membrane.
B = T.J. Engineering 97 H 32 spiral-wound module, cellulose acetate membrane.
C = Abcor TM 5-14 module, tubular configuration; cellulose acetate membrane.

Source: Reference 10.

The membrane performance is affected by feed constituent concentrations, operating pressure, operating temperature, flow rate, and pH. In tests performed by Abcor, the effects of pH were evaluated by comparing results of corrosive test solutions with neutralized test solutions.[10] Flux and rejection data were not affected by changes in pH, but extreme pH values were found to decrease the membrane life. Additional tests were performed to evaluate the effects of feed concentrations on flux and %-rejection. Corrosive waste rinse streams that were tested included: chromic acid, cadmium cyanide, zinc cyanide, and copper cyanide. In order to calculate these values, samples of feed, concentrate, and permeate were analyzed for TDS, temperature, pH, and chemical composition. The analytical test methods used are presented in Table 5.3.3. A summary of the operating parameters and results for these tests is presented in Tables 5.5.4 through 5.5.7. As can be seen from these tables, both flux and rejection decrease with increasing feed concentration.

Rejection and flux results were satisfactory for the chromic acid test runs, but hydrolysis and degradation of the membranes occurred. Three types of membranes were tested on the chromic acid test solution: hollow-fiber polyamide membrane, spiral-wound cellulose acetate membrane, and tubular cellulose acetate membrane. Although the tubular module showed the best results, large floor space requirements and high capital costs make this alternative less cost-effective. Only polyamide membranes could be used for testing of the cyanide plating rinses because the cellulose acetate membranes are rapidly degraded in the high pH range. Good results were observed for rejection values.

Additional tests were conducted on a pilot-scale level at American Electroplating Company in Worcester, MA to recover a zinc cyanide plating rinse.[13] Since the system operated at room temperature, an evaporator was used in conjunction with the RO unit in order to achieve the level of concentration necessary for closed-loop operation.

A schematic of the pilot system used to recover the zinc cyanide plating rinse is presented in Figure 5.5.3. Feed was pumped from the first plating rinse tank to a series of cartridge filters in parallel. Filters ranging from 1 to 20 m were used during the testing. After filtration, the feed was pressurized to 700 psi using a multi-stage centrifugal feed pump before being passed through spiral-wound RO modules arranged in series. The permeate from

TABLE 5.5.3. ANALYTICAL TEST METHODS USED DURING PRELIMINARY INVESTIGATIONS OF REVERSE OSMOSIS APPLICATIONS CONDUCTED BY WALDEN DIVISION OF ABCOR, INC.

Constituent	Test method	Procedure Reference No.
Cadmium	Atomic absorption	129
Total chrome	Atomic absorption	129
Hexavalent chrome	Colorimetric	108
Copper	Atomic absorption	129
Zinc	Atomic absorption	129
Cyanide	Titration	207B
Total organic carbon	Combustion-infrared	138A
Total dissolved solids	Gravimetric	148B

Source: Reference 10.

TABLE 5.5.4. ANALYTICAL RESULTS FOR REVERSE OSMOSIS TREATMENT OF SPENT CHROMIC ACID PLATING RINSE

Membrane module[a]	Waste feed		Operating conditions				% - Rejection	
	%-TDS	% of bath	Pressure (psig)	Temp. (°C)	pH of Feed	Flux[b]	Basis: TDS	Basis: Cr+6
Hollow fiber			400			2.59	84	97
spiral	0.40	1.5	600	29	1.9	15.3	97	96
tubular			800			10.0	99	98
Hollow fiber			400			1.97	95	87
spiral	1.83	6.7	600	29	1.2	13.2	94	86
tubular			800			8.58	97	91
Hollow fiber			400			1.20	90	91
spiral	4.11	15	600	29	1.2	10.6	92	92
tubular			800			7.31	95	7
Hollow fiber			400			leak	leak	leak
spiral	9.43	34	600	28	0.9	leak	leak	leak
tubular			800			6.60	94	97

[a]Three commercially-available membrane modules were tested:

DuPont B-9 hollow-fiber module (polyamide membrane);
T.J. Engineering 97H32 spiral-wound module (cellulose acetate membrane); and
Abcor, Inc., TM5-14 tubular module (cellulose acetate membrane).

[b]Gallon/minute/single DuPont B-9 permeator size 0440-035.
Gallon/day/ft^2 for spiral-wound and tubular modules.

Source: Reference 10.

TABLE 5.5.5. ANALYTICAL RESULTS FOR REVERSE OSMOSIS TREATMENT OF SPENT COPPER CYANIDE PLATING RINSE

Waste feed		Operating conditions				% - Rejection		
%-TDS	% of bath	Pressure (psig)	Temp. (°C)	pH of Feed	Flux[a]	TDS	Cu^+	CN^-
1.93	5.2	400	26	12.2	1.20	98	99+	99+
3.71	10	400	26	12.5	0.62	97	99+	99+
7.98	22	430	27	12.9	0.076	77	84	92

[a]Gallon/minute permeator. Hollow fiber membrane module (polyamide membrane).
Source: Reference 10.

TABLE 5.5.6. ANALYTICAL RESULTS FOR REVERSE OSMOSIS TREATMENT OF SPENT CADMIUM CYANIDE PLATING RINSE

Waste feed		Operating conditions				% - Rejection		
%-TDS	% of bath	Pressure (psig)	Temp. (°C)	pH of Feed	Flux[a]	TDS	Cd^+	CN^-
2.43	9	400	27	12.5	0.67	97	99	95
3.12	12	400	27	12.5	0.24	96	99	92
9.82	37	430	27	12.9	0.028	9.2	78	10

[a]Gallon/minute permeator. Hollow fiber membrane module (polyamide membrane).
Source: Reference 10.

TABLE 5.5.7. ANALYTICAL RESULTS FOR REVERSE OSMOSIS TREATMENT OF SPENT ZINC CYANIDE PLATING RINSE

Waste feed		Operating conditions				% - Rejection		
%-TDS	% of bath	Pressure (psig)	Temp. (°C)	pH of Feed	Flux[a]	TDS	Zn	CN^-
2.43	9	400	27	12.5	0.67	97	99	95
3.12	12	400	27	12.5	0.24	96	99	92
9.82	37	430	27	12.9	0.028	9.2	78	10

[a]Gallon/minute permeator. Hollow fiber membrane module (polyamide membrane).
Source: Reference 10.

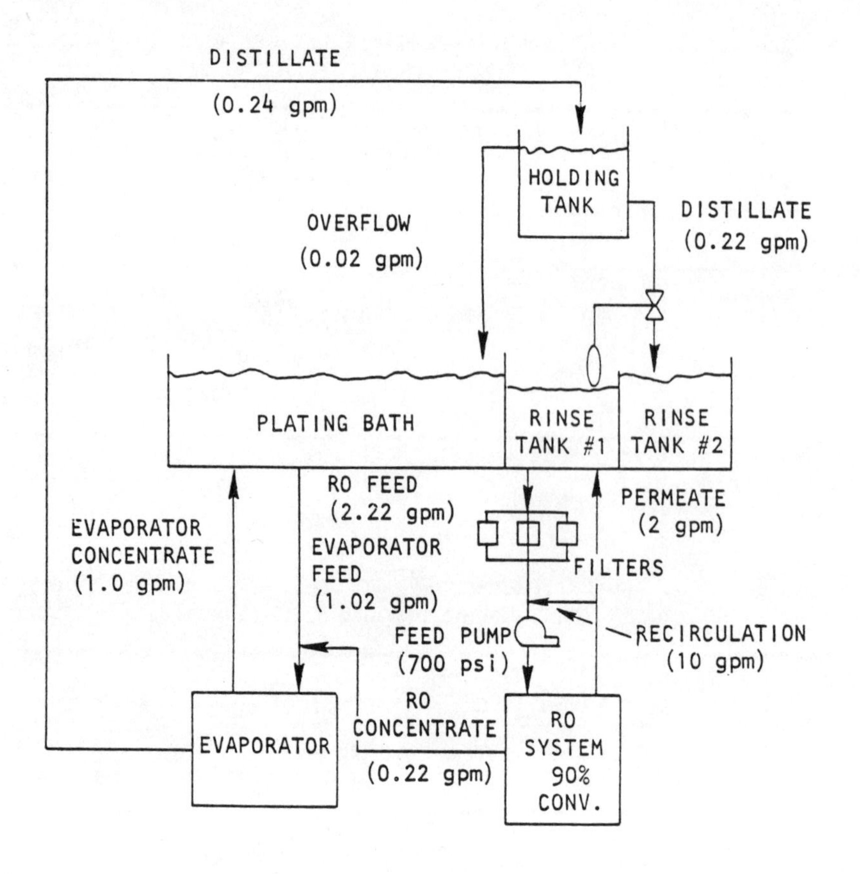

Figure 5.5.3. Schematic of reverse osmosis/evaporation system used to recover zinc cyanide plating solution at American Electroplating Company.
Source: Reference No. 13.

the RO system was returned to the first rinse tank, and the concentrate from the last RO module was sent to the evaporator. The distillate from the evaporator was pumped back to the second rinse tank, and the concentrate was metered back to the plating tank. Operating parameters for the system are presented in Table 5.5.8.

The typical composition of the plating bath is presented in Table 5.5.9. Samples were collected and analyzed for zinc, free cyanide, total solids, conductivity, and pH.[13] The system was operated at a fixed feed concentration in order to evaluate any trends in the flux and rejection data. The system was operated for a total of 1,000 hours. However, the membranes were exposed to the highly alkaline (pH 13) rinsewater for a total of 4,200 hours because the membranes remained in contact with the plating rinse during downtime. Total exposure time is a good measure of the chemical resistance of the membranes whereas total operating time is a better indicator of the effect of membrane plugging due to particulates present in the feed.

Both flux and rejection values gradually declined during the operating period as a result of membrane plugging. However, after cleaning the RO modules with a 2 percent citric acid solution, rejection and flux values improved within 24 hours. Periodic cleaning of the modules was also found to extend their lifetime.

Although RO membrane systems are currently available which can be used with corrosive wastes, lifetimes for these membrane systems are a limiting factor in their cost-effective application.[6] Research is currently being conducted to develop more chemically-resistant membranes.[8,14]

5.5.3 Process Costs

The economics of RO systems varies with the operating parameters, membrane type, waste feed characteristics, and desired product quality.

Capital costs include costs for RO modules, evaporation system, pumps, and associated piping and instrumentation. Typical capital costs average $25,000 for a RO system, and approximately $40,000 for the evaporation system.[5]

TABLE 5.5.8. TYPICAL OPERATING PARAMETERS DURING TESTING OF REVERSE OSMOSIS SYSTEM AT AMERICAN ELECTROPLATING COMPANY FOR THE RECOVERY OF ZINC CYANIDE PLATING RINSE

Item	Parameter
Feed pressure	700 psi
Recirculation flow rate	10 gpm
Temperature	70 to 90°F
Concentrate flow rate	0.2 gpm
Evaporator vacuum	26 to 27 in. Hg
Evaporator temperature	100 to 110°F
Evaporator steam pressure	5 psi
Evaporator concentrate flow rate	1 gpm

Source: Reference 13.

TABLE 5.5.9. TYPICAL COMPOSITION OF ZINC CYANIDE PLATING BATH AT AMERICAN ELECTROPLATING COMPANY

Constituent	Concentration
Zn (as metal)	20,000 mg/L (2.7 oz/gal)
CN (as NaCN)	60,000 mg/L (8.0 oz/gal)
Caustic	75,000 mg/L (10.0 oz/gal)
Brightener	4 mg/L (4 gal/1,000 gal)
Polysulfide carbonates	155,000 mg/L
Total solids	350,000 mg/L (35% by weight)

Source: Reference 13.

Capital and operating costs will increase proportionally with system size. System size will be determined by the desired conversion, the feed concentrations, operating temperatures and pressures, and membrane type. Table 5.5.10 presents estimates of capital costs for various RO system conversions.

Operating costs will include membrane replacement costs, filter replacement costs, energy costs for the pumps, and evaporator, operating labor (minimal) and maintenance costs for cleaning and replacing membranes and filters. Table 5.5.11 summarizes the operating cost requirements. Figure 5.5.4 shows that minimum total costs using an RO/evaporation system combination occur when the RO system conversion is 90 percent. Typical operating costs for an RO/evaporation system designed for 90 percent conversion are $12,000/year.[5]

Savings will be realized in neutralization and disposal costs, chemical purchase costs, and process water requirements. An EPA study of the economics of a reverse osmosis/evaporation system for the recovery of zinc cyanide from rinsewaters showed only a $10,000 savings/year for rinsewater treatment, water and makeup chemical costs, which was not deemed to be cost-effective.[5,13] However, it is anticipated that the savings will increase as costs for neutralization and disposal increase.

In summary, the cost-effectiveness of the reverse osmosis technique is dependent upon the following factors: production rate, type and concentration of rinsewater constitutents, water supply, wastewater disposal costs, and useful lifetime of RO membrane.[5] As more chemically resistant membranes are developed RO systems will have more cost-effective applications for corrosive waters. Also, with the implementation of the land disposal ban and the resulting rise in sludge disposal costs, reverse osmosis will become a more cost-effective alternative to conventional neutralization practices.

5.5.4 Process Status

Reverse osmosis systems have been widely applied in wastewater desalination and are also currently available for recovering corrosive wastewater streams. Research has focused on the recovery of corrosive plating

TABLE 5.5.10. CAPITAL COSTS FOR VARIOUS RO SYSTEM CONVERSIONS

Item	RO system conversion (%)				
	0	70	80	90	95
Required permeate flow (gpm)	0	2.575	3.625	5.85	8.91
Required membrane area (sq ft)[a]	0	371	522	842	1.283
Required number of modules[b]	0	6	6	12	18
Membrane module cost ($)[c]	0	3,780	5,040	7,560	11,340
Housing cost ($)[d]	0	1,700	2,559	3,400	5,100
Total cost for RO system ($)[e]	0	21,780	23,890	28,560	34,040
Required evaporator capacity (gph)[f]	120	66.2	54.4	39.0	28.1
Total cost for evaporator ($)	44,129	39,199	39,199	33,880	33,880
Total system	44,129	60,979	63,089	62,440	67,920

[a]Design flux - 10 g/f/d.
[b]Based on area of 70 ft^2/module.
[c]Based on $630/module.
[d]Based on $850/3-module housing.
[e]Based on system cost of $1,500 for system without modules, housings, and high-pressure pump. Pump/motor cost = $1,300 for 4 gpm permeate; $2,600 for 4 gpm permeate.
[f]Double-effect evaporator with cooling tower package. Based on rated capacities of 200 gph ($44,129), 100 gph ($39,199), and 50 gph ($33,880).

Source: References 8, 13, and 15.

TABLE 5.5.11. OPERATING COSTS FOR A REVERSE OSMOSIS SYSTEM

Operating requirement	Cost basis	Unit cost
Steam heat requirements (for evaporator)	No. 4 fuel oil at $0.393/gal; Heating value of $140,000 BTU/gal; 80% boiler efficiency	$3.50/100 lbs
Electrical	To run pressure pump and evaporator	$0.055/KWH
Cartridge filters	4 cartridges/month (average)	$4.68/cartridge
Cleaning chemicals	3 cleanings/year with citric acid	$0.82/lb

Source: References 13 and 16.

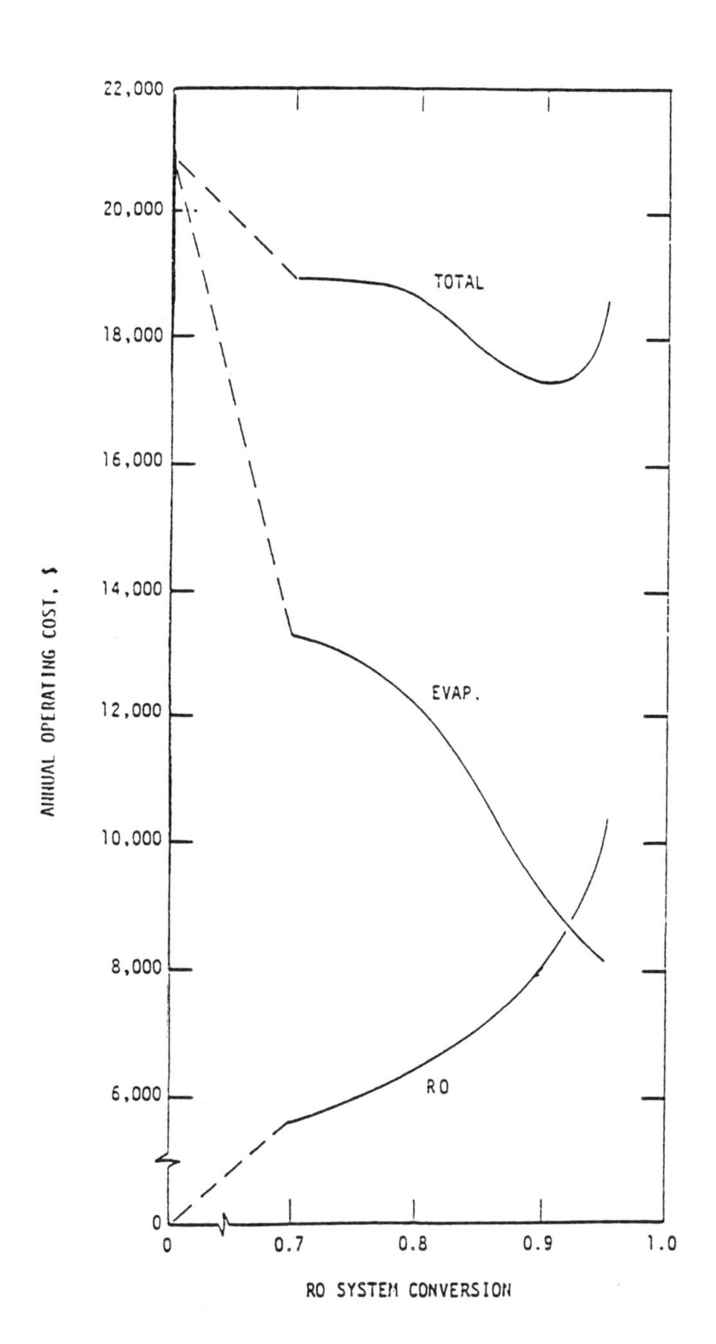

Note: System costs shown in this figure include direct operating costs and capital amortization costs (based on straight-line depreciation over 10 years with zero salvage value).

Figure 5.5.4. Annual operating costs for various RO system conversions.

Source: Reference 13.

rinses. However, cost-effective use of RO systems for this application is generally limited due to the shortened lifetime of membranes exposed to corrosive wastes and the associated high costs for membrane cleaning and replacement. However, future development of membranes which are able to withstand corrosive and oxidizing solutions is expected.[5,17]

If membrane resistance is increased, reverse osmosis would be a very cost-effective alternative to conventional treatment technologies. Excluding membrane cleaning and replacement costs, the only significant operating cost is the electricity required for operation of the pump. However, current systems are limited in the degree of attainable concentration of the reject stream which frequently requires the use of an evaporator in conjunction with the RO unit. The use of a combination reverse osmosis/evaporation treatment system can be more cost-effective than evaporation alone.

Five systems designed by Osmonics, Inc. (Minnetaka , Minnesota) are currently in operation for the treatment of corrosive wastes.[8] Although detailed performance data is not available for these systems, performance has reportedly been satisfactory.[8] Fluid Systems Division of UOP, Inc. and Desalination Systems, Inc. (both in San Diego, CA) currently market RO membranes that are capable of being used in the corrosive pH ranges.[6,7] However, membrane replacement is more frequent compared to use with more neutral solutions. Ionics, Inc. (Watertown, MA) is currently conducting research to develop a more chemically resistant membrane.[14] However, preliminary test results are not available at this time.

REFERENCES

1. Metcalf & Eddy, Inc. Wastewater Engineering: Collection, Treatment, and Disposal. McGraw-Hill Book Company, NY. 1972.

2. U.S. EPA. Treatability Manual, Volume III: Technologies for Control/Removal for Pollutants. EPA-600/8-80-042c. July 1980.

3. U.S. EPA, Office of Research and Development, Cincinnati, OH. Handbook for Remedial Action at Waste Disposal Sites. EPA-625/6-85-006. October 1985.

4. U.S. EPA. Sources and Treatment of Wastewater in the Nonferrous Metals Industry. Prepared by Radian Corporation for the U.S. EPA, Industrial Environmental Research Laboratory, Cincinnati, OH, under EPA Contract No. 68-02-2068. EPA-600/2-80-074. April 1980.

5. Higgins, T.E., CH_2M HILL. Industrial Processes to Reduce Generation of Hazardous Waste at DOD Facilities - Phase 2 Report, Evaluation of 18 Case Studies. Prepared for the DOD Environmental Leadership Project and the U.S. Army Corps of Engineers. July 1985.

6. Comstock, D. Desalination Systems, Inc. Telephone conversation with L. Wilk, GCA Technology Division, Inc. September 19, 1986.

7. Filtwell, J. Fluid Systems Division of UOP, Inc., San Diego, CA. Telephone conversation with L. Wilk, GCA Technology Division, Inc. September 19, 1986.

8. Osmotics, Inc., Minnetaka, MN. Telephone conversation with L. Wilk, GCA Technology Division, Inc. September 19, 1986.

9. Sundstrom, D.W., and H.E. Klei. Wastewater Treatment. Prentice-Hall Inc., Englewood Cliffs, NJ. 1979.

10. Donnelly, R.G., R.L. Goldsmith, K.J. McNulty, and M. Tan. Reverse Osmosis Treatment of Electroplating Wastes. Plating. May 1974.

11. Crampton, P., and R. Wilmoth. Reverse Osmosis in the Metal Finishing Industry. Metal Finishing. March 1982.

12. Cushnie, G.C. Centec Corporation, Reston, VA. Navy Electroplating Pollution Control Technology Assessment Manual. Final Report prepared for the Naval Facilities Engineering Command under Contract No. F08635-81-C-0258. NCEL-CR-84-019. February 1984.

13. McNulty, K.J., and J.W. Kubarewicz. Field Demonstration of Closed-Loop Recovery of Zinc Cyanide Rinsewater Using Reverse Osmosis and Evaporation. In: Proceedings of the 2nd Conference on Advanced Pollution Control for the Metal Finishing Industry, Co-sponsored by the American Electroplaters Society and the U.S. EPA Industrial Environmental Research Laboratory, Kissimmee, FL, February 5-7, 1979. EPA-660/8-79-014. May 1979.

14. Jain, S.M., Ionics, Inc., Watertown, MA. Telephone conversation with Jon Spielman, GCA Technology Division, Inc. August 12, 1986.

15. Mahoney, F., F. Mahoney Co./Permutit Systems, Hingham, MA. Telephone conversation with L. Wilk, GCA Technology Division, Inc. September 19, 1986.

16. Chemical Marketing Reporter. Chemical Prices for Week Ending July 18, 1986. 230(3): 32-40. July 21, 1986.

17. McNulty, K.J. Walden Division of Abcor, Inc., Wilmington, MA. Telephone conversation with Jon Spielman, GCA Technology Division, Inc. August 4, 1986.

5.6 DONNAN DIALYSIS & COUPLED TRANSPORT

5.6.1 Process Description

Other membrane technologies currently being researched include Donnan dialysis and coupled transport. Neither of these technologies is presently commercially available. However, both show potential for cost-effective applications in the recovery of corrosive plating solutions.

Donnan dialysis and coupled transport are similar processes in that both employ a concentration gradient to drive ions from a spent solution across a membrane into a stripping solution. These processes utilize a cell consisting of a membrane separating two compartments. The waste is fed to one compartment and the stripping solution (into which the concentrate will be extracted) is fed to the other. As with electrodialysis, the cells are generally arranged in stacks or multi-cells containing many membrane-spacer units which allows treatment of larger quantities of waste at a faster rate. Figure 5.6.1 diagrams a typical stack unit.

Unlike other membrane technologies (i.e., electrodialysis and reverse osmosis), energy requirements are minimal for donnan dialysis and coupled transport. The only energy required is the energy to pump the feed and stripping solutions across the cell.[2] The large hydraulic pressures required for reverse osmosis and the large electric current flow required by electrodialysis, are not required for these technologies.[2]

The major difference between these processes is the type of membrane employed, and the transport mechanisms involved. The coupled transport membrane is highly selective and therefore has more specific process applications, whereas the Donnan dialysis membrane has application to a wider variety of solution constituents. However, greater purity can be achieved using the coupled transport membrane.[3]

Donnan dialysis uses an anion- or cation- selective membrane, which functions similarly to ion exchange resins. For an anion exchange membrane, cations in both solutions (on each side of the membrane) are prevented from diffusing across the membrane, but anions will redistribute themselves across the membrane until equilibrium is reached; i.e., ratios of all similarly charged anions are equal. With a cation-exchange membrane, cations will

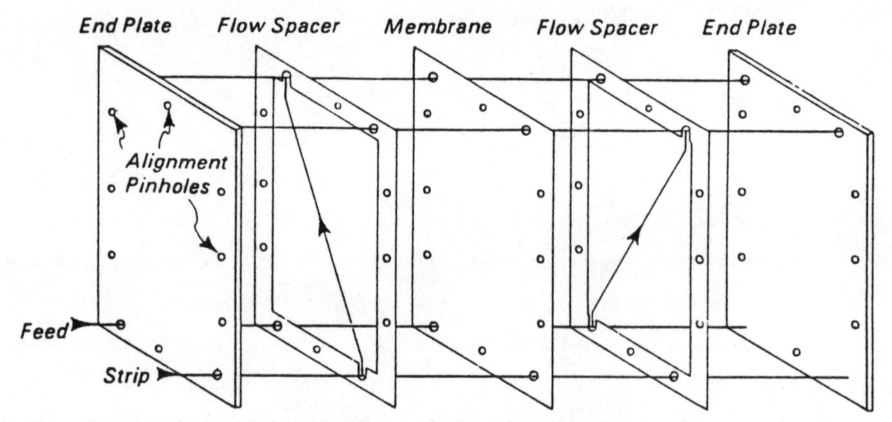

Figure 5.6.1. Membrane stack unit.

Source: Reference 1.

diffuse across the membrane and movement of anions will be restricted. The driving force for these exchange reactions is the potential created by the displacement of the system from the equilibrium ratios which can be controlled by adjusting solution concentrations.[2]

The microporous membrane used in the coupled transport process contains a metal-complexing agent within its pores.[3,4] Ions combine with the complexing agent and are removed from the spent solution. On the other side of the membrane, the ion solubility is favored over that of the complex in the stripping solution. Thus, a transfer of the ion across the membrane occurs due to the coupling of these two complexation reactions.[4] The membrane used in the coupled transport is selective to the transport of one metal ion over other ions; i.e., it is highly specific. Therefore, a different membrane is required for each application.[3] However, the greater selectivity of the membrane allows more complete removal of the ion from the spent solution.[3]

The two types of coupled transport reactions which may occur are cotransport and countertransport.[4] With cotransport, ions are only tranferred in one direction across the membrane. A typical cotransport process is diagrammed in Figure 5.6.2. In countertransport processes, the transport of one ion is balanced by the flow of another ion in the opposite direction. A generalized flow diagram of a countertransport system for recovering metal ions is shown in Figure 5.6.3.

Operating Parameters--

The Donnan dialysis operation is based upon the Donnan equilibrium principle, which can be described by the following equation:[2,6]

$$\left[\frac{(C_i)_l}{(C_i)_r}\right]^{1/Z} = \text{Constant}$$

Where:
- C = activity, approximately equal to concentration,
- i = any mobile ionic species,
- l = left-hand side of membrane,
- r = right-hand side of membrane,
- Z = valence of the ionic species.

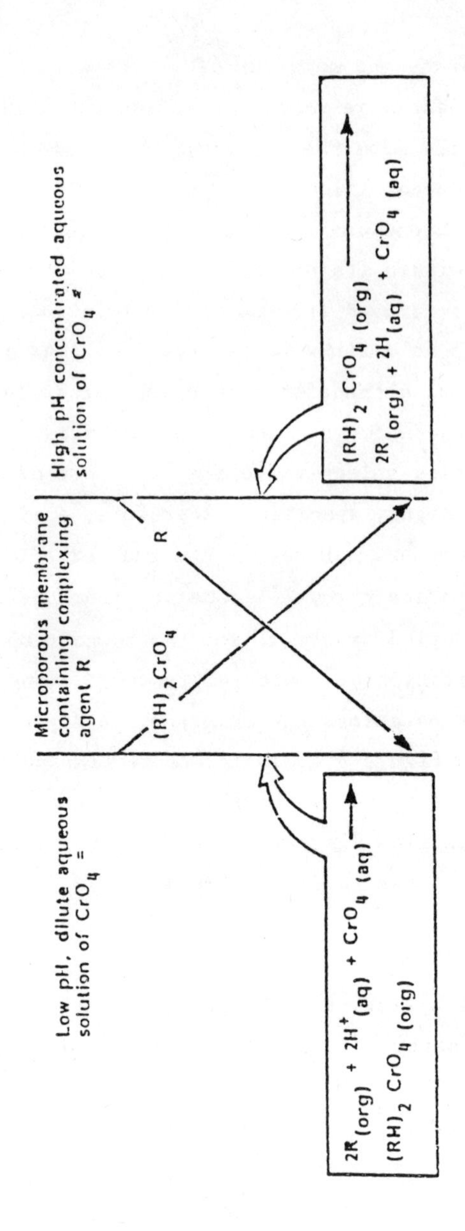

Figure 5.6.2. Co-transport process for chromate recovery.

Source: Reference 5.

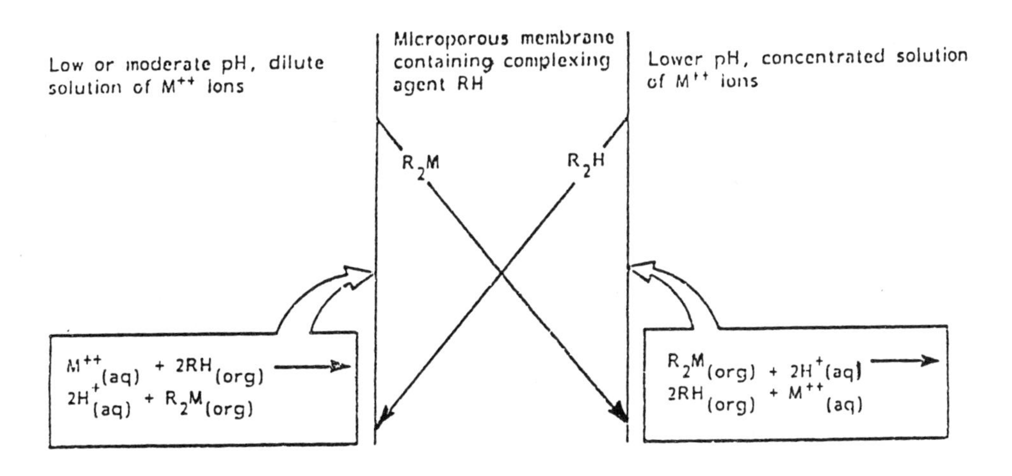

Figure 5.6.3. Counter-transport process for metal cation recovery.

Source: Reference 5.

The driving force which causes the ions to move from one side of the membrane to the other is the concentration potential resulting from the differences in concentration between similarly charged ions (i.e., both positive or both negative) on each side of the membrane. The degree of separation can be increased by any of the following changes in the system:[6]

- Increasing initial concentrations on one side of the membrane; i.e., in the spent solution;
- Decreasing initial concentrations on the other side of the membrane; i.e., stripping solution;
- Adding a complexing agent to remove ions from the solution on one side of the membrane; and/or
- Using a countercurrent flow system.

This equilibrium principle is also in effect during a coupled transport process. However a complexing agent is incorporated into the membrane which increases the quantity of a specific ion that can be transported across the membrane resulting in improved separation.

The operation of both of these processes is most significantly affected by the membrane characteristics. Important membrane properties include: equilibrium water content, ion-exchange capacity, osmotic water transport rate, metal complex anion transport rate, chemical resistance, and physical resistance.

The equilibrium water content is the amount of water which the membrane holds when the system has reached Donnan equilibrium. It is generally expressed as the grams of water absorbed per dry gram of membrane. This property measures the hydrophilicity of the membrane.[2]

The ion exchange capacity of the membrane can be expressed as the number of equivalents (eq) per dry gram of membrane, where an equivalent is equal to the molecular weight of the ion in grams divided by its electrical charge or valence.[7] It indicates the quantity of ions that can be exchanged with the given exchange medium. For example, a membrane with an exchange capacity of 1 eq/dry g could remove 37.5 g of divalent zinc from solution per dry gram of membrane ($Zn+2$, molecular weight = 65, 65/2 = 37.5). A Donnan membrane is non-specific in that it will transfer all anions (anion-exchange membrane) or all cations (cation-exchange membrane). However, a coupled transport membrane is highly specific and will only transport one type of metal ion.

The osmotic water transport rate is the amount of water that can be transported across the membrane in a given time period. It is typically expressed as a rate per unit of membrane area; i.e., volume/unit time/membrane area.

The metal complex ion transport rates can be determined using the following first order rate equation:[2]

$$k = \frac{\ln (C_o/C)}{t}$$

Where: k = rate constant,
ln = natural log,
t = residence time in the cell,
Co = initial concentration of ion in cell inlet, and
C = concentration of ion in cell outlet.

The residence time (t) can be calculated by dividing the cell volume by the feed flow rate. The metal ion transport rates will be an important factor in selecting the membrane area required for a specific application. The rate of transport across the membrane will vary with changes in pH and solution concentrations.[3,8]

The physical and chemical resistance of the membrane will determine the membrane stability. Lower operating costs will be incurred with more resistant membranes.

Pretreatment--

Pretreatment in the form of filtration may be necessary to remove any suspended particulates present in the spent solution. Suspended particulates can clog the membrane and reduce membrane efficiencies.

Post-treatment--

Post-treatment requirements can not be evaluated for Donnan dialysis at this time due to the lack of available performance data. Post-treatment requirements for coupled transport will vary with the specific application. In some cases, the recovered product will not be suitable for direct return to the plating bath; e.g., chrome recovery as discussed below. In these cases, it may be necessary to employ an ion exchange treatment process to convert the recovered product to a reusable form.

5.6.2 Process Performance

Limited performance data is available for these membrane technologies. Donnan dialysis has only been tested in laboratory-scale experiments. Attempts to test the process on a pilot-scale level were hindered by mechanical problems. Coupled transported has been both laboratory and field tested. The information presented in this section is a summary of available findings to date.

Donnan Dialysis--

Although the Donnan dialysis process has been in existence for quite some time, only laboratory-scale demonstrations have been performed.[2,6,9] In addition, the Southwest Research Institute (San Antonio, Texas) attempted to develop a pilot-scale system for field testing of recovery applications during studies conducted in 1984 and 1985.[1] Initial testing was conducted to develop an appropriate membrane for use with metal finishing waste solutions. Several membranes were evaluated for metal ion transport. A schematic of the test system is shown in Figure 5.6.4.

Initially, the membranes were tested with relatively low feed concentrations of about 50 ppm metal ions. Table 5.6.1 presents the results of these tests. As expected, increased ion exchange capacity correlated positively with increased metal ion transport rates. In addition, ion exchange capacity increased with increasing osmotic flow rates.[1]

The four membranes which demonstrated the highest metal ion transport rates during the preliminary tests were subjected to further testing with higher metal ion concentrations in the feed (approximately 500 ppm). Table 5.6.2 presents the results of these tests, which show that increasing the feed concentration lowers the metal ion transport rate.

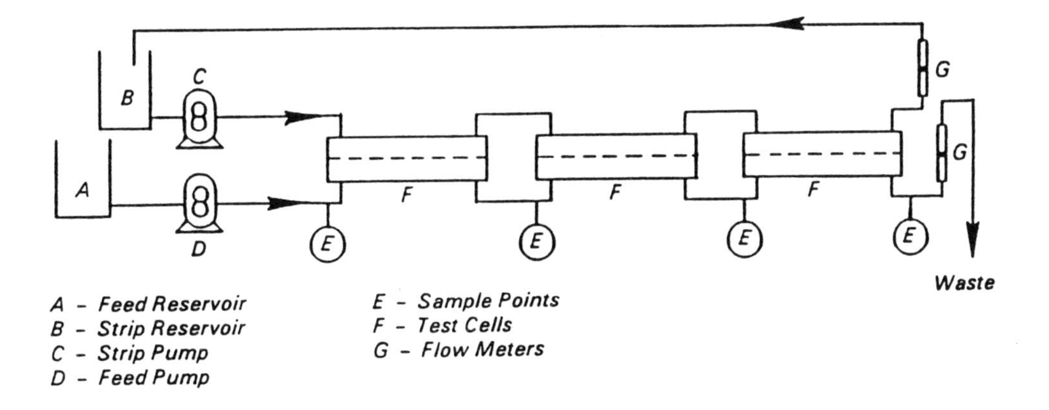

Figure 5.6.4. Membrane test system.

Source: Reference 1.

TABLE 5.6.1 RESULTS OF PRELIMINARY MEMBRANE TESTING PERFORMED BY THE SOUTHWEST RESEARCH INSTITUTE

Membrane No.	Ion exchange capacity (meq/dry g)	Osmotic water[a] flow rate (ml/hr/cm^2)	Metal complex anion transport rate constant (min^{-1},[b])		
			Cu	Cd	Zn
E11Q4	2.5	0.040	3.1	2.8	1.9
E12Q4	2.6	0.053	2.3	2.2	2.9
E12Q5	1.1	0.010	0	-	-
E14Q4	0.7	0.010	1.1[c]	0.3	0.2
E15Q1	1.1	0.012	1.0	0.4	0.1
E16Q1	2.2	0.102	2.5	-	-
E16Q2	1.2	0.016	0	-	-
E18Q1	2.2	0.054	2.2	1.4	0.6
E18Q2	1.4	0.009	0	-	-

[a]0.2N NaCl Versus Deionized H_2O - cell effective area = 122 cm^2.

[b]Feed-50 ppm of metal; stripping solution 0.2N NaCl.

[c]Duplicate determination of the rate constant also showed 1.1/min.

Source: Reference 1.

TABLE 5.6.2. RESULTS OF TESTS TO DETERMINE THE EFFECT OF METAL ION CONCENTRATION ON TRANSPORT RATE

Membrane No.	Membrane type	Equilibrium water content g H_2O/g	Ion exchange capacity (meq/dry g)	Osmotic[a] water flow rate ml/hr/cm^2	Metal complex anion transport rate (1/min)					
					Low conc. (50 ppm)			High conc. (5,000 ppm)		
					Cu	Cd	Zn	Cu	Cd	Zn
E11Q4	Vinylpyridine-grafted polyethylene (4-VP) CH3I	1.08	2.5	0.040	3.1	2.8	1.9	1.5	1.5	1.5
E12Q4	Vinylpyridine-grafted polyethylene (4-VP/N-VP) CH3I	2.45	2.6	0.053	2.3	2.2	2.9	0.7	0.8	0.6
E16Q1	Vinylbenzyl chloride-grafted polyethylene (VBC) (CH3) 3N	1.52	2.2	0.102	2.5	-	-	0.6	0.9	b
E16Q1	Vinylbenzyl chloride-grafted polyethylene (VBC/N-VP) (CH3) 3N	1.00	2.2	0.054	2.2	1.4	0.6	1.41	0.6	0.4

[a]0.2N NaCl versus deionized H_2O - cell effective area = 122 cm^2.

[b]Membrane ruptured.

Source: Reference 1.

An important factor in the performance of the system is the stability of the membrane. During the testing conducted by the Southwest Research Institute, it was observed that the metal ion transport rate decreased over time. The decline in the transport rate was attributed to problems in membrane stability, the precise cause of which was not determined. Due to time constraints, the best available membrane was used in the construction of a prototype Donnan dialysis stack.[1]

Although the Southwest Research Institute intended to test the prototype Donnan dialysis system for the recovery of various plating solutions, the project was never completed due to leakage problems. Future studies will focus on developing membranes with higher metal ion transport rates and also on making mechanical improvements to the dialysis stack.[8]

In summary, the performance of the Donnan dialysis system can not be realistically evaluated at this time due to lack of available performance data. However, preliminary testing indicated that a technically feasible system could be developed.

Coupled Transport--

The coupled transport process has been laboratory-tested on several types of metal-containing solutions (primarily metal finishing wastes). The most developed application is in the removal of hexavalent chromium from plating rinses.[3] Field tests of this application have been performed, but limited performance data is currently available.

The hexavalent chromium application works most effectively with solutions having a pH ranging from 0 to 4.[3,10] The process is capable of reducing the level of hexavalent chrome in the rinsewater to less than 1 ppm. However, unlike other membrane technologies, the product can not be returned directly to the plating bath.[11,12] The product is a 5 to 10 weight-percent solution of sodium chromate, which can generally be used for other in-house processes. Alternatively, the sodium chromate product can be passed through a cation-exchange column to recover approximately 5 weight-percent chromic acid for return to the bath.[3] The process is generally most effective for treating solutions with concentrations greater than 50 ppm; however, more dilute solutions can be treated.[3]

5.6.3 Process Costs

Realistic costs for the Donnan dialysis process can not be developed at this time due to the lack of commercial-scale testing. It is expected that the primary cost would be for the membrane unit. Operating costs would be expected to be minimal, and savings would be realized in reduced disposal costs and reduced purchase costs for recovered chemicals.

Preliminary cost estimates for coupled transport hexavalent chromium treatment systems have been prepared by the developer (Bend Research Corporation) based on pilot-scale testing.

Capital costs for the coupled transport process can vary widely with the application.[3] Capital costs for pilot-scale units developed by Bend Research Corporation have ranged from $500 to $1,000,000.[3] Standardized process units are not available since each unit has to be custom-tailored for a specific application. However, currently the most widespread application for the coupled transport process is for the recovery of chrome in plating shops. For a typical plating shop, using three countercurrent rinse tanks (1,000 gallons each), the capital equipment cost would be approximately $20,000.[3] The total membrane area for this recovery system would be 500 sq ft.[3] An increase in solution volume to be handled (at the same pH and concentration) would require a proportional increase in membrane area.

Operating costs would include periodic maintenance and membrane replacement. The membrane performance gradually deteriorates over time. Approximately every 6 months, the coupled transport unit needs to be drained so that the membrane can be regenerated with a proprietary regenerant solution. Bend Research Corporation currently performs this service for the pilot-scale units that they have installed. Costs for regeneration are approximately $2/sq ft, plus labor. The estimated lifetime of the membrane (with cleanings every 6 months) is approximately 2 years.[3] However, this estimate is conservative since membranes have not been tested for periods longer than 2 years[3]. The approximate unit cost for the membrane is estimated to be $20/sq ft.[3]

The process requires minimal day-to-day operating maintenance. If solution conditions are changed (e.g., pH or concentration changes), then modifications, such as increases in membrane area, may be required.

Savings will be realized in reduced disposal costs, and benefits from recovered chemicals and recovered process waters. The estimated payback period for the chrome recovery system is approximately 2 years.[3]

5.6.4 Process Status

Donnan dialysis has not yet been tested on a pilot-scale. Much of the research that has been performed to date with Donnan dialysis has concentrated on membrane development. The Southwest Research Institute (San Antonio, TX) is currently doing research for purposes of developing a pilot scale unit.[8] Although preliminary research has demonstrated the technical feasibility of the process, pilot-scale testing is needed to determine if sufficient solution concentrations can be achieved. The Donnan dialysis process could prove to be a cost-effective alternative to conventional treatment practices because of its minimal operating requirements.

Bend Research Corporation (Bend, Oregon) has done most of the development work for the coupled transport technology and has a patent pending on the process.[10] Although the process is applicable to the treatment of several metal-containing solutions, the most developed application is for the treatment of hexavalent chromium in plating rinses. The process was recently licensed to Concept Membrane, Inc. for marketing and sales purposes. However, commercial units are not currently available for purchase.[3,4]

Although the coupled transport process is more developed than Donnan dialysis, both of these membrane technologies show good potential as cost-effective alternatives to conventional neutralization and disposal practices for metal-containing corrosive wastes.

REFERENCES

1. Hamil, H.F. Southwest Research Institute, San Antonio, TX. Project Summary: Fabrication and Pilot Scale Testing of a Prototype Donnan Dialyzer for the Removal of Toxic Metals from Electroplating Rinse Waters. EPA/600/s2-85-080. August 1985.

2. Hamil, H.F. Southwest Research Institute, San Antonio, TX. Fabrication and Pilot Scale Testing of A Prototype Donnan Dialyzer for the Removal of Toxic Metals from Electroplating Rinse Waters. Prepared for U.S. EPA-ORD, Water Engineering Research Laboratory, Cincinnati, OH, under EPA Cooperative Agreements CR-807476 and CR-809761. January 1985.

3. Friesen, D. Bend Research Corporation, Bend, Oregon. Telephone conversation with Lisa Wilk, GCA Technology Division, Inc. September 25, 1986.

4. Higgins, T.E., CH2M Hill. Industrial Processes to Reduce Generation of Hazardous Wastes at DOD Facilities - Phase 2 Report, Evaluation of 18 Case Studies. Prepared for the DOD Environmental Leadership Project and the U.S. Army Corps of Engineers. July 1985.

5. Cushnie, G.C. Centec Corporation, Reston, VA. Navy Electroplating Pollution Control Technology Assessment Manual. Prepared for the Naval Facilities Engineering Command, Alexandria, Virginia. NCEL-CR-84.019. February 1984.

6. Davis, T.A., Wu, J.S., and B.L. Baker. Use of Donnan Equilibrium Principle to Concentrate Uranyl Ions By an Ion-Exchange Membrane Process. AICHE Journal, 17(4): 1006-1008. July 1971.

7. U.S. EPA. Coupled Transport Systems for Control of Heavy Metal Pollutants. EPA-600/2-79-181. August 1979.

8. Hamil, H.F. Southwest Research Institute, San Antonio, TX. Telephone conversation with Lisa Wilk, GCA Technology Division, Inc. September 24, 1986.

9. Wen, C.P., and H.F. Hamil. Metal Counterion Transport in Donnan Dialysis. Journal of Membrane Science, 8: 51-68. 1981.

10. Martin, M., Bend Research Corporation, Bend, OR. Telephone conversation with Lisa Wilk, GCA Technology Division, Inc. September 24, 1986.

11. Babcock, W.C. Industrial Water Reuse with Coupled Transport Membranes. Prepared for the U.S. Dept. of the Interior, Dept. of Reclamation. Report No. OWR/RU-83-3. May 1983.

12. Babcock, W.C., et al. Renovation of Electroplating Rinse Waters with Coupled Transport Membranes. In: Proceedings of the Fourth Annual Conference on Advanced Pollution Control for the Metal Finishing Industry. EPA-600/9-82-022. December 1982.

5.7 SOLVENT EXTRACTION

5.7.1 Process Description

Solvent extraction is widely used as an analytical chemistry technique and for the recovery of metals in the field of hydrometallurgy.[1,2] In addition, research has shown applications for solvent extraction in the recovery of spent nitric/hydrofluoric acid pickling liquors generated by the steel industry.[3,4]

Solvent extraction is a separation technique which utilizes the differential distribution of constituents between the aqueous phase and the extractant (organic phase) to separate constituents from a mixed solution. A generalized flow diagram of a solvent extraction process is shown in Figure 5.7.1. The equipment used in the extraction step can consist of a single-stage mixing and settling unit, several single-stage units connected in series, or in a single unit, multi-stage unit operating by countercurrent flows in a column or differential centrifuge.[5] However, in order for the process to be cost-effective, it is usually necessary to recover the extracting solvent for reuse.

With a single-stage system, the extracting solvent is mixed with the solution, allowed to settle, and separated. The extracted solute can then be recovered by stripping from the extractant. The process may be operated as a batch or continuous technique, although with the latter, different vessels are required for mixing/settling and decanting.[5]

With a multi-stage system, the extracting solvent and the solution flow countercurrently through a vertical tower, which may contain internal devices to influence the flow pattern and provide intimate contact between the streams. The flow through the tower may be stage-wise or continuous-contact type.[5]

Although the solvent extraction process is a separation technique, it can also be used as a recovery technique. Use of solvent extraction for recovery involves the following general steps:[4]

1. Extraction - Constituents are transferred from aqueous phase to organic phase using an organic solvent as an extractant.

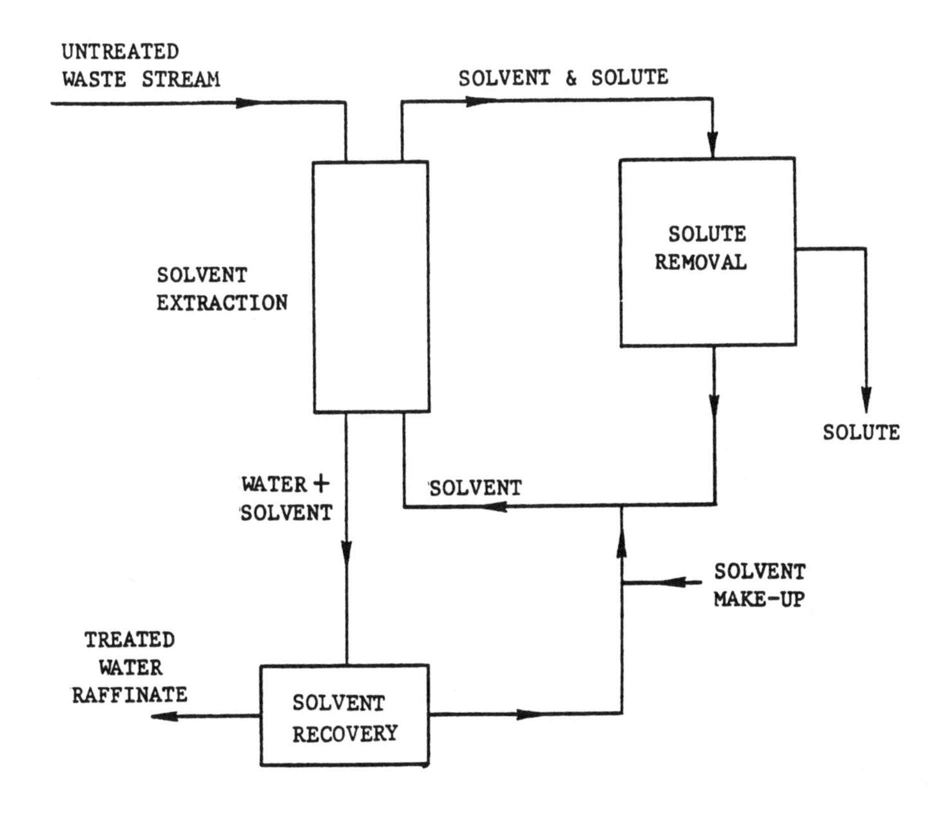

Figure 5.7.1. Generalized flow diagram of solvent extraction process.

Source: Reference 5.

2. Scrubbing - Impurities (e.g., metals) which are co-extracted with the desired constituents are transferred back to the aqueous phase.

3. Back-Extraction/Stripping - The constituent to be recovered (e.g., nitric/hydrofluoric acid) is transferred from the organic phase to a concentrated aqueous phase.

Various methods may be used to accomplish the recovery of acid pickling liquors including: the Republic Steel process, the AX process (also known as the Stora process), the Nisshan process, the Solex process, and the Kawasaki process.

The Republic Steel process was the first application of solvent extraction to the recovery of corrosive spent sulfuric acid pickling liquor. Since iron was the main contaminant in the pickling liquor, solvent extraction was used to remove the iron so that the sulfuric acid could be recovered for reuse. The process involves oxidation of ferrous ions, complexing ferric ions with a cyanide complex, extracting the iron complex with a solution consisting of 25 percent tributylphosphate (TBP) and 75 percent kerosene, and stripping the iron with ammonia.[4] Both sulfuric acid and ferric oxide (a marketable product) are recovered. However, this process was never applied on a commercial-scale.[4]

The AX process was developed in Sweden for the purpose of recovering spent hydrochloric-nitric acid pickling liquors. A flow diagram of the process is presented in Figure 5.7.2. Sulfuric acid is initially added to the spent liquor in order to liberate the nitrate and fluoride from their soluble metal salts.[3] The spent liquor then flows to the top of a column where it is extracted countercurrently with a solution of TBP in kerosene. Nitric and hydrofluoric acids are removed as kerosene-soluble adduct complexes and are subsequently added to recover the nitric and hydrofluoric acids from the TBP acid complex. Stripped through addition of water, an activated carbon filter is used to remove traces of TBP prior to returning the acids to the pickling tank. The solvent extraction produces a metal sulfate waste stream which requires further treatment prior to disposal.[3]

The Nisshan extraction process is similar to the AX process, except that it uses hydrochloric acid instead of sulfuric acid to convert the metal nitrates and fluorides to nitric and hydrofluoric acids in the scrubbing

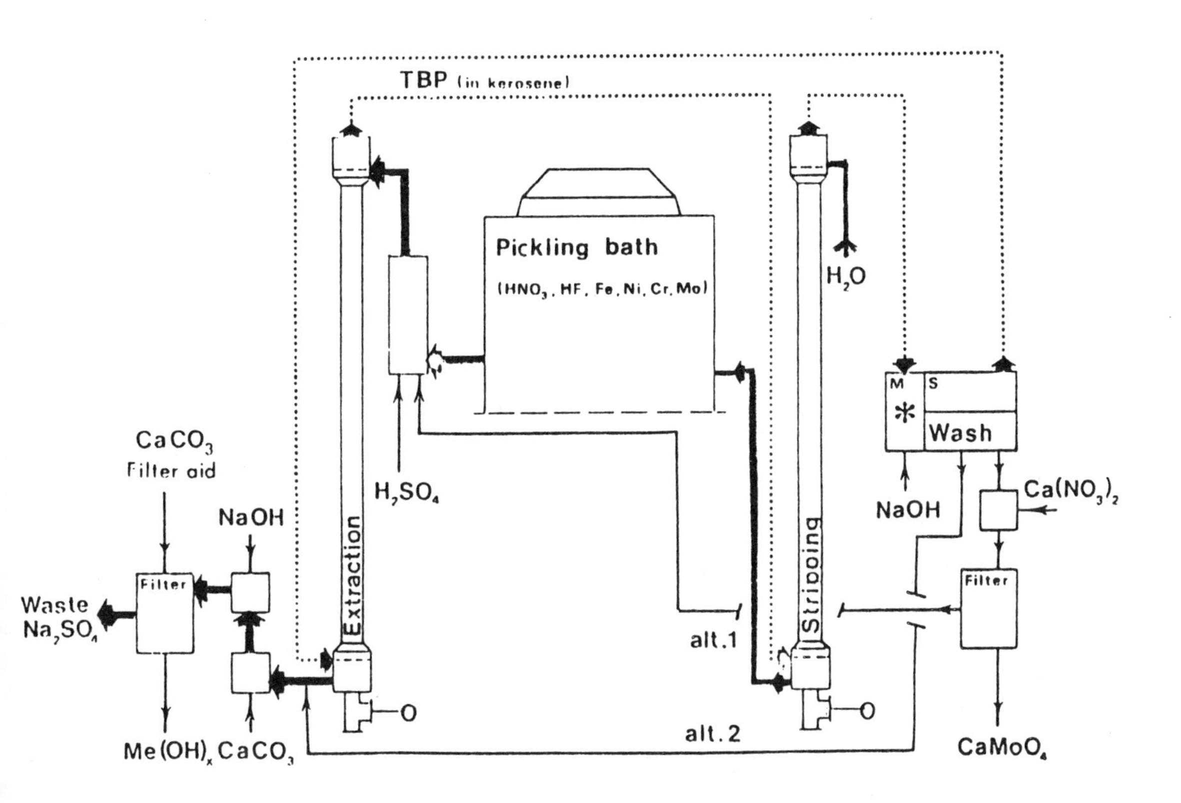

Figure 5.7.2. Flow diagram of Ax process.
Source: Reference 6.

step. The use of hydrochloric acid in this step results in a sludge containing iron, chromate, and nickel, which can be reused as raw material for steelmaking.[4] Additionally, the Nisshan process makes use of the different affinities of nitric and hydrofluoric acid for TBP by fractionally recovering these products in the water stripping step.[4]

With the Solex process, metal ions are removed prior to extracting the acids with the organic solvent. The Solex process uses a dialkylphosphoric acid (e.g., diethyl hexyl phosphoric acid, D_2EHPA) in an organic solution containing a small amount of carboxylic acid and active hydrogen atoms to extract iron from the spent pickling liquor.[3] The iron is then reduced to its divalent state by the addition of a sodium chloride solution containing either sodium sulfate, sodium nitrate, or ammonia. The iron complex is then stripped from the organic extractant with a strong solution of hydrochloric acid and the regenerated D_2EHPA is recycled.

Following removal of iron,the Solex process proceeds in similar fashion to the AX and Nisshan processes. HCl or H_2SO_4 is added to the spent pickle liquor to convert any remaining metal ions to chlorides or sulfates. A TBP solution is used to extract the nitrate and fluoride ions after which nitric and hydrofluoric acids are stripped from the organic extractant using water. Since the second extraction primarily recovers nitric acid, this waste is again subjected to TBP extraction/water stripping to recover additional hydrofluoric acid.[7]

Kawasaki Steel modified the Solex process for application in their Chiba Works plant in Japan.[4] The Kawasaki process shown in Figure 5.7.3, also extracts iron prior to extracting the acids. A solution containing 30 percent D_2EHPA and 70 percent n-paraffin is used to extract the ferric ions from the spent acid. The ferric iron is then complexed and stripped from the extractant using an ammonium fluoride solution. The organic extractant is recycled to the iron extraction process while the ferric ammonium fluoride complex is precipitated as crystals. A thermal decomposition process (discussed in Section 5.8) is used to convert the iron to ferric oxide, a usable product in the steel process.

After iron removal, hydrochloric acid is added to the waste acid solution to convert any remaining metal salts to metal chlorides. The waste acid solution is then subjected to a second extraction with a 70 percent

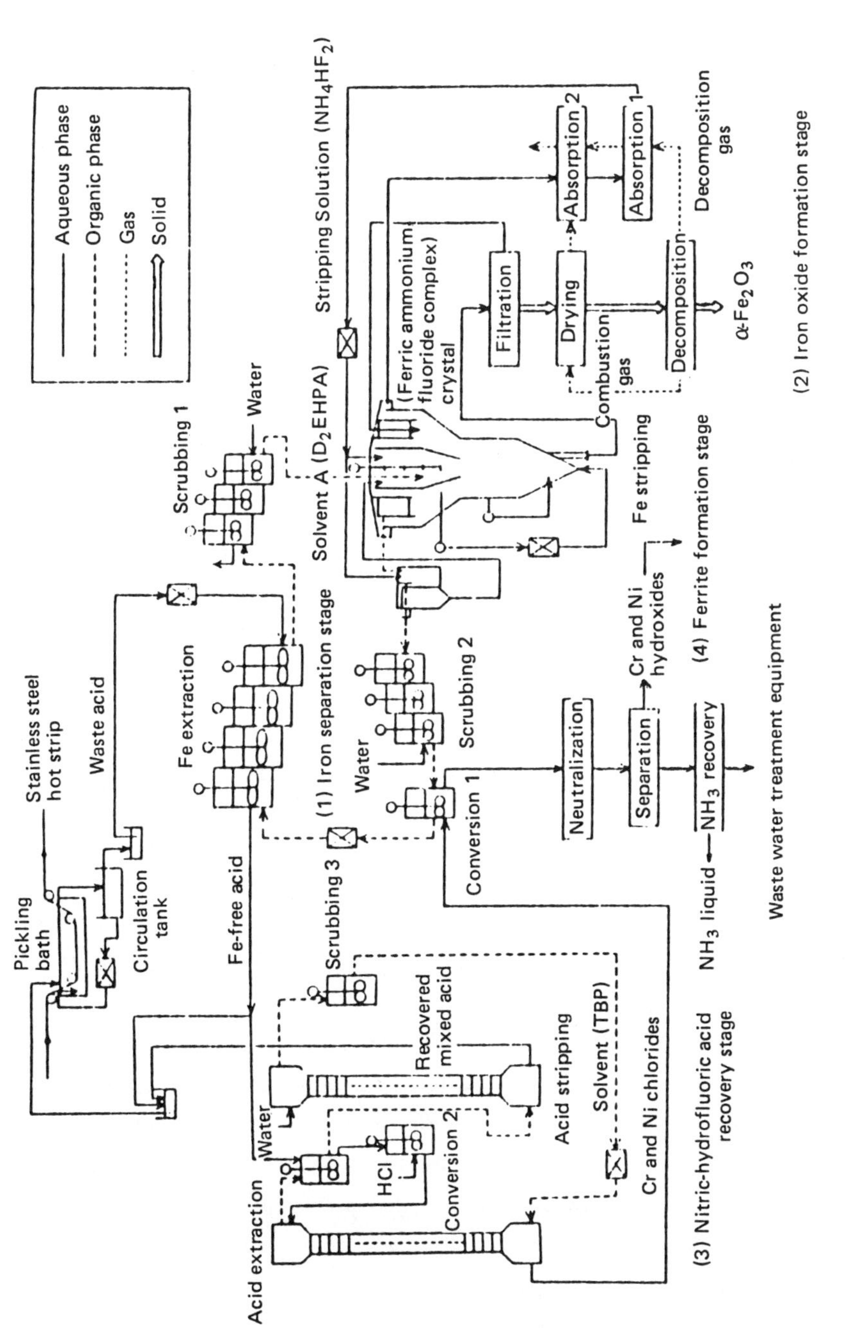

Figure 5.7.3. Flow diagram of Kawasaki process.

Source: Reference 4.

TBP, 30 percent n-paraffin solution to remove the nitrate and fluoride ions. The remaining aqueous solution contains metal chlorides which are further treated to form metal oxides for reuse in the steel process. The organic phase is stripped with water to recover nitric and hydrofluoric acids for return to the pickling bath and TBP/n-paraffin extractant which is recycled to the second-stage extractor.

Operating Parameters--

Important process parameters include the distribution ratios of nitric and hydrofluoric acids (i.e., organic phase concentration divided by aqueous phase concentration), and the stability constants (B-values) of the acids and metals in the aqueous phase.[6] Both parameters vary with the ionic medium, ionic strength, and the temperature of the system.[6] The distribution ratio will also be effected by the solution concentration.[4,6] Additional factors to consider include: the flow ratio of organic phase to aqueous phase, residence times of the two phases, and number of extraction stages required. These factors are discussed in detail below in terms of their interrelationships and effects on extraction efficiency.

During the extraction process, it is desirable to have high distribution ratios. Tributyl phosphate (TBP), used as the acid extractant, exhibits a strong affinity for uncharged molecules. The preferential order of acid extraction using TBP is as follows:

Nitric Acid > Hydrofluoric Acid > Hydrochloric Acid > Sulfuric Acid

The distribution ratios of nitric acid and hydrofluoric acid increase linearly with TBP concentration with optimal concentration for extraction being 75 percent TBP in kerosene.[6] As shown in Figure 5.7.4, the extraction of nitric acid is significantly improved by the addition of sulfuric acid. However, increasing the sulfuric acid additions will increase the post-treatment (neutralization) costs. The recommended optimum excess sulfuric acid concentration is 20 percent.[6] Alternatively, post-treatment costs can be lowered by using hydrochloric acid in this step instead of sulfuric acid since only stoichiometric amounts of hydrochloric acid are required for stripping.[7] Figure 5.7.5 shows that the distribution ratio for

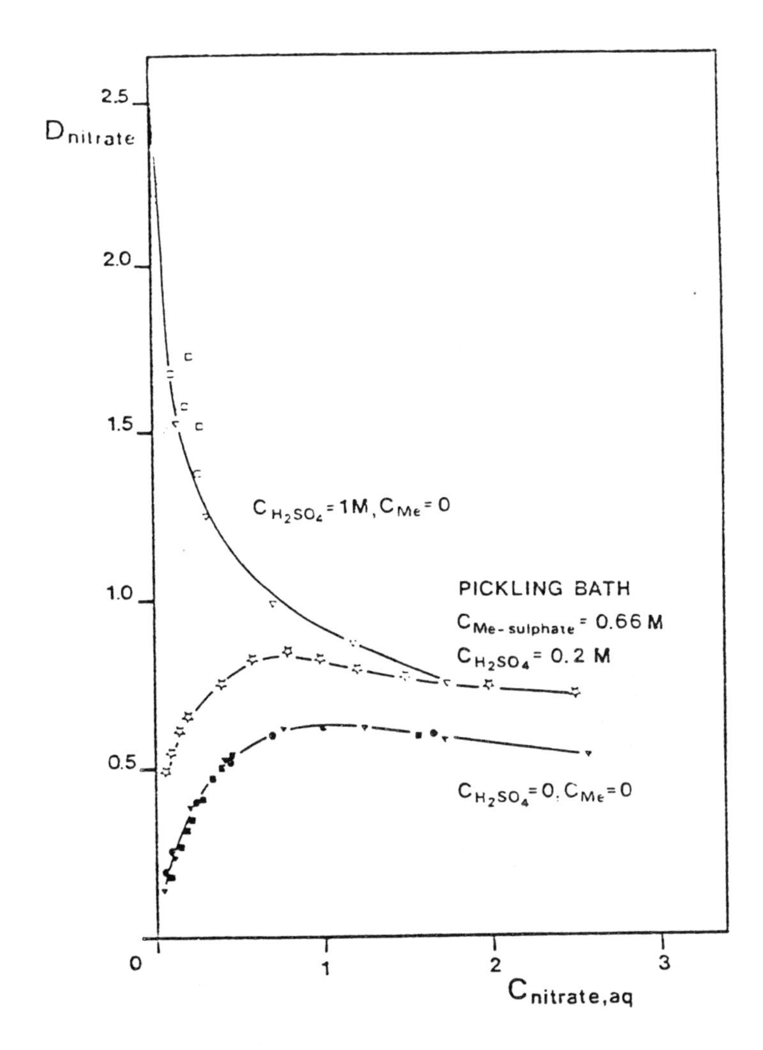

Figure 5.7.4. Distribution of nitrate between 75% TBP in kerosene and water as a function of concentration of total nitrate in the aqueous phase.

Notes: C = concentration
D = distribution ratio
Me = metal
aq = aqueous

Source: Reference 6.

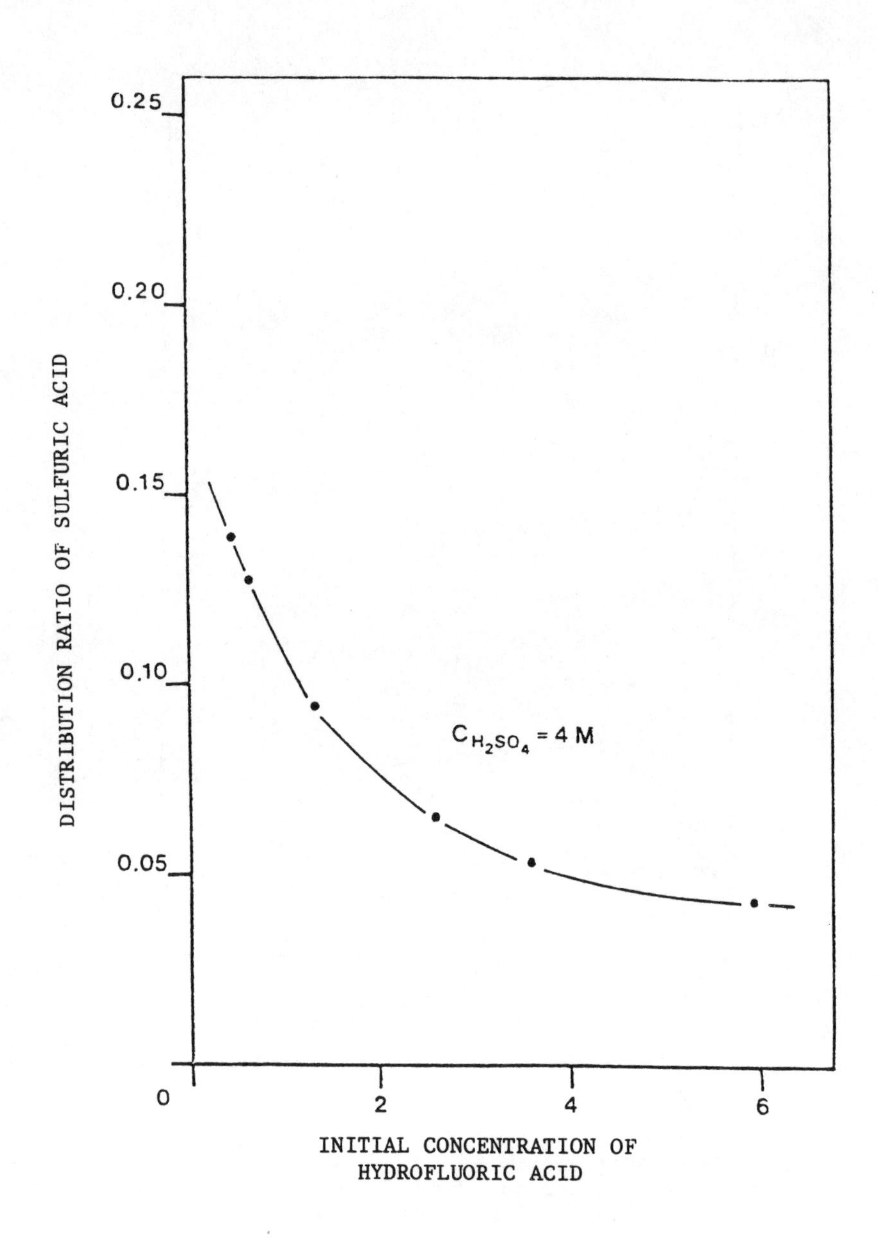

Figure 5.7.5. Distribution of H_2SO_4 between 100% TBP and water as a function of total added HF concentration.

Notes: C = concentration
D = distribution ratio
aq = aqueous

Source: Reference 6.

hydrofluoric acid decreases with increasing concentration. The data plotted in Figure 5.7.6 demonstrate that the quantity of sulfuric acid extracted is dependent on the hydrofluoric acid concentration. Similar results are observed when hydrochloric acid is used instead of sulfuric acid.[7]

In a mixed solution containing both nitric and hydrofluoric acids, the extraction efficiency for nitric acid will be better than that for hydrofluoric acid, although the extent to which this is true is dependent upon their relative concentrations.[6] As can be seen from Figure 5.7.7, at high fluoride concentrations, the distribution ratio of fluoride decreases with increasing nitrate concentration, but at low fluoride concentrations this effect is minimal or even reversed. In addition, at low nitric and hydrofluoric acid concentrations, increasing concentrations of sulfuric acid will increase the distribution ratio of fluoride.[6] Thus, as in the Solex process, the majority of hydrofluoric acid is extracted after the nitric acid has been removed.[6]

During the stripping process, it is more advantageous to have lower distribution ratios to ensure that most of the species is removed with the aqueous phase. As shown in Figure 5.7.8, nitric acid is readily stripped with water resulting in nearly complete recovery. However, low hydrofloric acid concentrations results in a high distribution ratio and, therefore, reduces the effectiveness of stripping.[6] However, the stripping efficiency for hydrofluoric acid can be increased by using an additional mixer-settler containing nitric acid, without sulfuric acid.[4,6]

The distribution ratios decrease with increasing solution temperature.[6] Therefore, more effective recovery is achieved if the solution is extracted at a lower temperature (approximately 10°C) and stripped at a higher temperature (above 60°C), but this would require additional equipment and energy costs for heating and cooling the feed.[6]

The organic flow in both extraction and stripping stages will be the same. The water flow used during the stripping stage should not be greater than 1.5 times the pickle liquor flow.[6] Since a high distribution ratio is good for extraction, but not for stripping, it is important to optimize the flow ratios according to the effects of concentration changes on distribution ratios.[6] For a unit capable of treating 70,000 tons/year of pickle liquor, the optimum organic:aqueous phase ratios were calculated to be 2.3 for extraction and 1.6 for stripping.[6]

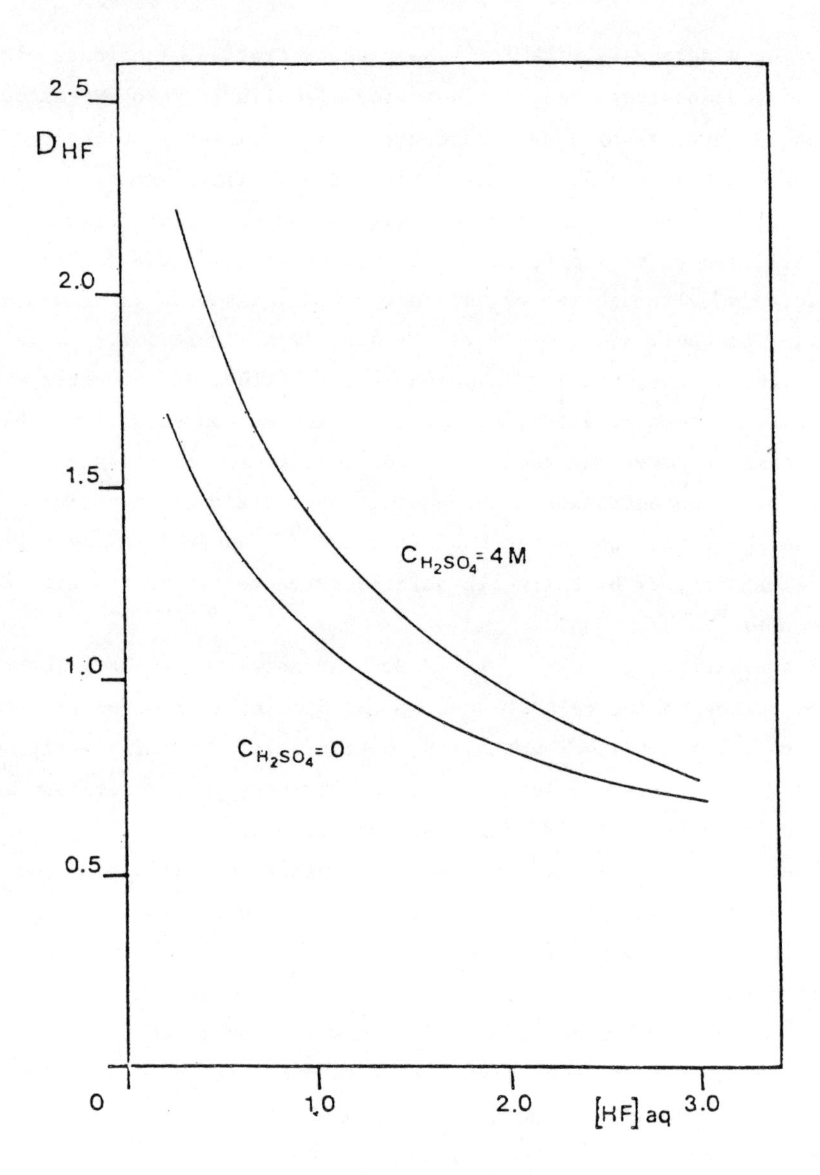

Figure 5.7.6. Distribution of HF between 100% TBP and water as a function of added concentration of HF and H_2SO_4.

Notes: C = concentration
D = distribution ratio
aq = aqueous

Source: Reference 6.

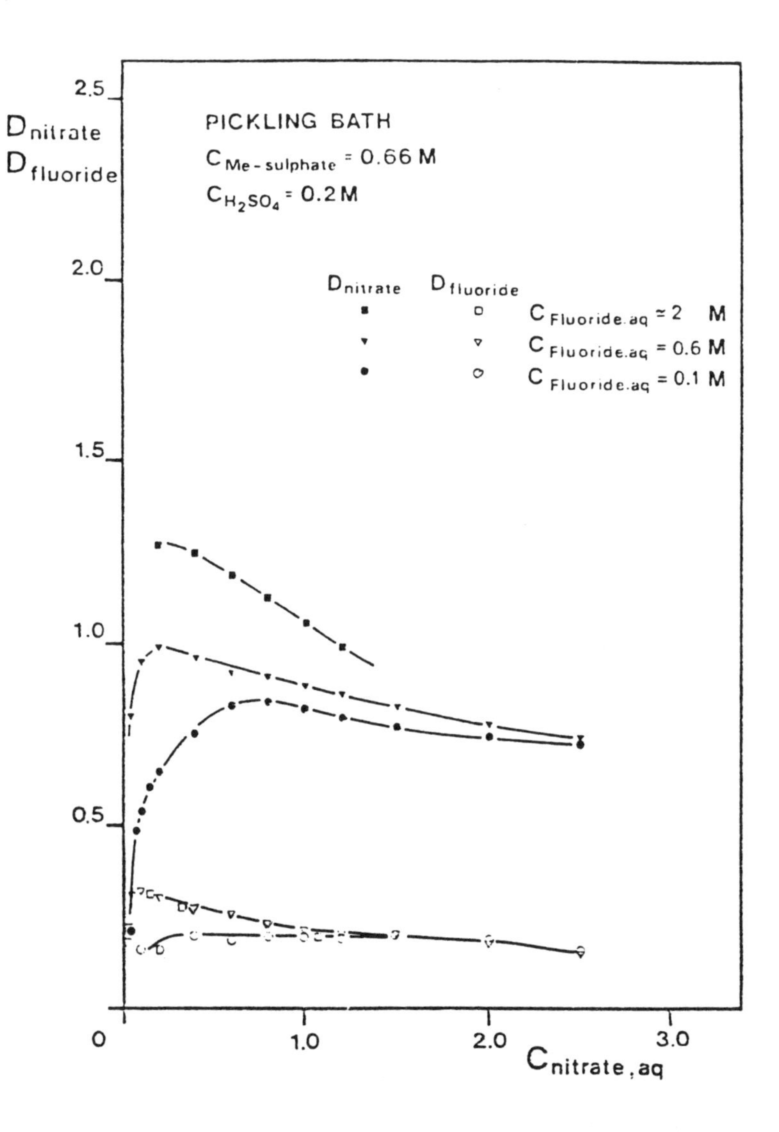

Figure 5.7.7. Extraction of HNO_3 and HF by 75% TBP in kerosene from aqueous solutions 0.33 M in Me_2 $(SO_4)_3$ as a function of total aqueous nitrate and fluoride concentration.

Notes: C = concentration
D = distribution ratio
Me = metal
aq = aqueous

Source: Reference 6.

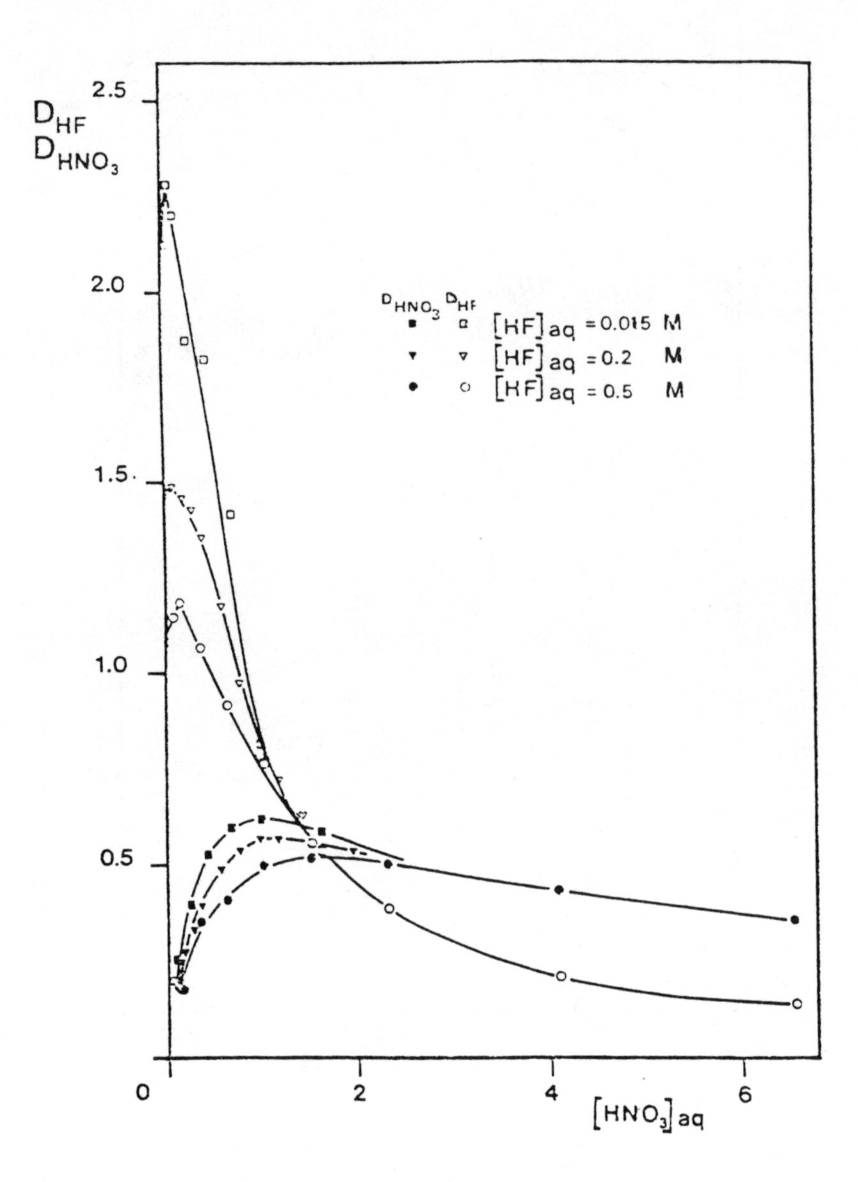

Figure 5.7.8. The distribution of HNO_3 and HF between 75% TBP in kerosene and water as a function of aqueous concentration of HNO_3 and HF.

Notes: C = concentration
D = distribution ratio
aq = aqueous

Source: Reference 6.

Pre-Treatment--

Removing iron prior to extracting nitric and hydrofluoric acids, as in the Solex and Kawasaki processes, permits greater recovery of hydrofluoric acid since the iron complex has a greater distribution ratio than hydrofluoric acid in the TBP extractant. The order of extraction is as follows:[7]

$$HNO_3 > HFeCl_4 > HF > HCl$$

As shown in Figures 5.7.9 and 5.7.10, the extraction efficiencies are adversely affected to a significant degree by the presence of iron in the solution.

Post-Treatment--

The solvent extraction process will generate a metal sludge that will require further handling. The sludge can either be neutralized and disposed, or recovered for reuse using thermal decomposition techniques as in the Kawasaki process. Metals recovery is facilitated if hydrochloric acid is used instead of sulfuric acid in the scrubbing step.

The spent scrubbing acid used to free the nitrate and fluoride ions will also require neutralization. The use of hydrochloric acid (Kawasaki process) instead of sulfuric acid (AX process) to free the nitrate and fluoride reduces sludge generation because a caustic soda neutralization can be used instead of lime.[7]

Extracting solutions can be recovered and reused in the solvent extraction process. The organic solvent exiting the stripping stage will contain small concentrations of acids (mostly HF), calcium, and molybdenum. It can be recovered by washing with a 1-M solution of sodium hydroxide in a mixer-settler prior to reuse in the extractor.[6] The wash solution can be treated by adding calcium nitrate to precipitate calcium fluoride, calcium molybdate, and a sodium nitrate filtrate which is recycled to the spent pickling bath feed to recover the extractable nitrate.

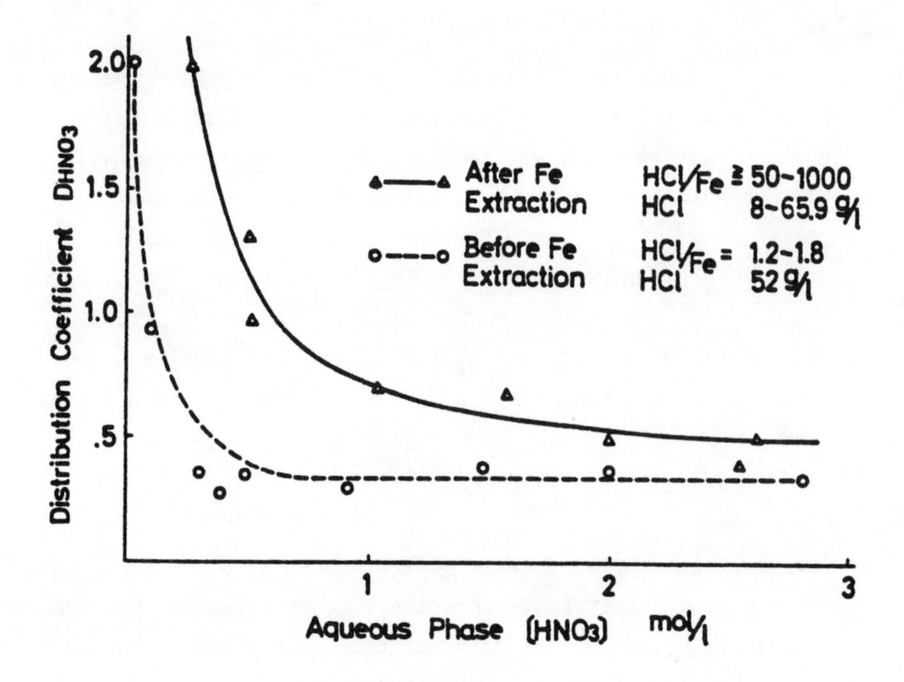

Figure 5.7.9. Extraction of nitric acid with 75% TBP in kerosene as a function of iron content.

Source: Reference 7.

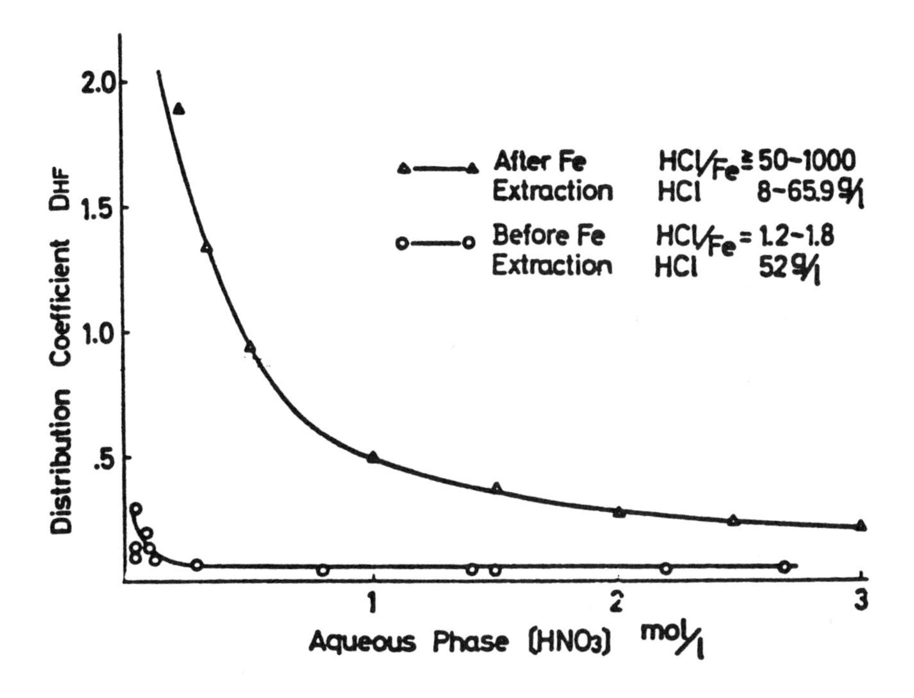

Figure 5.7.10. Extraction of hydrofluoric acid with 75% TBP in kerosene as a function of iron content.

Source: Reference 7.

5.7.2 Process Performance

Although research on the application of solvent extraction to acid recovery was performed in 1962 by the Republic Steel Company, the first commercial-scale application of the process was demonstrated at Stora Kopparbergs Bergslags in Sweden in 1973.[4] The AX process was developed utilizing countercurrent multiple-stage extractions in a pulsed column mode and a multi-stage mixer-settler arrangement.[6] Operating and design parameters for these two systems are presented in Tables 5.7.1 and 5.7.2, respectively. Results of tests conducted in each of these operational modes are summarized in Table 5.7.3. The results were successful in that they demonstrated the potential for recovering nitric and hydrofluoric acids while maintaining a continuous pickling process. High sludge generation, caused by the addition of sulfuric acid (which forms metal sulfates) to free the nitrate and fluoride ions for organic extraction, reduces the cost-effectiveness of this solvent extraction technique.

Nisshan Steel Co., Ltd. in Japan made modifications to the AX process and conducted commercial-scale testing of the modified process at Nisshan Steel, Shunan Works, in Japan.[4] Hydrochloric acid was added to the spent nitric-hydrofluoric acid to free the nitrate and fluoride ions. The nitrate and fluoride ions were then extracted with an organic solvent consisting of 75 percent TBP and 25 percent aromatic hydrocarbons. Nitric acid was used to strip the acids from the organic phase into the aqueous phase. The Nisshan process was able to recover 94 percent of the nitric acid and 16 percent of the hydrofluoric acid.[4] The lower hydrofluoric acid recovery was attributed to the fact that ferric ions form complex ions with fluoride ions.[4]

Further research led to the development of the Solex process which was modified for commercial application by Kawasaki Steel Corp. in Japan.[4] The typical composition of the spent nitric-hydrofluoric acid pickling liquor at this point is presented in Table 5.7.4. As described in Section 5.7.1, the process consisted of four stages: iron extraction, iron oxide formation, a second extraction and back-extraction (stripping) to recover nitric-hydrofluoric acid, and ferrite formation. The operating and design parameters for the recovery system are presented in Table 5.7.5. As shown, greater acid recoveries are achieved with the Kawasaki process. In addition, iron is also recovered.

TABLE 5.7.1. OPERATING AND DESIGN PARAMETERS FOR THE MULTI-STAGE PULSED MODE COLUMN

Item	Parameter
Treatment capacity	600 L/hr
Pulse amplitude	0 to 40 mm
Frequency amplitude	20 to 120 cycles/min
Number of stages	
Extraction	10
Stripping	5
Extraction column dimensions	
Height	9 m
Diameter	25 cm
Stripping column dimensions	
Height	3 m
Diameter	90 mm

Source: Reference 6.

TABLE 5.7.2. OPERATING AND DESIGN PARAMETERS FOR THE MULTI-STAGE MIXER-SETTLERS

Item	Parameter
Treatment capacity	600 L/hr
Number of stages	
Extraction	5
Stripping	10
Stripping column dimensions	
Height	8 m
Diameter	35 m

Source: Reference 6.

TABLE 5.7.3. RESULTS OF TESTS CONDUCTED AT STORA IN SWEDEN USING THE AX PROCESS[a]

Parameter	Extraction			Stripping		
	Pulsed column Test 1	Pulsed column Test 2	Mixer-settler	Pulsed column Test 1	Pulsed column Test 2	Mixer-settler
Flow rate						
Aqueous (mL/m)	985	1,070	49	915	1,000	53
Organic (mL/m)	2,860	2,530	127	1,500	1,789	102
Organic:aqueous	2.9	2.4	2.6	1.6	1.8	1.9
Pickling bath concentration (aqueous feed)						
HNO_3 (M)	1.68	2.25	2.25	0.54	0.91	1.10
HF (M)	1.68	1.87	1.87	0.41	0.35	0.39
Iron (g/L)	33	44	44	-	-	-
Chromium (g/L)	5	7.2	7.2	-	-	-
Nickel (g/L)	8	11.5	11.5	-	-	-
Molybdenum (g/L)	0.7	1.13	1.13	0.36	0.48	-
Concentration of exiting organic stream						
HNO_3 (M)	0.54	0.91	1.07	0.04	0.11	0.23
HF (M)	0.41	0.35	0.42	0.17	0.09	0.16
Iron (g/L)	-	-	-	-	-	-
Chromium (g/L)	-	-	-	-	-	-
Nickel (g/L)	-	-	-	-	-	-
Molybdenum (g/L)	0.36	0.48	-	0.17	0.25	-
Concentration of exiting aqueous stream						
HNO_3 (M)	0.02	0.01	0.01	0.92	1.38	1.39
HF (M)	0.47	1.05	0.84	0.42	0.38	0.39
Iron (g/L)	33	-	-	-	-	-
Chromium (g/L)	5	-	-	-	-	-
Nickel (g/L)	8	-	-	-	-	-
Molybdenum (g/L)	0.01	0.02	-	0.26	0.39	-
Percent extraction						
Nitric acid	99	99	99	93	88	79
Hydrofluoric acid	72	44	55	59	74	59

[a] Not analyzed.

Source: Reference 6.

TABLE 5.7.4. TYPICAL COMPOSITION OF SPENT PICKLING LIQUOR AT KAWASAKI STEEL CHIBA WORKS IN JAPAN

Constituent	Concentration (g/L)
Nitric acid	180 to 200
Hydrofluoric acid	40 to 45
Iron	28 to 30 (35 maximum)
Chromium	10 to 15
Nickel	5 to 10

Source: Reference 4.

TABLE 5.7.5. TYPICAL OPERATING PARAMETERS AND RESULTS FOR KAWASAKI PROCESS

Parameter	Iron separation stage	Nitric-Hydrofluoric acid recovery stage
Waste liquor constituents	Waste acid	Iron-free acid
Treatment capacity	1.0 m^3/hr (24 m^3/day)	0.3 m^3/hr (7.2 m^3/day)
Extraction or recovery percentage		
Nitric acid	98.5	95.7
Hydrofluoric acid	75.3	79.5
Iron	95	-

Source: Reference 4.

5.7.3 Process Costs

It is difficult to evaluate the economics of solvent extraction due to the limited number of facilities using the process. Currently, there are no commercial-scale solvent extraction systems being used in the United States.

Capital costs would include the equipment for the solvent extraction process and the thermal decomposition equipment (see Section 5.9) if ferric oxide recovery is judged to be cost effective. Operating costs would include the costs for solvent (TBP, kerosene, D_2EHPA), sulfuric or hydrochloric acid used in scrubbing, labor, and maintenance. Solvents and acids can be recycled to reduce acid purchase requirements, reduce neutralization and disposal costs, and generate savings due to metal oxide recovery.

5.7.4 Process Status

Solvent extraction has been demonstrated to be effective in the recovery of spent nitric-hydrofluoric acid pickling liquors generated by the steel industry. Commercial-scale systems have been tested and are in use in both Sweden and Japan. However, no commercial-scale solvent extraction systems have yet been employed in the United States.

Of the four solvent extraction processes developed for application to acid wastes, the Kawasaki (or Solex) process has shown the most promising results. Commercial-scale testing of the Kawasaki process has demonstrated 95 percent recovery of nitric acid, and 70 percent recovery of hydrofluoric acid.[4] In addition, 95 percent of the iron was recovered for reuse.[4] Kawasaki intends to eventually market the process.[4,8]

Although the process has been demonstrated to be technically feasible with good recovery results, limited data are available to assess its economic viability. However, with further demonstrations this process could prove to be an effective alternative to conventional neutralization and disposal practices.

REFERENCES

1. Clevenger, T.E., and J.T. Novak. Recovery of Metals from Electroplating Wastes Using Liquid-Liquid Extraction. Journal of the Water Pollution Control Federation, 55(7): 984-989. July 1983.

2. McDonald, C.W. Removal of Toxic Metals from Metal Finishing Wastewater by Solvent Extraction. Prepared for the U.S. EPA, Industrial Environmental Research Laboratory, Cincinnati, Ohio. EPA-600/2-78-011. February 1978.

3. Stephenson, J.B., Hogan, J.C., and R.S. Kaplan. Recycling and Metal Recovery Technology for Stainless Steel Pickling Liquors. Environmental Progress, 3(1): 50-53. February 1984.

4. Watanabe, T., Hoshimo, M., Uchino, K., and Y. Nakazato. A New Acid and Iron Recovery Process in Stainless Steel Annealing and Pickling Line. Kawasaki Steel Technical Report No. 14, pp. 72-82. March 1986.

5. U.S. EPA. Treatability Manual, Volume III: Technologies for Control/Removal of Pollutants. EPA-600/8-80-042c. July 1980.

6. Rydberg, J., Reinhardt, H., Lunden, B., and P. Haglund. Recovery of Metals and Acids from Stainless Steel Pickling Bath. In: Proceedings of the 2nd Annual International Symposium on Hydro-Metal, Chicago, Illinois. February 25 throuth March 1, 1973.

7. Watanabe, A., and S. Nishimura. Process for Treating an Acid Waste Liquid. Assigned to Solex Research Corporation, Osaka, Japan. U.S. Patent No. 4,166,098. August 28, 1979.

8. Japan Metal Bulletin. Kawasaki Recovers Iron and Acids from Pickling Line. No. 4530. May 19, 1984.

5.8 THERMAL DECOMPOSITION

5.8.1 Process Description

Thermal decomposition is a capital intensive, regeneration process which can be used to recover acid wastes, including sulfuric and hydrochloric acid pickling liquors and sulfuric acid wastes generated by the organics industry. Currently, the most significant area of application for this technology is the recovery of hydrochloric acid from spent pickling liquors generated by the steel industry. By using thermal decomposition processes, both free and bound acids can be recovered.

The hydrochloric acid regenerating process involves precipitating hydrous ferrous chloride and decomposing it in a roaster or furnace to form iron oxide and hydrochloric acid.[1] A general flow diagram of a HCl regeneration process is presented in Figure 5.8.1. The spent pickle liquor is preconcentrated in a venturi scrubber by heat exchange with the hot gases emerging from the reactor. The preconcentration step removes excess water from the spent pickling liquor which reduces the energy requirements for the subsequent roasting step.

Following preconcentration, the acid stream is fed to the roaster. Free water and hydrochloric acid evaporate in the upper regions of the roaster while hydrolysis takes place in the lower regions, resulting in the formation of iron oxide and hydrochloric acid. The iron oxide is discharged as a free-flowing powder from the base of the reactor. At the high reactor temperatures, hydrochloric acid vaporizes and exits through the top of the reactor with the combustion gases.

Hot gases exiting the reactor will contain hydrochloric acid gas, superheated water vapor, and inert combustion products. These gases pass through the venturi scrubber where a direct heat exchange occurs between the hot gases and the spent pickle liquor, such that preconcentrating the liquor and cooling the hot gases to 100°C occurs simultaneously. The cooled gases exit the venturi scrubber and are then directed under negative pressure to an absorber. This is charged with make-up water or water from the fume scrubber to recover hydrochloric acid at an approximate concentration of 20 percent.

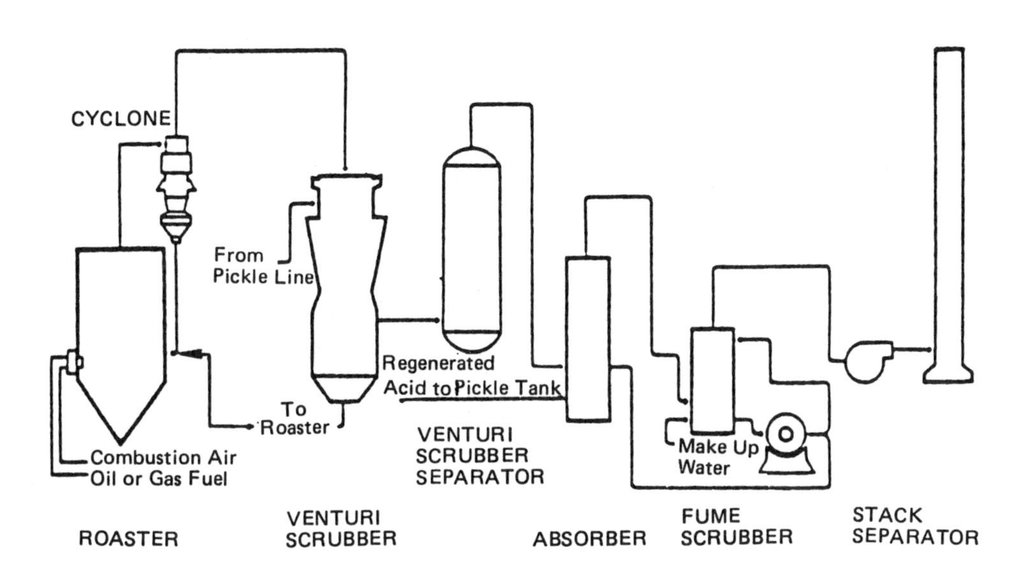

Figure 5.8.1 General flow diagram of HCl regeneration by thermal decomposition.

Source: Reference 2.

The remaining vapors from the absorber are discharged to the fume scrubber and exhausted to the atmosphere.[3] Several types of reactors may be used for the roasting step, including the following:

- Fluidized-Bed Reactor - Concentrated pickle liquid is evaporated in a fluidized-bed of granular iron oxide. Iron oxide product will adhere to the fluidized grains, and must be discharged at a continuous rate to maintain desired bed volume.
- Spray Roaster - Concentrated pickle liquor is atomized and evaporated at high temperatures in the reactor.
- Sliding Bed Reactor - Concentrated pickle liquor is sprayed onto a continuously circulating bed of hot iron oxide. Iron oxide is recirculated through the system using a surge bin and bucket elevator.

A spray roasting thermal decomposition process has recently been developed for the regeneration of waste sulfuric acid from the production of titanium dioxide pigments.[4] In this case, the spent solution contains iron and other metal sulfates. As with the HCl regeneration process, free sulfuric acid is vaporized and exits with the combustion gases to a condensor/absorber, where the free sulfuric acid can be recovered. However, unlike HCl regeneration, the initial roasting reaction does not regenerate the sulfuric acid directly; i.e., through hydrolysis of iron chloride. Instead, ferrous and other metal sulfates are oxidized to sulfur dioxide and an anhydrous metal salt precipitate. The metal sulfate salts can be further roasted to convert nearly all the sulfates to oxides and sulfur dioxide. The sulfur dioxide resulting from these roasting reactions is converted to sulfuric acid via conventional vanadium-catalyzed processes.[5,6]

Currently, research is underway for a thermal decomposition system which is capable of recovering sulfuric acid from wastes which contain a much broader range of constituents, such as those generated by organics industries.[7] Waste acids generated in the manufacture of herbicides, dyes, and bactericides may contain sulphonated polynuclear aromatics, chlorinated hydrocarbons, and nitrates. The conventional thermal decomposition techniques mentioned previously cannot be used for these wastes because of the hazardous air emissions that would be generated. Waste acids and acid tars with metal impurities cannot be regenerated with conventional thermal decomposition systems because of slag formation on the refractory wall lining which impedes the performance.[7]

A process developed by the German Ministry of Research and Technology (BMFT) uses a rotating furnace, an intermediate chamber, two additional combustion chambers, and a waste heat boiler to recover organic contaminated sulfuric acid streams.[7] Figure 5.8.2 presents a flow diagram of the process. Waste acid is fed with combustion air and sulfur to the rotating furnace which contains a coke bed, that serves as a heat reservoir and collects ashes. The sulfur, hydrocarbons, and hydrogen are oxidized in the furnace . Partial coking of the incoming hydrocarbons occurs which maintains a continuous bed of coke, and also reduces the nitrates and sulfur trioxide in the decomposition gas. The decomposition gas then flows to the intermediate chamber, where additional air is mixed with the decomposition gases before directing the flow to the secondary combustion chambers. In these units, liquefaction of the ash particles in the gas stream occurs, forming a slag bed which causes a 1 to 2 percent excess of oxygen in the decomposition gas mixture and results in a temperature drop. In subsequent steps, the decomposition gas temperature is further reduced in a heat exchanger, the gas is washed, and sulfur dioxide is condensed and recovered. The sulfur dioxide is then converted to sulfuric acid via vanadium-catalyzed processes.[5,7]

Operating Parameters--

Operating parameters which will affect the efficiency of thermal recovery systems include: heat requirements, process water requirements, and volume of material processed. These parameters also determine the size of the system required for regeneration, which in turn will affect the capital and operating costs. Typical operating parameters for HCl regeneration systems are presented in Table 5.8.1.

Indirect heating of the pickling tank, using either in-tank heaters or external circulation heaters, reduces the overall volume of waste acid to be treated by approximately 30 percent.[10] By reducing the volume of waste acid to be processed, a smaller size regeneration system can be employed with fewer capital costs.

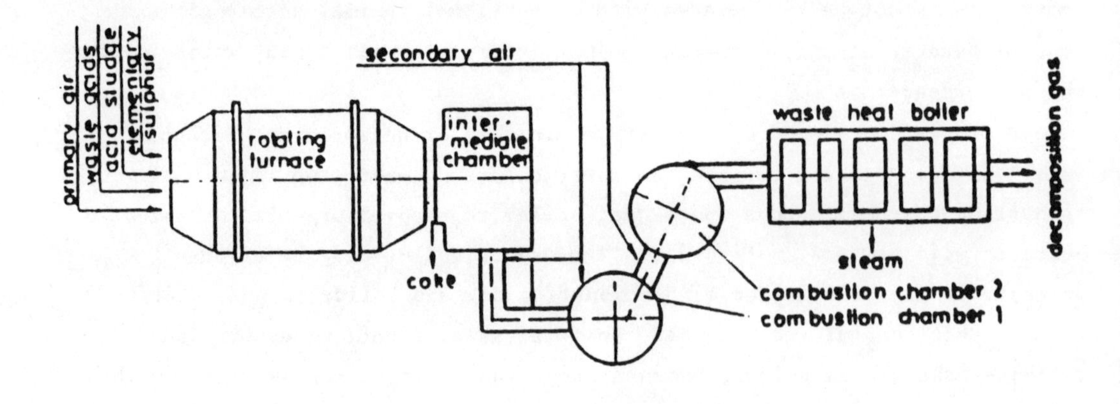

Figure 5.8.2. Flow diagram of a thermal decomposition process using a rotating furnace.

Source: Reference 7.

TABLE 5.8.1. TYPICAL OPERATING PARAMETERS FOR THERMAL DECOMPOSITION ACID REGENERATION SYSTEM

Parameter	Result
Reactor temperature	800 to 100°C
Temperature of gases exiting reactor	~400°C
Temperature of gases exiting venturi scrubber	100°C
Average cycle time	~10 hours
Specific heat comsumption	5,000 kcal/kg Fe_2O_3
Specific energy consumption	0.3 to 0.4 kwh/kg Fe_2O_3

Source: References 1, 2, 3, 6, 8, 9, and 10.

In order to reduce the volume of rinse water required for the process, a cascade rinsing system can be employed. Good rinsing can be achieved with 7 to 10 gallons of rinse water per ton of steel pickled by using four or five stages of cascade rinsing.[10] By reducing the volume of rinse water generated, treatment costs are reduced.

5.8.2 Process Performance

The performance of a thermal decomposition system can be evaluated on the basis of percent acid recovered, the purity (concentration) of recovered acid and oxide products, processing time, and byproduct generation.

Several companies currently offer thermal decomposition systems for the regeneration of hydrochloric acid from spent pickling liquors. The most commonly used reactor is the spray roaster. Typical results for these system are presented in Table 5.8.2. Unlike other recovery processes which are limited to recovery of free acid, thermal decomposition processes are able to recover more than 99 percent of the total HCl equivalent in the spent pickling waste.[10] In addition, a high quality iron oxide byproduct is generated, which can be reused within the plant or marketed for use in ferrite magnets, pigments, molding sands, glass and other industries.[10] Table 5.8.3 lists the typical composition and purity of the iron oxide byproduct formed.

The sulfuric acid thermal regeneration process for titanium dioxide production waste has not been tested at the commerical-scale level. However, pilot tests have demonstrated the potential to recover 93 to 96 percent of the sulfuric acid equivalent in the spent pickling waste.[4] The sulfuric acid regeneration process will generate a waste stream of unreacted sulfates which will require neutralization. However, this amount will be significantly less than the amount that is required if no recovery process is employed. For example, a regeneration system for a 50,000 metric-ton/year titanium dioxide production plant will generate 19,600 to 46,000 metric-tons/year of waste requiring neutralization as compared to the 100,000 to 200,000 metric-tons/year of waste requiring neutralization without the regeneration process.[4]

TABLE 5.8.2. TYPICAL PERFORMANCE OF A THERMAL DECOMPOSITION SYSTEM FOR REGENERATION OF HYDROCHLORIC ACID

Parameter	Result
Range of available regeneration capacities	5 to 75 gpm
Composition of spent pickling liquor	
Hydrochloric acid	0.5 to 5.0%
Ferric chloride	20 to 25%
Composition of regenerated acid	
Hydrochloric acid	20%
Iron	0.25%
Regeneration efficiency (percent of HCl equivalents recovered from waste)	>99%
Purity of iron oxide byproduct	98.5 to 99.4%

Source: Reference 10.

TABLE 5.8.3. TYPICAL COMPOSITION OF IRON OXIDE BYPRODUCT GENERATED BY THERMAL DECOMPOSITION PROCESS

Constituent	Percent composition
Ferrous oxide	98.5 to 99.4
Aluminum oxide	0.05 to 0.12
Manganese oxide	0.35 to 0.45
Chromium oxide	0.02 to 0.06
Silicon oxide	0.01 to 0.06
Calcium oxide	0.01 to 0.06
Magnesium oxide	0.01 to 0.04
Sodium oxide	0.01 to 0.06
Potassium oxide	0.01 to 0.03
Nickel oxide	0.01 to 0.05
Copper oxide	0.01 to 0.05
Chloride	0.05 to 0.20
Sulfur trioxide	0.02 to 0.04
Water solubles	0.10 to 0.70
Ignition loss	0.25 to 1.00

Source: Reference 10.

The rotary furnace thermal decomposition technique has also only been demonstrated at the pilot-scale. Unlike the previously mentioned decomposition systems, the rotary furnace is designed to handle waste acids with a variety of constituents. The German Ministry of Research and Technology (BFMT) tested 18 different waste sulfuric acids with varying concentrations, composition, and consistency. Table 5.8.4 lists the approximate compositions of these waste feeds. The performance evaluation results are summarized in Table 5.8.5. In all cases, it was found that sulfur dioxide could be obtained of sufficient quality to permit use in sulfuric acid production. Performance was not affected by changing waste viscosity, carbon content, or chemical forms of sulfur.

5.8.3 Process Costs

The economics of using thermal decomposition systems to regenerate waste hydrochloric acids from spent pickling liquors can be evaluated on the basis of the following factors:[2]

- Cost of recovered acid if purchased;
- Cost of treatment and disposal of waste acid by other technologies;
- Pickling waste quantity (which determines regeneration system size requirements and costs); and
- Quality and market value of the byproduct iron oxide.

These costs will vary with the particular application, but in all cases the capital costs will be high. Typically, capital costs for an installed system can range from $1 to 7 million for regeneration systems with capacities ranging from 5 to 75 gpm.[10,11,12] In some cases, the value of the recovered hydrochloric acid may be less than the costs for regeneration. However, with increasing disposal costs, and a developing market for the iron oxide byproduct, this situation could be reversed.[6]

Table 5.8.6 presents an economic evaluation of a small, medium, and large HCl regeneration system. As can be seen from the table, thermal decomposition is most cost-effective for plants generating large quantities of spent acid.

TABLE 5.8.4. COMPOSITION OF WASTES FED TO THE ROTARY FURNACE DURING TESTING CONDUCTED BY THE BMFT

	Approximate percent composition					
Waste acid type no.	Sulfuric acid	Organic carbon	Water	Ash	Chlorine	Difficult impurities
1	60	2	30			Ammonia
2	76	3-4	17.5	0.05		Brominated polynuclear aromatics
3	41-48	0.1-1.0	51-57	0.06	0.02	Aliphatics, Polynuclear aromatics
4	50-55	0.1-2.0	40-45	0.06		Ammonia
5	43	0.1-0.4	54	0.005	0.6	Aliphatics, Chlorinated aromatics
6	35	0.4	56	0.8	0.07	Ammonia
7	66-70	15-22	10-14	0.1		Aromatics (alkylbenzenes)
8	90	4-5	4-6	0.01		Aromatics
9	50-53	1.2-5.3	42-45	0.05		Aliphatics, Aromatics, Ketones
10	65	2	32	0.05	0.1	Aromatic sulfonic acids
11	26	0.2	73	0.07	0.02	Sulfonated aromatics, Phosphates
12	50	0.1	50	0.1	20 ppm	Chlorinated aromatics

(continued)

TABLE 5.8.4 (continued)

Waste acid type no.	Approximate percent composition: Sulfuric acid	Organic carbon	Water	Ash	Chlorine	Difficult impurities
13	75-59		1.0-3.6	19-21	0.4-4.4	Aliphatics, Aromatics
14	90-92	0.2-0.5	7.5-9.5	0.02		Aromatics, Heterocycles
15	77	5.3	18	0.002		Sulfonated branched unsaturated alkanes
16	68	19	10	0.05		Sulfonated branched unsaturated alkanes
17	30	0.3	69	0.01	0.02	Nitric acid, nitrated aromatics, Ketones
18	30-35	30	30-33	6		Chlorinated aromatics, Alkalis, Alkaline earth

Source: Reference 7.

TABLE 5.8.5. SUMMARY OF ANALYTICAL RESULTS OF THE ROTARY FURNACE PILOT TESTS CONDUCTED BY THE BMFT

Parameter	Result
Throughput Rates	
Waste sulfuric acid	1,000 to 2,300 kg/hr
Acid tar	700 to 1,050 kg/hr
Sulfur	800 to 1,200 kg/hr
Average Composition of Decomposition of Gas	
Oxygen	0.8 to 2.0 %-vol.
Sulfur dioxide	10.0 to 15.0%-vol.
Carbon dioxide	6.0 to 10.0%-vol.
Composition of Processed Decomposition Gas (mg/m^3)	
Sulfur trioxide	220 to 640*
Elemental sulfur	2 to 45
Ammonia	0 to 20
NO_x	0 to 4
NO_2	ND
Hydrogen sulfide	ND
Hydrocarbons	ND
Average sulfuric acid output rate	170 tons/day

*The SO_3 can be converted to sulfuric acid via vanadium catalyzed reactions.

Source: Reference 7.

TABLE 5.8.6. ECONOMIC EVALUATION OF HYDROCHLORIC ACID REGENERATION USING THERMAL DECOMPOSITION

Item	Cost basis	System size (gpd)		
		10,000	100,000	200,000
Capital Costs	TIC*	$3,907,000	$14,974,000	$23,487,000
Operating Costs				
Labor	1.8% of TIC	69,000	266,000	427,000
Maintenance	3% of TIC	120,000	460,000	720,000
Fuel	12,000 Btu/gal/waste	59,000	225,000	363,000
Electricity	0.10 kwh/gal waste	11,000	40,000	65,000
Water	1 gal/gal waste	6,000	24,000	39,000
	Total	$265,000	$1,015,000	$1,614,000
Cost Savings				
Acid value	50% of PMV**	457,000	4,570,000	9,133,000
Iron oxide value	$100/ton	187,500	1,875,000	3,750,000
Treatment and disposal costs	Caustic soda	12,000	120,000	240,000
	Total	$656,500	$6,565,000	$13,123,000
Net Annual Savings	Savings - operating	$391,500	$5,550,000	$11,509,000
Payback Period (years)	Capital÷Net savings	10.0	2.7	2.0

* TIC = Total installed cost
**PMV = Present market value

Source: References 9, 10, and 13. Costs updated to July 1986 dollars.

Costs for the two sulfuric acid regeneration systems can not be realistically determined at this time, since both processes have not been developed beyond the pilot-scale level. Cost savings would be realized by the recovered acids and the reduced disposal requirements. Significant quantities of these waste acids are currently neutralized and disposed at high costs. For example, a typical 50,000 metric ton/year pigment plant using lime or limestone neutralization can produce up to 360,000 metric tons/year of gypsum, while consuming over 130,000 tons of limestone.[4] Significant savings in materials would be realized if these acids could be recovered. However, capital costs for these systems are expected to be the same or slightly higher than the costs for HCl regeneration systems. In addition, according to current market prices, the value of the recovered sulfuric acid would be less than the value of the same quality of hydrochloric acid. Therefore, a sulfuric acid regeneration system would not be cost-effective except for large quantity generators or waste disposal facilities.

5.8.4 Process Status

Thermal decomposition is an effective but capital intensive process for the recovery of hydrochloric acid from spent pickling liquors. It is a demonstrated technology and is currently being used by several steel manufacturers. There are currently no commercial-scale applications of thermal decomposition for wastes other than spent hydrochloric acid pickling liquors. However, research has demonstrated the technical feasibility of using thermal decomposition to regenerate waste sulfuric acid effluents from spent sulfuric acid pickling liquors and from organic industry waste acids.

The thermal decomposition process has the advantage of being able to recover bound as well as free acid from waste, which distinguishes it from the previously mentioned recovery technologies. More than 99 percent of the hydrochloric acid equivalents in waste pickle liquor can be regenerated, and an estimated 93 to 96 percent of sulfuric acid equivalents can potentially be regenerated by thermal decomposition.[4,11,12]

However, capital costs for thermal decomposition will be prohibitive for small volume generators. Although the total quantity of waste sulfuric acid generated by the steel industry is greater than the amount of hydrochloric acid generated, the latter is generated by a small number of large quantity generators. Combined with the higher purchase price of hydrochloric acid, HCl regeneration systems may be more implementable than sulfuric acid regeneration systems in the steel industry. However, large quantities of waste sulfuric acid are generated by individual organic chemical manufacturing plants, and therefore acid regeneration may have a wider application for this industry.

For large quantity generators, the use of thermal decomposition will realize a net savings in reduced acid purchase, waste treatment, and disposal costs. In addition, the burden to the environment will also be significantly reduced. With increasing costs for disposal, and increasing development of the technology for other waste types, thermal regeneration systems are likely to find wider application in corrosive waste treatment.

REFERENCES

1. Ruthner, Michael, and Othmar Ruthner. Twenty-five (25) Years of Process Development in HCl Pickling and Acid Regeneration. Iron and Steel Engineer. November 1979.

2. Wadhawan, Satish C. Economics of Acid Regeneration - Present and Future. Iron and Steel Engineer. October 1978.

3. Bierbach, Herbert, and Klaus Hohmann. Continuous Regeneration of Hydrochloric Acid Pickle Liquors and Other Metal Chlorides According to the Lurgi Process. Wire World International, 15(5):161-163. September/October 1973.

4. Smith, Ian, Gordon M. Cameron, and Howard C. Peterson. Acid Recovery Cuts Waste Output. Chemical Engineering. February 3, 1986.

5. Franklin Associates. Industrial Resource Practices: Petroleum Refineries and Related Industries. Prepared for U.S. EPA, Office of Solid Waste, Washington, DC, under EPA Contract No. 68-01-6000 (1982B). 1982.

6. Versar, Inc. National Profiles Report for Recycling - A Preliminary Assessment. Draft Report Prepared for the U.S. EPA, Waste Treatment Branch, under EPA Contract No. 68-01-7053 (Work Assignment No. 17). July 8, 1985.

7. Driemal, Klaus, Norbert Lowicki, and Joachim Wolf. Harmless Disposal of Waste Sulfuric Acids Containing Polynuclear Sulphonated Aromatics In A Rotating Furnace with Special Further Combustion Stages. Project of the German Ministry of Research and Technology (BMFT). In: Recycling International, 4th edition, pp. 1078-1083. Karl J. Thome-Kozmiensky, editor. Published by EF-Verlag fur Energie- and Umwelttechnik. ISBN 3-924511-05-5. 1984.

8. Elliot, A.C. Regeneration of Steelworks Hydrochloric Acid Pickle Liquor. Effluent and Water Treatment Journal. July 1970.

9. Camp, Dresser, and McKee, Inc. (CDM). Technical Assessment of Treatment Alternatives for Wastes Containing Corrosives. Prepared for the U.S. EPA, Office of Solid Waste, under EPA Contract No. 68-01-6403 (Work Assignment No. 39). September 1984.

10. Perox, Inc. Brochure: Hydrochloric Acid Regeneration and Iron Oxide Production. Received August 1986.

11. Wadhawan, Satish C. Perox, Inc., Pittsburgh, Pennsylvania. Telephone Conversation with Jon Spielman, GCA Technology, Inc. August 6, 1986.

12. Wadhawan, Satish C. Perox, Inc., Pittsburgh, Pennsylvania. Letter to Jon Spielman, GCA Technology, Inc. August 7, 1986.

13. Chemical Monitoring Reporter, 230(3):32-40. July 21, 1986.

5.9 WASTE EXCHANGE

5.9.1 Description

Waste exchange involves the transfer of an unwanted waste material to a company which is capable of using it in its industrial process. Corrosive wastes can be exchanged and reused for neutralization (see Section 4.1), pH adjustments, and cleaning solutions. Corrosive wastes which are most commonly exchanged are listed in Table 5.9.1.

Waste exchanges are typically initiated by a third party, who uses either passive or active means to effect the transfer. The passive approach occurs through the use of an information clearinghouse, which is typically sponsored by governmental agencies (e.g., regional chambers of commerce) or industry associations. Active waste transfers are handled by materials exchange services which actually buy, treat, process and/or store wastes, and solicit potential users for them.[1,2]

The clearinghouse approach is the most prevalent waste exchange technique. Clearinghouses generally issue a catalog in which generators list the wastes they wish to transfer, and potential users list the wastes they desire. Information typically contained in the catalogue includes: waste description, quantity generated, availability, and general location of the waste.[2] Waste materials are assigned a code number so that the identity of the listing company can be kept confidential. An interested user will send a letter of inquiry to the clearinghouse, who will then forward the letter to the generator. The actual waste transfer will be handled by the generator and the user.

Clearinghouses exchange information only. They do not actively seek out customers, negotiate transfers, set values, process materials, transport materials, or provide legal advice; these functions are left to the participating waste generator and waste user.[1,3] In order to be successful, clearinghouses must be responsive to industry needs. They typically provide:[1]

- Support of plant managers and engineers faced with disposal problems;
- Industry endorsement and assistance; and
- Confidentiality of industry identities and waste generation data.

TABLE 5.9.1. LIST OF MOST COMMONLY EXCHANGED CORROSIVE WASTES

Acids	Alkalis
Hydrochloric acid	Sodium hydroxide
Hydrofluoric acid[a]	Sodium carbonate
Nitric acid[a]	Acetylene sludge
Phosphoric acid[a]	
Sulfuric acid[a]	
Picric acid[a]	
Pickle liquors ($FeCl_2$ or $FeSO_4$)[a]	

[a]In concentrations exceeding 15 percent.

Source: References 1 and 2.

A materials exchange takes a more active role in the waste transfer by actually identifying a potential match and assisting in consumating a contract between the waste generator and the waste user.[1] The material exchange acts as an agent or broker, seeking a buyer or seller for the waste material.[3] In some cases, the materials exchange will actually take possession of the wastes, charge a fee for handling the transaction, process the waste (to make it marketable), and sell the waste to a potential user.[2] In other cases, a materials exchange will use various information sources to match generators with potential users. Typical sources of information include: waste exchange catalogs, personal contacts with industry, trade associations, trade journals, and referrals from other industries and government agencies. The materials exchange will then actively assist in the arrangements between the generator and potential user.[2]

By taking a more active role in the waste transfer, a materials exchange incurs greater legal risks than a clearinghouse. The legal liability for a materials exchange is the same as that for other companies involved in the hauling, treating, and handling of hazardous waste materials. Legal accountability for parties involved in a waste exchange encompasses the following areas:[1]

- Public Liability - Generator is responsible for packaging, handling, and transportation of waste while under the control of a transfer agent or user.
- Liability to Third Party - Personal or property injury resulting from a waste in transit between generator and user.
- Contractual Liability - Responsibilities to users with regard to content of the waste.

Although a materials exchange incurs greater risks and costs, it is generally more effective than a clearinghouse in recycling industrial waste.

5.9.2 Application

Important factors in the success of waste exchange operations include: the compatibility of generator waste material with user processes, the purity of the waste material, the quantity of specific materials, variability in waste characteristics or availability, and the amount of processing required.[1]

With a clearinghouse, the responsibility of ensuring compatibility and purity of the waste lies with the waste generator and the waste user. However, in taking a more active role, a materials exchange also assumes this responsibility. Laboratory analyses are performed to determine the constituents of the waste material and the level of impurities present.

Generally, successful results are achieved when wastes are transferred from industries having high purity requirements to those with lower purity requirements. Both caustic and acidic wastes with relatively high quantities of impurities can be reused as cleaning solutions (lower purity requirements). For example, sulfuric acid wastes from pharmaceutical manufacturers can be reused by iron and steel plants in cleaning rolling steel.[11] Greater purity is generally required when the corrosive wastes are reused in a manufacturing process. Although purity is not as critical when corrosive wastes are reused for neutralization or pH adjustments, compatibility is very important.

Most users require large amounts of waste material that is available on a regular basis. However, there are occasions when small-volume, one-time-only wastes will be required. A clearinghouse will generally list all volumes of waste, whereas a materials exchange will only list those wastes available in sufficient quantity to guarantee a transfer. Often, industries producing large quantities of a specific waste type will find it more economical to process the waste in-house and reuse it, or send it to a recovery facility. Medium and small-volume generators typically cannot afford in-house processing equipment and are, therefore, more likely candidates for waste transfer.

Large volumes of waste materials are generated as sludge from a variety of industries. The large volume, complex composition, and physical form of sludge makes their transfer impractical.[1,2,3] However, if waste streams are segregated and processed near the point of origin (i.e., prior to sludge generation), valuable resources (e.g., acids, alkalies, and metals) may be recovered more cost-effectively.

Waste exchanges are more successful when the area served by the waste exchange includes a larger, more diversified industrial base. Industries which represent the majority of both potential users and generators that could benefit from the use of waste exchanges are listed in Table 5.9.2.

TABLE 5.9.2. INDUSTRIES REPRESENTING THE MAJORITY OF GENERATORS AND POTENTIAL USERS BENEFITING FROM WASTE EXCHANGE

SIC code	Industry description	Percent of wastes transferrable
2911	Petroleum refining	63
2865, 2669	Organic chemicals	22
2831, 2833, 2834	Pharmaceuticals	17
35XX	Industrial machinery	17

Source: Reference 1.

5.9.3 Costs

Costs of waste exchange will vary with the type of service employed, the amount of waste to be transferred, the distance between exchanging companies, and reprocessing costs. In order for a waste exchange to be a cost-effective alternative for users and generators, the transfer must meet the following requirements:[1,4,5]

- User's net costs for waste material must be lower than purchase costs for feedstock;
- Generator's net costs for waste transfer must be lower than waste disposal costs; and
- User's and generator's net gain from the waste transfer will adequately compensate any transportation and/or reprocessing costs.

The costs to waste generators and potential waste users for using the services of a clearinghouse are minimal. The prevailing practice by clearinghouses is to charge a flat service fee, regardless of the waste quantity listed. This fee generally ranges from $5 to $20 per listing.[1] Clearinghouses, which are not associated with an organization or trade association, may also charge a subscribers fee for their waste catalogs.

If a waste generator or potential waste user chooses to use the clearinghouse form of waste exchange, they will also incur additional costs for laboratory analysis of the waste to be exchanged, transportation costs for transferring the waste, and costs for any reprocessing required. For the waste generator, these costs can be offset by reduced disposal costs and/or revenues from the sale of the waste material. Costs incurred by the waste user will be offset by reduced purchase costs for feedstock material. A transfer over a distance greater than 50 miles is generally not economical if the waste is valued at less than $0.13/lb.[1,6,7]

Costs for a materials exchange service would vary with the particular waste to be transferred. These costs would depend upon the market potential of the waste material, transportation costs, laboratory analytical costs, and reprocessing costs.[1] Laboratory analysis, reprocessing, and transportation may sometimes be subcontracted to other companies by the material exchange.[3] Although costs for use of a materials exchange service are higher, the generator and user do not have to handle the waste transfer and, therefore, incur fewer costs and risks during the actual waste transfer. Examples of cost-effective materials exchanges are given in Table 5.9.3.

5.9.4 Status

A number of clearinghouses and material exchanges are currently operating in Europe, Canada, and the United States. Tables 5.9.4 and 5.9.5 list the clearinghouses and material exchanges currently operating in North America. Waste exchange can serve the following functions:[1,4]

- Saves valuable raw materials;
- Saves energy by not having to process raw materials for disposal;
- Saves purchase costs for raw materials; and
- Benefits health and environmental quality by decreasing the procurement of raw materials and the disposal of waste.

With rising raw materials costs, and increasing restrictions on land disposal, waste exchange is becoming an increasingly attractive alternative to conventional practices for handling corrosive wastes. Under the HSWA of 1984, small generators will have increased requirements for waste handling. Waste exchange may be a cost-effective alternative for these small generators, who do not have the resources for inhouse recycling/reuse.

TABLE 5.9.3. EXAMPLES OF ECONOMIC USES OF WASTE EXCHANGES FOR CORROSIVE WASTES

User/generator	Tons generated	Waste description	Savings over disposal costs ($)	Savings over raw material purchase($)	Net value of exchange ($)
Storage facility	0.11	Sulfuric acid	350	25	375
Milling company	12.10	Soda ash, liquids	1,350	1,320	2,670
Manufacturer	1.20	Sodium hydroxide	NA	1,600	1,600
Steel processor	480.00	Sulfuric acid	55,200	NA	55,200
Manufacturer	144.00	Sodium hydroxide	35,200	168,000	203,200
County agency	0.44	Oxalic acid	900	350	1,250
Castings mfgr.	Ongoing	Sodium hydroxide	*	*	6,060
Metal products mfgr.	Ongoing	Sodium hydroxide	*	9,582	9,582

Notes: NA = Not applicable.
* = Not available.

Source: References 2, 8, 9, and 10.

TABLE 5.9.4. CLEARINGHOUSE (INFORMATION) WASTE EXCHANGES IN NORTH AMERICA

Clearinghouse	Location	Contact	Telephone No.
Alberta Waste Materials Exchange	Edmonton, Alberta, CAN	Charles Wood	(403)436-6303
California Waste Exchange	Sacramento, CA	Robert McCormick	(916)324-1818
Canadian Inventory Exchange[a]	Ste-Adle, Quebec, CAN	Philippe LaRoche	(514)229-6511
Canadian Waste Materials Exchange	Mississauga, Ontario, CAN	Robert Laughlin	(416)822-4111
Enkarn Research Corporation[a]	Albany, NY	J.T. Engster	(518)436-9684
Georgia Waste Exchange[a]	Marietta, GA	Michael Weeks	(404)363-3022
Great Lakes Regional Waste Exchange	Grand Rapids, MI	William Stough	(616)451-8992
Industrial Materials Exchange Service	Springfield, IL	Margo Ferguson	(217)782-0450
Industrial Waste Information Exchange	Newark, NJ	William E. Payne	(201)623-7070
Manitoba Waste Exchange	Winnipeg, Manitoba, CAN	Rod McCormick	(204)257-3891
Midwest Industrial Waste Exchange	St. Louis, MO	Clyde H. Wiseman	(314)231-5555
Montana Industrial Waste Exchange	Helena, MT	Buck Boles	(406)442-2405
Northeast Industrial Waste Exchange	Syracuse, NY	Lewis Cutler	(315)422-6572
Ontario Waste Exchange	Mississauga, Ontario, CAN	Brian Forrestal	(416)822-4111
Piedmont Waste Exchange	Charlotte, NC	Mary McDaniel	(704)597-2307
Southern Waste Information Exchange	Tallahassee, FL	Roy Herndon	(904)644-5516
Tennessee Waste Exchange	Nashville, TN	Sharon Bell	(615)256-5141
Wastelink, Div. of Tencon Assoc.[a]	Cincinnati, OH	Mary E. Malotke	(513)248-0012
Western Waste Exchange	Tempe, AZ	Nicholas Hild	(602)965-2975

[a]Operates for profit.

Source: References 3 and 11.

TABLE 5.9.5. MATERIAL EXCHANGES IN NORTH AMERICA

Material exchange	Location	Contact	Telephone No.
Resource Recovery of America, Inc.	Mulberry, FL	Robert Kincart	(813)425-1084
Zero Waste Systems, Inc.	Oakland, CA	Trevor Pitts	(415)893-8257
NYC Environmental Facilities Corp.	Albany, NY	Peter A. Marini	(518)457-4132

Source: References 3 and 12.

REFERENCES

1. GCA Corporation, Technology Division. Industrial Waste Management Alternatives Assessment for the State of Illinois, Volume IV: Industrial Waste Management Alternatives and Their Associated Technologies/Processes. Final Report prepared for the Illinois Environmental Protection Agency, Division of Land Pollution Control. GCA-TR-80-80-G. February 1981.

2. New York State Environmental Facilities Corporation (EFC). Industrial Materials Recycling Act Program, 4th Annual Report. 1985.

3. Jones, E.B., and W. Banning. The Role of Waste Exchanges in Waste Minization and Reclamation Efforts. In: Proceedings of the Hazardous and Solid Waste Minimization Seminar, sponsored by Government Institutes, Inc., Washington, D.C. May 8-9, 1986.

4. Terry, R.C., et al. Arthur D. Little, Inc. Waste Clearinghouses and Exchanges: New Ways for Identifying and Transferring Reusable Industrial Process Waste. EPA/SW-130C. October 1976.

5. Terry, R.C., Berkowitz, J.B., and C.H. Porter. Waste Clearinghouses and Exchanges. Chemical Engineering Progress. December 1976.

6. Laughlin, R.G.W., Golomb, A., and H. Mooij. Waste Materials Exchanges for Environmental Protection and Resource Conservation. In: Proceedings of the 24th Ontario Industrial Waste Exchange Conference, Toronto, Ontario. June 1, 1977.

7. Laughlin, R.G.W. Canadian Waste Exchange Program: Successes and Failures. In: Proceedings of the National Conference on Hazardous and Toxic Wastes Management, New Jersey Institute of Technology. June 5, 1980.

8. New York State Environmental Facilities Corporation (EFC). Industrial Materials Recycling Act Program, 5th Annual Report. 1986.

9. New York State Environmental Facilities Corporation (EFC). Industrial Materials Recycling Act Program, 3rd Annual Report. 1984.

10. New York State Environmental Facilities Corporation (EFC). Industrial Materials Recycling Act Program, 2nd Annual Report. 1983.

11. Versar, Inc. National Profiles Report for Recycling - A Preliminary Assessment. Draft Report prepared for the U.S. EPA, Waste Treatment Branch, under EPA Contract No. 68-01-7053 (Work Assignment No. 17). July 8, 1985.

12. Marini, P.A. New York State Environmental Facilities Corporation. Telephone conversation with J. Spielman, GCA Technology Division, Inc. July 29, 1986.

6. Considerations for System Selection

6.1 GENERAL CONSIDERATIONS

Waste management options consist of four basic alternatives: source reduction, waste exchange, recycling/reuse, use of a treatment (i.e., neutralization) or disposal processing system or some combination of these waste handling practices (see Figure 6.1.1). Recovery, treatment, and disposal may be performed onsite in new or existing processes or through contract with a licensed offsite firm which is responsible for the final disposition of the waste. Selection of the optimal waste management alternative will ultimately be a function of regulatory compliance and economics, with additional consideration given to factors such as safety, public and employee acceptance, liability, and uncertainties in meeting cost and treatment objectives.

Many of the technologies discussed in previous sections can be utilized to meet land disposal ban requirements or to achieve adequate waste recovery rates. However, practicality will limit applications to waste streams possessing specific characteristics. Since many processes yield large economies of scale, waste volume will be a primary determinant in system selection. The physical and chemical nature of the waste stream and pertinent properties of its constituents will also determine the applicability of waste treatment processes. Treatment will often involve neutralization and the use of other technologies in a system designed to progressively recover or destroy hazardous constituents in the most economical manner. Incremental costs of contaminant (i.e., toxic organics and heavy metals) removal will increase rapidly as low concentrations are attained.[1]

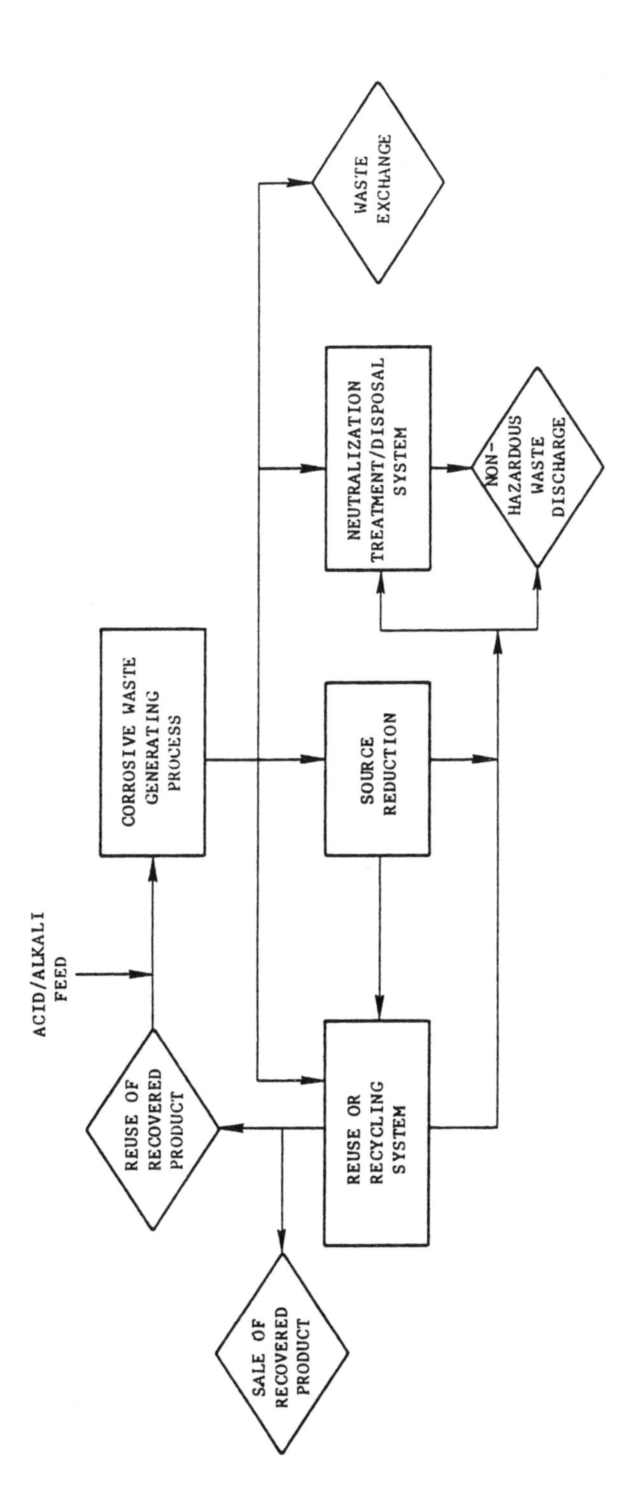

Figure 6.1.1. Corrosive waste management options.

6.2 WASTE MANAGEMENT PROCESS SELECTION

All generators of hazardous corrosive wastes will be required to undertake certain steps to characterize regulated waste streams and to identify potential treatment options. Treatment process selection should involve the following fundamental steps:

1. Characterize the source, flow, and physical/chemical properties of the waste.

2. Evaluate the potential for source reduction.

3. Evaluate the potential for waste exchange.

4. Evaluate the potential for reuse or sale of recycled acid/alkali streams and other valuable waste stream constituents; e.g., recovered metals or solvents.

5. Identify potential treatment and disposal options based on technical feasibility of meeting the restrictions imposed by the land disposal ban. Give consideration to waste stream residuals and fugitive emissions to air.

6. Determine the availability of potential options. This includes the use of offsite services, access to markets for recovered products or waste exchange, and availability of commercial equipment and existing onsite systems.

7. Estimate total system cost for various options, including costs of residual treatment and/or disposal and value of recovered products. Cost will be a function of Items 1 through 5.

8. Screen candidate management options based on preliminary cost estimates.

9. Use mathematical process modeling techniques and laboratory/pilot-scale testing as needed to determine detailed waste management system design characteristics and process performance capabilities. The latter will define product and residual properties and identify need for subsequent processing.

10. Perform process trials of recovered products and wastes available for exchange in their anticipated end use applications. Alternatively, determine marketability based on stream characteristics.

11. Calculate detailed cost analysis based on modeling and performance results.

12. Perform final system selection based on relative cost and other considerations; e.g., safety, acceptance, liability, and risks associated with data uncertainties.

Key system selection steps are discussed in more detail below.

6.2.1 Waste Characterization

The first step in identifying appropriate waste management alternatives to land disposal involves characterizing the origin, flow, and quality of generated wastes. An understanding of the processing or operational practices which result in generation of the waste forms the basis for evaluating waste minimization options. Waste flow characteristics include quantity and rate. Waste quantity has a direct impact on unit waste management costs due to economies of scale in processing costs and marketability of recovered products. Flow can be continuous, periodic, or incidental (e.g., spills) and can be at a relatively constant or variable rate. This will have a direct impact on storage requirements and waste management process design; e.g., continuous or batch flow.

Waste physical and chemical characteristics are generally the primary determinant of waste management process selection for significant volume wastes. Of particular concern is whether the waste is pumpable, inorganic or organic, and whether it contains recoverable materials or constituents which may interfere with processing equipment or process performance. Waste properties such as degree of corrosivity, reactivity, ignitability, heating value, viscosity, concentrations of toxic organic chemical constituents, biological and chemical oxygen demand, and solids, oil, grease, metals and ash content need to be determined to evaluate applicability of certain waste management processes. Individual consistuent properties such as solubility (affected by the presence of chelating compounds), vapor pressure, partition coefficients, reactivity, reaction products generated with various biological and chemical (e.g., neutralizing, oxidizing, and reducing) reagents, and adsorption coefficients are similarly required to assess treatability.

The presence or absence of buffers will affect neutralization reagent and pH control system requirements. Chelators will enhance metal solubility, requiring over neutralization to alkaline pH to effect metal precipitation. Cyanides and chromium will require treatment through chlorination and reduction, respectively, prior to combined neutralization with other corrosive wastes.

Finally, variability in waste stream characteristics will necessitate overly conservative process design and additional process controls, thereby increasing costs. Marketability of recovered products or materials offered for waste exchange will also be adversely affected by variability in waste characteristics.

6.2.2 Source Reduction Potential

Source reduction potential is highly site specific, reflecting the variability of industrial waste generating processes and product requirements. Source reduction alternatives which should be investigated include raw material substitution, product reformulation, process redesign and waste segregation. The latter may result in additional handling and storage requirements, while viability of other waste reduction alternatives may be more dependant on differential processing cost and impact on product quality.

Many opportunities exist for firms to achieve waste minimization through implementation of simple, low-cost methodologies currently proven in successful programs.[2] Lack of available techniques has been less of an impediment to increased implementation than perception that these methods are not available.[2] Historically, management has favored end-of-pipe treatment and has been reluctant to institute waste reduction and reuse practices. This reluctance is primarily due to potential for process upsets or adverse impacts on product quality. Other risks of installing waste reduction methods include uncertain investment returns and production downtime required for installation. However, in the wake of increasing waste disposal and liability costs, source reduction has repeatedly proven to be cost effective, while at the same time providing for minimal adverse health and environment impact.[2] Thus, source reduction should be considered a highly desireable waste management alternative.

6.2.3 Waste Exchange Potential

As discussed in Section 5.8, corrosive wastes have significant potential for being managed through waste exchange. Wastes will be good candidates for exchange if: 1) contaminant concentrations are low, consistent, and at levels which are compatible with user processes; 2) processing requirements are minimal; and 3) the waste is available in sufficient volumes on a regular basis.[3] Wastes generated from processes with high purity requirements may be used directly in processes with lower specifications; e.g., paint stripping or equipment cleaning. Another reuse method with high exchange potential is acid/alkali waste combination for mutual neutralization. Economics are particularly favorable when these individual wastes would have required separate post-treatment for metal precipitation or organic removal. Finally, waste exchange may prove to be the least cost management option for firms with wastes that have high recovery potential, but lack the waste volume or capital to make onsite recovery viable.[3]

Potential for waste exchange is reduced when industries are faced with concerns about liability, confidentiality, and quality of the waste. Additionally, transportation costs are frequently a limiting factor in the exchange of high volume, low concentration wastes.[2]

6.2.4 Recovery Potential

As part of the waste characterization step, the presence of potentially valuable waste constituents should be determined. In the case of concentrated corrosive solutions, the bulk of the waste typically has recycle potential. Economic benefits from recovery and isolation of this or other materials may result if they can be reused in onsite applications or marketed as saleable products. In the former case, economic benefits result from decreased consumption of virgin raw materials. This must be balanced against possible adverse effects on process equipment or product quality resulting from buildup or presence of undesirable contaminants. Market potential is limited by the lower value of available quantity or demand. Market potential will be enhanced with improved product purity, availability, quantity, and consistency.

Onsite reuse has several advantages relative to marketing for offsite use including reduced liability and more favorable economics. Offsite sale is less profitable due to transportation costs and the reduced purchase price which can typically be charged offsite users as a result of uncertainties in product quality. Thus, economics and liability combine with factors such as concerns about confidentiality to encourage onsite reuse whenever possible.

In practice, recycling of corrosive wastes has been limited to recovery of highly concentrated solutions, particularly those which are not amenable to management in low-cost wastewater treatment systems. Recycling options have been discussed in detail in Section 5.0. These are summarized in Table 6.2.1 with information provided on current applications, residuals generated, and availability.

6.2.5 Identifying Potential Treatment and Disposal Options

Following an assessment of the potential for source reduction and recycling, the generator should evaluate treatment systems which are technically capable of meeting the necessary degree of neutralization and hazardous constituent removal or destruction. Guideline considerations for the investigation of treatment technologies are summarized in Table 6.2.2.

The treatment objectives for a waste stream at a given stage of treatment will define the universe of candidate technologies. Treatment objectives must ultimately assure that the waste meets appropriate specifications for surface water discharge (NPDES, POTW), land disposal (pH, Extraction Procedure Toxicity Characteristic leaching test for metals and organics, treatment standards for priority organics), or reuse in an industrial process.

Restrictive waste characteristics (e.g., concentration range, flow, interfering compounds) and technological limitations of candidate treatment processes will reduce the list of candidate technologies to a list of potential applications for a specific waste. Consideration must be given to pretreatment options for eliminating restrictive waste characteristics (e.g., cyanide or chromium pretreatment), process emissions, residuals and their required treatment, and opportunities for recovery. System design will be based on neutralization requirements and the most difficult compound to recover, remove or destroy.

TABLE 6.2.1. SUMMARY OF RECOVERY/REUSE TECHNOLOGIES FOR CORROSIVE WASTES

Process	Applicable waste streams	Stage of development	Performance	Residuals generated	Cost
Evaporation/ distillation	Metal plating rinses; acid pickling liquors	Well-established for treating plating rinses.	Plating solution recovered for reuse in plating bath. Rinse water can be reused.	Impurities will be concentrated, therefore, crystallization/ filtration system may be required.	Can be cost-effective for recovering corrosive plating solutions from rinse waters.
Crystallization	H_2SO_4 pickling liquors; HNO_3/HF pickling liquors; caustic aluminum etch solutions.	20 to 25 systems currently in operation (fewer applications for caustic recovery).	97-98% recovery for H_2SO_4 (80-85% metal removal).	Ferrous sulfate heptahydeate crystals (can be traded or sold).	Cost-effective if treating large quantities of waste.
			99% HNO_3 and 50% HF recovered.	Metal fluoride crystals (can recover additional HF by thermal decomposition).	
			80% recovery of NaOH.	Aluminum hydroxide crystals (can be traded or sold).	
Ion exchange	Plating rinses; acid pickling baths; aluminum etching solutions; H_2SO_4 anodizing solutions; rack-stripping solutions (HF/HNO_3).	Several RFIE units in operation for treatment of corrosives.	Cocurrent systems not technically feasible for direct treatment of corrosives; can be used in conjunction with neutralization technologies to lower overall costs.	Cocurrent process generates spent regenerant, which is also corrosive.	RFIE and APU are cost-effective.
		Units for direct treatment of acid bath only available from ECO-TEC, Ltd.	RFIE units show good results. Conventional RFIE performs best with dilute solutions. APU performs best with high metal concentration (30 to 100 g/L).	Recovered metals which can be reused, treated, disposed, or marketed.	
Electrodialysis	Recovery of chromic/ sulfuric acid etching solutions.	Units currently being sold, but limited area of application. 5 in operation.	85% recovery of etching solution. 45% copper removal; 30% zinc removal.	Metals which can be treated, disposed, or regenerated for reuse.	Cost-effective for specific applications (chromic/sulfate acid etchants).
	Recovery of plating rinses (particularly chromic acid rinse water).	Several in operation.	Works best when copper concentrations are in the 2 to 4 oz/gal usage.	Chromic acid can be returned to plating bath; rinse water can be reused.	Low capital investment; cost-effective for specific application (chromic acid rinses).
	Recovery of HNO_3/HF pickling liquors.	Marketed, none in operation to date.	3 M HF/HNO_3 recorded.	2 M KOH Soln which can be recycled back to the pretreatment step for this ED application.	Cost-effective for large quantity generator.

(continued)

TABLE 6.2.1 (continued)

Process	Applicable waste streams	Stage of development	Performance	Residuals generated	Cost
Reverse osmosis	Plating rinses.	Corrosive waste membranes marketed by four companies. RD module systems applicable to corrosives available from two companies.	90% conversion achieved with cyanide plating rinses.	Recovered plating solution returned to plating bath (after being concentrated by an evaporator). Rinsewater reused.	Cost-effective for limited applications. Development of a more chemically resistant membrane would make it very cost-effective for a wider area of application.
Donnan dialysis/ coupled transport	Plating rinses; potentially applicable to acid baths.	Donnan analysis only lab-scale tested.	Data not available for Donnan analysis (further testing required).	Data not available for Donnan analysis.	No cost data available for Donnan analysis.
		Coupled transport lab and field tested. Coupled transport system is currently being marketed.	Coupled transport has demonstrated 99% recovery of chromate from plating rinses. Other plating rinses should be applicable, but not fully tested.	For chromate plating rinse applications, sodium chromate is generated; can be used elsewhere in plant or subjected to ion exchange to recover chromic acid for recycle to plating solution.	Average capital cost for plating shop is $20,000. Can be cost-effective for specific applications.
Solvent extraction	HNO_3/HF pickling liquors.	Commercial-scale systems installed for development purposes in Europe and Japan. No commercial-scale installations in U.S.	95% recovery of HNO_3; 70% recovery of HF.	Metal sludge (95% iron can be recovered by thermal decomposition).	Not available.
Thermal decomposition	Acid wastes.	Well-established for recovering spent pickle liquors generated by steel industry. Pilot-scale stage for organic wastes.	99% regeneration efficiency for pickling liquors.	98-99% purity iron oxide which can be reused, traded, or marketed.	Expensive capital investment. Only cost-effective for large quantity waste acid generators.

TABLE 6.2.2. GUIDELINE CONSIDERATIONS FOR THE INVESTIGATION OF WASTE TREATMENT, RECOVERY, AND DISPOSAL TECHNOLOGIES

A. Objectives of treatment:

- Primary function (pretreatment, treatment, mutual neutralization, residuals treatment)
- Primary mechanisms (neutralization, destruction, removal, conversion, separation)
- Recover waste for reuse
- Recovery of specific chemicals, group of chemicals (acids, alkalis, metals, solvents, other organics)
- Polishing for effluent discharge (NPDES, POTW)
- Immobilization or encapsulation to reduce migration (inorganic sludge)
- Overall volume reduction of waste
- Selective concentration of constituents (acids, alkalis, metals, solvents, other organics)
- Detoxification of hazardous constituents

B. Waste applicability and restrictive waste characteristics:

- Acceptable concentration range of primary and restrictive waste constituents
- Acceptable range in flow parameters
- Chemical and physical interferences (compatibility with neutralization reagent)

C. Process operation and design:

- Batch versus continuous process design
- Fixed versus mobile process design
- Equipment design and process control complexity (pH control)
- Variability in system designs and applicability
- Spatial requirements or restrictions
- Estimated operation time (equipment down-time)
- Feed mechanisms (wastes and reagents; solids, liquids, sludges, slurries)
- Specific operating temperature and flow
- Sensitivity to fluctuations in feed characteristics
- Residuals removal mechanisms
- Reagent selection and requirements
- Ancillary equipment requirements (tanks, pumps, piping, heat transfer equipment)
- Utility requirements (electricity, fuel and cooling, process and make-up water)

(continued)

TABLE 6.2.2 (Continued)

D. Reactions and theoretical considerations:

- Waste/reagent reaction (neutralization, destruction, conversion, oxidation, reduction)
- Competition or suppressive reactions (buffers)
- Enhancing conditions (specify chemicals)
- Fluid mechanics limitations (mass and heat transfer)
- Reaction kinetics (temperature and concentration effects)
- Reactions thermodynamics (endothermic/exothermic/catalytic)

E. Process efficiency:

- Anticipated overall process efficiency
- Sensitivity of process efficiency to:
  - feed concentration fluctuations
  - reagent concentration fluctuations
  - process temperature fluctuations
  - toxic constituents (biosystems)
  - physical form of the waste
  - other waste characteristics

F. Emissions and residuals management:

- Extent of fugitive and process emissions and potential sources (processing equipment, storage, handling)
- Ability (and frequency) of equipment to be "enclosed"
- Availability of emissions and residuals data/risk calculations
- Products of incomplete reaction
- Relationship of process efficiency to emissions or residuals generation
- Air pollution control device requirements
- Process residuals (fugitive/residual reagents, recovered products, filter cakes, sludges, incinerator scrubber water and ash)
- Residual constituent concentrations and leachability
- Delisting potential

G. Safety considerations:

- Safety of storing and handling corrosive wastes, reagents, products and residuals
- Special materials of construction for storage and process equipment
- Frequency and need for use of personnel protection equipment
- Requirements for extensive operator training
- Hazardous emissions of wastes or reagents
- Minimization of operator contact with wastes or reagents
- Frequency of maintenance of equipment containing hazardous materials
- High operating temperatures
- Difficult to control temperatures
- Resistance to flows or residuals buildup
- Dangerously reactive wastes/reagents
- Dangerously volatile wastes/reagents

A key consideration in the choice of a corrosive wastewater neutralization system is reagent selection. Potential reagents and their associated advantages and disadvantage with respect to handling, processing and sludge generation, are summarized in Table 6.2.3. Sludge characteristics and volume will have a significant impact on ultimate reagent selection since disposal of this material constitutes a large percentage of total treatment costs. The presence of toxic organics in corrosive wastewaters will also significantly add to post-treatment costs. These will increase with organic concentration and decrease with reactivity, volatility, and biodegradability.[4]

Concentrated organics and sludges will typically be treated through neutralization prior to handling in other equipment. Organic wastes will then be amenable to recovery, treatment, or disposal (e.g., incineration) technologies as summarized in Section 4.1 and discussed in detail in the literature.[4,5] Sludges will be neutralized and dewatered prior to aqueous treatment of the supernatent. Inorganic dewatered residues can be solidified/ encapsulated prior to land disposal and organic solids can be incinerated or otherwise thermally destroyed (Section 4.1).

Potential reagent and processing system equipment designs and configurations have been summarized in Section 4.0. Ultimately, the selection of a specific treatment system from the list of potentially applicable processes will depend on cost, availability, and site specific factors. These considerations are discussed below.

6.2.6 Availability of Potential Management Options

The availability of each component of a waste management system will affect its overall applicability. Existing available onsite treatment process capacity (e.g., wastewater treatment system), ancillary equipment, labor, physical space, and utilities will have a significant impact on the economic viability of a treatment system. Purchased equipment must be available in sizes and processing capabilities which meet the specific needs of the facility. Offsite disposal, recovery, and treatment facilities and companies using exchanged materials or purchasing saleable products must be located within a reasonable distance of the generator to minimize transport costs. In

TABLE 6.2.3. SUMMARY OF NEUTRALIZATION TECHNOLOGIES

Process	Applicable waste streams	Stage of development	Performance	Residuals generated	Cost
Acid/alkali mutual neutralization	All acid/alkali compatible waste streams except cyanide.	Well developed.	Generally slower than comparable technologies due to dilute concentrations of reagents. May evolve hazardous constituents if incompatible wastes are mixed.	Variable, dependent on quantity of insolubles and products contained in each waste stream.	Least expensive of all neutralization technologies.
Limestone	Dilute acid waste streams of less than 5,000 mg/L mineral acid strength and containing low concentrations of acid salts.	Well developed.	Requires stone sizes of 0.074 mm or less. Requires 45 minutes or more of retention time. Can only neutralize acidic wastes to pH 6.0. Must be aerated to remove evolved CO_2.	Will generate voluminous sludge product when reacted with sulfate-containing wastes. Stones over 200 mesh will sulfonate, be rendered inactive, and add to sludge product.	Not cost-effective in treating concentrated wastes. May be cost-effective in treating dilute acidic wastes.
Lime	All acid wastes.	Well developed.	Requires 15 to 30 minutes of retention time. Must be slurried to a concentration of 10 to 35% solids prior to use. Can under- (below pH 7) or over- (above pH 7) neutralize.	Will generate voluminous sludge similar to limestone.	More expensive than crushed limestone (200 mesh).
Caustic soda	All acid wastes.	Well developed.	Requires 3 to 15 minutes of retention time. In liquid form, easy to handle and apply. Can under- or over-neutralize including pH 13 or higher.	Reaction products are generally soluble, however, sludges do not dewater as readily or as easily as lime or limestone.	Most expensive of all widely used alkaline reagents (five times the cost of lime).
Sulfuric acid	All alkaline wastes except cyanide.	Well developed.	Requires 15 to 30 minutes of retention time. In liquid form, but presents burn hazard. Highly reactive and widely available.	Will generate large quantities of gypsum sludge when reacted with calcium-based alkaline wastes.	Least expensive of all widely used acidic reagents
Hydrochloric acid	All acid wastes.	Well developed, but rarely applied due to high reagent cost.	Requires 5 to 20 minutes of retention time. Liquid form presents burn and fume hazard. More reactive than sulfuric.	Reaction products are generally soluble.	Approximately twice as expensive as sulfuric on a neutralization equivalent basis.
Carbonic acids, liquid carbon dioxide	All alkaline wastes except cyanide.	Emerging technology.	Retention time 1 to 1-1/2 minutes. In liquid form, must be vaporized prior to use. Can only neutralize alkaline wastes to pH 8.3 end point.	Will form calcium carbonate precipitate when reacted with calcium-based alkaline wastes.	Approximately 3 to 4 times as expensive as sulfuric. Therefore, limited to applications using more than 200 tons of reagents per year or with flow rate greater than 100,000 gpd.

addition, they must have available capacity for the waste type and volume generated. Finally, time constraints may eliminate certain treatment processes from consideration as a result of anticipated delays in procurement, permitting, installation, or start-up.

In general, neutralization systems are widely applied and readily available. However, several recovery systems and post-treatment systems for organic wastes have only recently been applied to corrosive waste treatment. Availability and uncertainty in expected performance will play a significant role in the decision to implement these technologies.

6.2.7 Management System Cost Estimation

The relative economic viability of candidate waste management systems will be the primary determinant of ultimate system selection. This must be evaluated on the basis of total system costs which includes the availability of onsite equipment, labor and utilities, net value of recovered products, and recovery/treatment/disposal processing costs. Capital equipment expenditures and financing constraints are frequently a limiting factor in system selection, particularly for processes which have higher uncertainties of success.

Costs for a given management system will be highly dependent on waste physical, chemical, and flow characteristics. Thus, real costs are very site-specific and limit the usefulness of generalizations. The reader is referred to the sections on specific technologies (Sections 4.0 and 5.0) for data on costs and their variability with respect to flow and waste characteristics. Costing methodologies have also been adequately described in the literature[6,7,8] and are available in software packages.[9] Major cost centers which should be considered are summarized in Table 6.2.4.

6.2.8 Modeling System Performance and Pilot-Scale Testing

Following this preliminary cost evaluation, which will enable the generator to narrow his choice of waste management options, steps must be taken to further finalize the selection process. These could involve the use of theoretical models to predict design and operating requirements. However,

TABLE 6.2.4. MAJOR COST CENTERS FOR WASTE MANAGEMENT ALTERNATIVES

A. Credits:

- Material/energy recovery resulting in decreased consumption of purchased raw materials
- Sales of waste products

B. Capital costs:[a]

- Processing equipment (reagent addition, reaction vessel, recovery apparatus, sludge and other residual handling equipment)
- Ancillary equipment (storage tanks, pumps, piping)
- Pollution control equipment
- Vehicles
- Buildings, land
- Site preparation, installation, start-up

C. Operating and maintenance costs:

- Overhead, operating, and maintenance labor
- Maintenance materials
- Utilities (electricity, fuel, water)
- Reagent materials
- Disposal, offsite recovery, and waste brokering fees
- Transportation
- Taxes, insurance, regulatory compliance, and administration

D. Indirect costs and benefits:

- Impacts on other facility operations; e.g., changes in product quality as a result of source reduction or use of recycled materials
- Use of processing equipment for management of other wastes

[a]Annual costs derived by using a capital recovery factor:

$$CRF = \frac{i(1+i)^n}{(1+i)^n - 1}$$

Where: i = interest rate and n = life of the investment. A CRF of 0.177 was used to prepare neutralization cost estimates in this document. This corresponds to an annual interest rate of 12 percent and an equipment life of 10 years.

models generally sacrifice accuracy for convenience and are often not sufficiently accurate for complex waste streams. Laboratory data, or pilot-plant and full-scale data, may ultimately be needed to confirm predicted performance. In fact, some data may be needed as model inputs for predicting system behavior.

In many cases, models are useful in predicting behavior and can be used in place of costly laboratory testing. Models are also useful in assessing relative performance, costs of various approaches to treatment, and the incremental costs of achieving increasingly stringent treatment concentration levels. Many suppliers of neutralization and recovery equipment use models to optimize design and operations parameters and to scale treatment processes. Equipment manufacturers are also often able to provide experimental equipment and models to establish process parameters and cost, including the costs required for disposal of residuals.

REFERENCES

1. Allen, C.C., and B. L. Blaney. Research Triangle Institute. Techniques for Treating Hazardous Waste to Remove Volatile Organics Constituents. Performed for U.S. EPA HWERL. EPA-600/2-85-127. March 1985.

2. Committee on Institutional Considerations in Reducing the Generation of Hazardous Industrial Wastes. Environmental Studies Board, National Research Council. Reducing Hazardous Waste Generation: An Evaluation and a Call for Action. National Academy Press, Washington, D.C. 1985.

3. GCA Technology Division, Inc. Industrial Waste Management Alternatives Assessment for the State of Illinois. Volume IV: Industrial Waste Management Alternatives and Their Associated Technologies/Processes. Final Report prepared for the Illinois Environmental Protection Agency, Division of Land Pollution Control. GCA-TR-80-80-G. February 1981.

4. Breton, M. et al. GCA Technology Division, Inc. Technical Resource Document: Treatment Technologies for Solvent Containing Wastes. Prepared for U.S. EPA HWERL under Contract No. 68-03-3243. August 1986.

5. Surprenant, N. GCA Technology Division, Inc. Technical Resource Document: Treatment Technologies for Halogenated Organic Wastes. Prepared for U.S. EPA HWERL under Contract No. 68-03-3243. October 1986.

6. Peters, M. S., and K. D. Timmerhaus. Plant Design and Economics for Chemical Engineers. 3rd Edition. Mc-Graw Hill Book Company, New York, NY. 1980.

7. U.S. EPA Design Manual: Dewatering Municipal Wastewater Sludges. U.S. EPA Municipal Environmental Research Laboratory, Cincinnati, OH. EPA-625/1-82-014. October 1982.

8. Mitre Corp. Manual of Practice for Wastewater Neutralization and Precipitation. EPA-600/2-81-148. August 1981.

9. Cunningham, V. L. et al. Smith, Kline & French Laboratories. Environmental Cost Analysis System. 1986.

Other Noyes Publications

TREATMENT TECHNOLOGIES FOR SOLVENT CONTAINING WASTES

by

M. Breton, P. Frillici, S. Palmer
C. Spears, M. Arienti, M. Kravett
A. Shayer, N. Surprenant
Alliance Technologies Corporation

Pollution Technology Review No. 149

This book provides technical information describing management options for solvent containing wastes. These options include treatment and disposal of waste streams as well as waste minimization procedures such as source reduction, reuse, and recycling.

Emphasis is placed on proven technologies such as incineration, use as fuel, distillation, steam stripping, biological treatment, and activated carbon adsorption; however, a full range of waste minimization processes and treatment recovery technologies which can be used to manage solvent wastes is covered in the book.

Potentially viable technologies are described in terms of performance in removal of regulated constituents, associated process residuals and emissions, and restrictive waste characteristics which affect the ability of a given technique to effectively treat the wastes under consideration.

Approaches to the selection of treatment/recovery options are reviewed, and pertinent properties of organic solvents which impact treatment technology/waste interactions are provided.

CONTENTS

ISBN 0-8155-1158-2 (1988) 753 pages

Other Noyes Publications

CORROSION INHIBITORS
An Industrial Guide

by

Ernest W. Flick

This volume describes approximately 750 corrosion inhibitors and rust preventives which are currently available for industrial usage. The book will be of value to industrial, technical and managerial personnel involved in the specification and use of these products. It has been produced from information received from numerous industrial companies and other organizations.

The data included represent selections from manufacturers' descriptions, made at no cost to, nor influence from, the makers or distributors of these materials. Only the most recent information has been included. All of the products listed here are currently available, which will be of interest to readers concerned with product discontinuances.

The book lists the following product information, as available, in the manufacturer's own words: (1) company name and product category, (2) trade name and product numbers, and (3) product description. Also included are a list of Suppliers' Addresses and a Trade Name Index.

The following companies are represented in the book:

Air Products and Chemicals
Akzo Chemie America
Albright & Wilson, Inc.
Allied-Kelite Division/ Witco Chemical Corp.
Alox Corp.
American Charcoal Company
Amoco Chemicals Corporation
Amoco Petroleum Additives Co.
Ampion Corporation
Anedco, Inc.
Angus Chemical Co.
Armite Laboratories
Baker Performance Chemicals
Bar's Leaks, Inc.
Betz Energy Chemicals, Inc.
Betz Laboratories, Inc.
Betz Process Chemicals, Inc.
Birchwood Casey
Bond Chemicals, Inc.
Bruce Products Corp.
Burmah-Castrol, Inc.
Butler Engineering Associates
Calgon Corp.
Chemical Technologies, Inc.
Chem-Pak
Ciba-Geigy Corp.
Cortec Division/ Sealed Air Corp.
Crown Technology, Inc.
Dow Corning Corp.
Du-Lite Corp.
E.I duPont de Nemours & Co.
Dykem Co.
Exxon Chemicals
FarBest Corp.
Filmite Oil Corp.
GAF Corp.
GOA Corporation
The B.F. Goodrich Co.
Frederick Gumm Chemical Co.
Halox Pigments
Percy Harms Corp.
Henkel Corp.
Hercules, Inc.
ICI Americas, Inc.
Ideas, Inc.
Jetco Chemicals, Inc.
Kano Laboratories, Inc.
Keil Chemical Division/ Ferro Corp.
King Industries
LaFrance Manufacturing Co.
Loctite Corp.
The Lubrizol Corp.
Maurer-Shumaker, Inc.
Mayco Oil & Chemical Co.
Mazer Chemicals, Inc.
Metalbrite Products Corp.
Metal Lubricants Co.
Harry Miller Corp.
Miranol Chemical Co., Inc.
Mitchell-Bradford Chemical Co.
Modern Industries, Inc.
Mogul Division/ The Dexter Corp.
Mona Industries, Inc.
The Must for Rust Co.
National Research & Chemical
New York Bronze Powder Co.
NL Baroid/NL Industries, Inc.
Northern Instruments Corp.
Oakite Products, Inc.
Olin Corporation
Park Chemical Company
Pennwalt Corporation
Penreco
Pillsbury Chemical & Oil, Inc.
Polar Chip
The PQ Corp.
Price-Driscoll Corp.
Process Research Products
Pro-Tech
Rawn Co., Inc.
Raybo Chemical Co.
Rustlick Products
Sandoz Colors & Chemicals
Henry E. Sanson & Sons, Inc.
Sharp Chemicals Co.
Sherex Chemical Co., Inc.
SKS Industries
Sonneborn Division/ Witco Corp.
Subox Division/ Carboline Company
Tennessee Chemical Co.
Three Bond of America, Inc.
3M/Adhesives, Coatings & Sealers Division
Arthur C. Trask Operation
Valvoline Oil Co.
R. T. Vanderbilt Co., Inc.
Van Straaten Corp.
Virginia Chemicals
Watcon, Inc.
Wayne Chemical Products Co.
Zip Aerosol Products

ISBN 0-8155-1126-4 (1987) **479 pages**